ANIMAL COGNITION

A Tribute to Donald A. Riley

Donald A. Riley

ANIMAL COGNITION

A Tribute to Donald A. Riley

Edited by
Thomas R. Zentall
University of Kentucky

Psychology Press
Taylor & Francis Group

NEW YORK AND HOVE

First published by Psychology Press

This edition published 2013 by Routledge

Psychology Press   Psychology Press
711 Third Avenue   27 Church Road
New York   Hove
NY 10017   East Sussex, BN3 2FA

Psychology Press is an imprint of the Taylor & Francis Group, an informa business

Reprinted 2009 by Psychology Press

Library of Congress Cataloging-in-Publication Data

Animal cognition : a tribute to Donald A. Riley / edited by
Thomas R. Zentall.
    p.   cm. — (Comparative cognition and neuroscience)
  "Based on a symposium held at Tolman Hall at the University of
California at Berkeley on November 21, 1991."
  Includes bibliographical references (p.) and index.
  ISBN 0-8058-1183-4 (cloth). — ISBN 0-8058-1184-2 (paper)
  1. Cognition in animals—Congresses. 2. Riley, Donald A.—
Congresses. I. Riley, Donald A. II. Zentall, Thomas R.
III. Series.
QL785.A723   1993
156'.3—dc20
                        92-38798
                            CIP

10 9 8 7 6 5 4 3 2 1

Contents

Preface

The chapters that appear in this volume were prepared in honor of Donald A. Riley, known to friends and colleagues as Al, who retired from the faculty of the University of California at Berkeley in the summer of 1991. The seeds for this volume were planted at the 1989 meeting of the Psychonomic Society in Atlanta by some of Al's former students (Mike Brown, Bob Cook, Bill Maki, Herb Roitblat, and Tom Zentall). At that time we considered a volume that would cover recent developments in all the areas of research to which Al had contributed. Once we realized how broad the volume would have to have been and how many chapters would have to have been included if the book were to have any coherence, we decided that a more narrow focus would be all that we could manage.

The essays that appear here are representative of a research area that has been classified loosely as animal cognition, although some of the contributors would probably have felt more comfortable with a more general and perhaps more traditional title such as animal learning. The decision was somewhat arbitrary. It was made, in part, because many of Al's students would describe their area of research as animal cognition. Furthermore, it reflects a functionalist philosophy that was prevalent in Al's laboratory and that many of Al's students absorbed. According to this philosophy, it is acceptable to hypothesize that an animal might engage in complex processing of information, as long as one can operationalize evidence for such a process and the hypothesis can be presented in the context of testable predictions that can differentiate it from other mechanisms. It represents a spirit that was passed on to us directly by Al and indirectly through Edward C. Tolman, the man whose name appears on the building in which we studied and conducted our research.

This volume is dedicated to Al Riley, a friend to all those who have contributed to this volume, a mentor to a fortunate few. For those of us trained under Al's guidance, the experience, although difficult to describe, had a profound effect on our approach to both research and teaching. Al's mentoring had a quality that can be described best as a cross between the Socratic and Rogerian methods. Asking Al a question often resulted in the response, "That is a very interesting question—what do *you* think it means?" But we never got the feeling from Al that he knew the answer and just wanted to see if we could figure it out for ourselves. Instead, he always gave us the impression (whether it was true or not) that he really didn't know the answer either, but he was sure that if we gave it enough thought, we could come up with a pretty good response.

Studying with Al was an experience full of challenges. It was not unlike Al to show up before a lab meeting with copies of a recent journal article he had just read, and with a puzzled but nonjudgmental look on his face, ask us what we thought of it. In our minds we wondered, had Al not understood the procedures or the conclusions, or had he thought there was some problem with the study but he just couldn't put his finger on it. After we had spent, sometimes, hours discussing the article, we often were able to list several problems with the study and had designed one or more experiments that we thought might resolve the problems raised. At the end of our lab meeting, we seldom knew whether the issues we discussed had anything to do with the reason Al introduced the paper for discussion. But it never mattered.

Al led lab discussions only in the most subtle way, asking for clarification of what at first appeared to be a minor detail of a proposed explanation, but what often turned out to be a major flaw in the argument. Had he seen it all along? Did he have some intuition that this was one of those important details? Or did it just happen to be important? We would never know.

Sometimes, when one of us proposed a particularly "creative" explanation for a phenomenon we were discussing, but the mechanisms we were proposing were hard to understand, Al would say (without the least bit of condescension or frustration), "write me a letter about that," as if to say, "this is just too complicated for me to understand without being able to see it spelled out on paper." It was a challenge for us to put it down in writing, where, more often than not, the cold, harsh, concrete words often made it very clear where our reasoning was flawed. Had Al seen the problem with our explanation when it was discussed, but wanted us to discover it on our own? Or had he really not understood the problem, and was he really just struggling to comprehend what was being said? Again, we never knew, but we always benefited intellectually from the "letter writing" experience.

Together with the challenge Al instilled in us, there was the faith that as bright graduate students we could come up with the answers to any questions raised. We always got the feeling from him that our ideas were as good as those of anyone doing research "out there." Those were some of the intangibles, the tools that allowed each of us to strike out in our own way, defining our own research programs, giving each of us the confidence and self-assurance that would enable us to prosper on our own. Most of all, Al provided us with a mirror that allowed us to see the best in ourselves. Al, for all this we are grateful.

—*Thomas R. Zentall*

xii

1. Douglas Grant
2. William Maki
3. Ruth Maki
4. Ervin Hafter
5. Diane Chatlosh
6. James Ison
7. Marvin Lamb

8. John S. Watson
9. David Olton
10. Edward Wasserman
11. Herbert Roitblat
12. William Roberts
13. Donald Riley
14. Sara Shettleworth

15. Ken Cheng
16. Cynthia Langley
17. Thomas Zentall
18. Norman Spear
19. Lyle Bourne
20. Werner Honig
21. Dennis Wright

22. Stewart Hulse
23. Robert Cook
24. Mark Rilling
25. Michael Brown
26. Anthony Wright
27. Janice Steirn
28. David Thomas
29. Pamela Jackson-Smith

Animal Cognition Conference, Berkeley, CA, November 21, 1991

xiii

COMPARATIVE COGNITION
AND NEUROSCIENCE

*Thomas G. Bever, David S. Olton,
and Herbert L. Roitblat, Senior Editors*

INTRODUCTION

1 Animal Cognition: An Approach to the Study of Animal Behavior

Thomas R. Zentall
University of Kentucky

The purpose of this introductory chapter is to provide a framework within which research in animal cognition can be better understood. It is quite fitting that some of the ideas presented here also can be found in a chapter by Riley, Brown, and Yeorg (1986). Although I had not read the chapter prior to writing this one, the ideas presented in it clearly formed a part of my training as an experimental psychologist in Al Riley's lab.

First, I argue that a cognitive approach to animal behavior can have value whether or not one can actually find evidence for cognitive processes in animals. I suggest that by adopting a cognitive approach one will carry out research that otherwise would not have been conducted. Specific, testable hypotheses generated by a particular cognitive account of an observed behavior can encourage the development of novel experimental designs. Thus, in addition to whatever evidence is found for cognitive functioning in animals, a cognitive approach has heuristic value.

Second, I argue that, in spite of the fact that many animals typically show little evidence of cognitive behavior when they are observed in their natural environment, it is not unreasonable to expect that evidence for cognitive behavior will be found in an "unnatural" laboratory setting.

Finally, I try to provide an overview of the scope of contributions to this book.

ANIMAL COGNITION: A WORKING DEFINITION

Animal cognition can be defined more readily by what it is not, than by what it is. Defined most conservatively, animal cognition consists of the learned behavior that is left after simple associative-learning explanations have been ruled out.

3

Such a definition by exclusion is especially appealing to those who were trained either as behaviorists or with behaviorism as the historical foundation of their research (see many of the chapters in the present volume).

Some of those who define themselves behaviorists (e.g., Tolman, 1932) might be willing to take this definition one step further and argue that behavior that cannot be accounted for by simple associative principles implies some kind of active central process (e.g., an attentional filter, the parameters of which change with experience, as proposed by Sutherland & Mackintosh, 1971).

On the other hand, others involved in animal cognition research feel quite comfortable adopting the language of human cognitive psychology (using terms such as information processing, problem solving, and reasoning) to describe hypothesized mechanisms underlying behavior. They may use terms such as these because the behavior displayed by animals sometimes appears similar in nature to behavior seen in humans under conditions that seem (at least by some) to be analogous. Their reasoning is, that although the use of those terms to describe processes underlying animal behavior may be premature, it puts the experimenter in a better position to evaluate the hypothesis that similar mechanisms are present.

More conservative researchers would argue that research in animal cognition would be viewed as more credible if one avoided terms that are only loosely based on data and that are difficult to operationalize. They propose that all learning and memory in animals is potentially accounted for without resorting to the internal or active processing of stimuli. The appeal of such an approach is great. The explanation of behavior strictly in terms of observables avoids postulation of hidden units or processors. On the other hand, even if one can explain a large proportion of the behavior of organisms in terms of simple reinforcement contingencies, together with stimulus generalization (along quantifiable dimensions of physical similarity) and overt behavioral mediation, there is some behavior that appears to be very difficult to explain without resorting to active internal processing of stimuli and the relations among them (i.e., cognitive processing).

I would like to propose that the ideal position to take is one that first asks to what extent a particular phenomenon lends itself to explanation in terms of simple contingencies, and then asks if a cognitive account provides a more reasonable explanation considering the total range of data that has been collected from that species (as well as related species). Although this research strategy may appear to be of a particularly nondirective or amorphous nature (because the decision-making rules cannot be clearly operationalized), there are three advantages to such an approach. First, it encourages the investigator to maintain relative objectivity. One need not be committed to (or biased in favor of) an explanation that views all behavior as either cognitive or noncognitive. We do not demand that humans always function at a cognitive level (humans who demonstrate complex cognitive functioning also show good evidence of simple reflexive responding, and learning that does not require anything more than simple conditioning principles). Why should we make such a demand of animals?

Second, although a cognitive account implies a more abstract or more complex underlying process, paradoxically, this research strategy may yield the most parsimonious explanation of behavior across contexts. If simple associative learning principles cannot explain the wide variety of data that has been collected to address a particular phenomenon, a well-articulated, cognitively based description may actually provide a "simpler" explanation.

Finally, allowing for the possibility of cognitive functioning by animals leaves one more open to benefit from serendipity. When unexpected results are found, more possible testable explanations may be generated.

RATIONALE FOR A COGNITIVE APPROACH TO ANIMAL BEHAVIOR

Objectivity. If one is not committed in advance to an S-R position one is more likely to design an experiment that both maximizes the likelihood of finding an effect and yet, involves the controls needed to rule out alternative explanations. Furthermore, one is more likely to be objective in interpreting the data from those experiments. For example, I have seen experiments in which investigators who are studying transfer effects (and who appear to favor a Stimulus-Response interpretation of all learning), focus entirely on the drop in performance relative to baseline levels rather than on the level of performance relative to chance levels (although the glass may be 80% empty, it is also 20% full).

Similarly, there are those who "hide behind" a lack of statistical significance (arguing for the absence of an effect) when the size of an effect comes close to, but fails to reach the .05 level of significance. This is especially misleading when the study involves a small number of subjects (low statistical power) and when other statistical tests (e.g., a nonparametric sign test) would have resulted in a significant outcome (see, e.g., D'Amato, Salmon, Loukas, & Tomie, 1985).

There also may be individual differences in cognitive capacity or at least in the magnitude of cognitive effects found, and procedures may not be optimal for their observation. But I would even take this argument one step further and propose that when dealing with issues of cognitive capacity, the failure to detect a real difference (a Type 2 error) may be just as serious an error as the failure to reject a difference attributable to sampling error (a Type 1 error).

Parsimony. One barrier to the acceptance of animal cognition research is that its results often are not considered to be parsimonious. We are better off being able to explain behavior in terms of the simple relations between observable stimuli and responses, than in terms of unobservable cognitions. For example, if after an animal has learned to discriminate between two stimuli, based on differential histories of reinforcement, it shows immediate differential responding to two new stimuli, it would be more parsimonious to explain this transfer in terms of stimulus generalization along dimensions of physical similarity. But what if one designs an experiment to rule out differential stimulus similarity as an explanation for the observed transfer effects? Furthermore, what if one designs the experiment in such a way that one would expect transfer effects based on notions of concept formation? If transfer effects were found under these conditions, the most parsimonious explanation might involve the animal's use of a concept.

To be more specific, let us say that we are interested in assessing the ability of pigeons to use an identity (or difference) relation. We train pigeons either on a match-to-sample or oddity-from-sample task involving stimuli that differ in the orientation of line (vertical vs. horizontal). Once these tasks have been learned, we transfer the pigeons to either a matching task or an oddity task involving stimuli that differ in hue (red vs. green). The fact that a vertical line is no more like a red field than a green field makes it quite unlikely that a theory based on simple associative learning together with stimulus generalization along definable dimensions of physical stimulus similarity can account for observed positive transfer when pigeons are trained with line matching and transferred to hue matching (or are trained with line oddity and transferred to hue oddity). Significant differences in performance between positive- and negative-transfer groups (birds trained with line matching and transferred to hue oddity, or trained with line oddity and transferred to hue matching) have been reported in a number of experiments (Zentall & Hogan, 1974, 1975, 1976, 1978, see also Lombardi, Fachinelli, & Delius, 1984; Chapter 3 of this volume). Taken as a whole, the difference between groups (or in some cases between conditions) is reliable by any definition (relative transfer, absolute transfer, statistical significance, or replicability).

On the other hand, the magnitude of the effect is often rather small (but see Zentall, Edwards, Moore, & Hogan, 1981; Zentall & Hogan, 1978, Phase 4). It is typically not as good as that shown by chimpanzees (80% correct) or humans (near 100% correct). One could argue, in fact, that the level of transfer is closer to 50% (i.e., zero transfer) than it is to terminal performance on the original task (90% correct, a criterion sometimes used). Thus, one could argue that, based on parsimony, the results are more in keeping with a stimulus-response chaining explanation than they are with the notion of concept use. But if one takes a more general view of concept use among species, one is confronted with the less parsimonious conclusion that humans and chimps can use concepts but pigeons cannot. To be sure, the pigeons' "glass" is not full, but neither is it empty. Clearly, it is more parsimonious to propose a continuum of the ability to use concepts, from extensive in humans to considerably less in pigeons, than it is to draw an arbitrary absence-of-ability-to-use-concepts line.

One should keep in mind, of course, that pigeons may be better able to use concepts than we currently have reason to believe. It may be that we just need to find better ways of "asking" them about their cognitive abilities. The more we find out about the physiology and behavior of these animals, and the more clever we become in designing appropriate experiments, the better we will be able to ask them about their intellectual capacities. It may be useful to give two concrete examples here of how our increasing knowledge about a species can allow us to better formulate research methodology.

The first example comes from research with pigeons. For some time it has been known that pigeons have two fields of vision, a frontal binocular field that they use, for example, for eating and a lateral monocular field that they use, for example, for flying. Remy and Emmerton (1991) recently reported that learning does not transfer well between the frontal field and the lateral field. Now, identity research with pigeons generally has involved presentation of stimuli along the horizontal dimension. A sample is presented on the center response key and the comparisons are presented to the left and right of the sample. When the pigeon pecks the sample, the visual stimulus most likely projects onto the frontal field. Although the pigeon is free to move its head, the first sight of the comparisons is likely to involve the lateral field. If there is little transfer between the two fields, then the two stimuli may not be perceived as identical. Would better concept transfer be found if the samples and comparisons were presented in a vertical display such that both samples and comparisons would be projected into the same visual field?

Research with learning set in animals provides another example of how methodological improvements can result in unexpected findings. Learning set, the improvement with practice in the rate of acquisition of simultaneous visual discriminations, is a phenomenon readily found with monkeys (see e.g., Harlow, 1949). However, findings of learning set with similar materials and procedures have not been reported with rats (Warren, 1965). The problem inherent in any conclusion based on these data is that it may not be appropriate to use similar materials and procedures with, for example, a monkey, a predominantly visual-tactile animal and a rat, a predominantly olfactory animal. Consistent with this hypothesis, when the task is modified to allow the rat to use its well-developed olfactory abilities, good evidence for the development of learning set has been reported (Slotnick & Katz, 1974). Thus, it is clear that the methodology used in assessing the capacities of an animal can determine whether one will find evidence of that behavior.

What should become clear from these examples is that parsimony may not be as helpful a rule of thumb as it seems. The complexity of behavior found may depend more on how clever the experimenter is in designing an experiment that will demonstrate the abilities of the organism being studied than it does on the inherent capacities of that animal. Furthermore, selecting the most parsimonious explanation may depend on one's frame of reference. In the limited context of a particular experiment, if two theories make similar predictions, the simpler one may be preferred. However, in the context of a series of experiments that suggest continuity of a more complex capacity across species, the more complex cognitive theory may, in fact, be more parsimonious.

Serendipity. It can be argued that a cognitive approach to research with animals has merit even if one finds no evidence for a particular underlying cognitive capacity. According to this view, the process of designing experiments to distinguish between cognitive and noncognitive theories (or among cognitive theories) often leads to unconventional experimental designs. And these designs often produce results that would not otherwise have been found. Unfortunately, reports of serendipity do not often appear in the literature. This omission is unfortunate because it gives the impression that our findings are always exactly what we set out to find, or at least, that our findings are consistent with one of the hypotheses stated in our introduction. For this reason, I would like to provide an example of research in which the original question asked did not get answered but the unexpected, yet interpretable, results that were obtained led to an unrelated line of research.

It has been proposed (Skinner, 1950) that when pigeons learn a conditional discrimination, they learn a stimulus-response chain (i.e., respond to one of the comparisons after having been stimulated by one of the samples; respond to the other comparison after having been stimulated by the other sample). In our matching-to-sample research, we asked, to what extent do pigeons learn not only about the correct comparison but also about the incorrect alternative? We asked this question by first training pigeons on either a matching-to-sample or oddity-from-sample task and then replacing either the incorrect or the correct comparison on a given trial with a stimulus that was familiar (in the context of a conditional discrimination) but that had never been seen before with that sample.

It was our expectation that for both tasks, replacing the correct comparison would disrupt performance greatly, whereas replacing the incorrect comparison would be minimally disruptive. Although the expected results were found for the matching task, the opposite results were observed for the oddity task (i.e., performance was greatly disrupted when the incorrect comparison was replaced but was minimally disrupted when the correct comparison was replaced; Zentall et al., 1981). Another way of describing the data is to note that for both tasks, performance was greatly disrupted whenever the comparison that matched the sample was replaced, but was minimally disrupted when the odd comparison was replaced.

The results suggested that in both tasks, the basis for performance was the matching comparison rather than the correct comparison. This conclusion was further supported by the finding that when neither of the comparisons matched the sample, as Skinner (1950) predicted, replacement of the correct comparison always produced considerable disruption of performance, whereas replacement of the incorrect comparison did not.

Thus, the question originally asked was to what extent comparison choice behavior in matching and oddity tasks was controlled by the incorrect comparison stimulus. The unexpected answer was, it depends on the task. Taken as a whole, the results indicate that performance of either task remains high when the nonmatching stimulus is removed, but performance of both tasks suffers when the matching stimulus is removed. These results suggest that the identity relation between the sample and the matching comparison stimulus serves as the basis for learning both tasks. (Our common coding research presented later in this volume, which resulted directly from questions of prospective memory in pigeons, provides another example of serendipity in animal cognition research.)

Of even more importance than the many cases of serendipitous findings that have resulted from the study of animal cognition is the contention that these results would not have been discovered had the original research not been conducted. Furthermore, the original research probably would not have been conducted if questions of cognition in animals had not been raised. Thus, even when no evidence for a uniquely cognitive process can be found, the research conducted with experimental designs aimed at testing for the presence of such effects can have heuristic value (i.e., they can result in unanticipated new findings). Of course, it is not necessary to hypothesize about the presence of cognitive processes in animals for one to generate (or to recognize the implications of) an unexpected outcome, but it does appear to help.

RATIONALE FOR THE EVOLUTION
OF COGNITION IN ANIMALS

With the advocacy of a shift toward more "relevant" research in psychology (e.g., research directed toward curing the social and emotional problems of our society) that became popular in the 1960s came the concern that research with animals on basic learning processes was not only irrelevant to an understanding of human behavior, but was also irrelevant to an understanding of the natural behavior of animals. What could we learn about animals by removing them from their natural environment and, for example, placing them in an artificial cubicle and requiring them to press an artificial lever in the presence of one artificial stimulus but not in the presence of another?

Some years ago Robert Boice (1973) published an article in which he described the rather limited and stereotypic "natural" behavior of feral rats. Boice's answer to the question, why study unnatural behavior in unnatural environments, was enlightening. He argued that if our ultimate interest was in understanding human behavior, then the study of animals under "artificial" conditions might be an excellent model for a better understanding of how humans have managed to adapt to having been placed in a similarly artificial environment (i.e., modern society). Apparently, humans, as well as other animals, have remarkable adaptive capacity that is typically not seen under natural conditions.

Once we accept the fact that both we and our experimental animals are being exposed to complex environments that are quite different from those in which we evolved, we are faced with the question of how human genes have allowed us to adapt to artificial environments that demand high-level cognitive capacities (e.g., reading, calculating, operating machinery).

In the course of the typical human hunter-gatherer's day I would guess that a minimum amount of cognitive capacity would be required. In fact, I would guess that the lifetime of the typical hunter-gatherer may pass with little need for any the higher level cognitive functioning of which humans are capable. So, how did it get there? I suggest that the capacity for cognitive functioning must not only evolve, but also must remain relatively stable in the absence of much selective pressure for its maintenance because in its absence, occasional (e.g., perhaps once every hundred generations) precipitous changes in the environment might result in elimination of the gene pool. It must be the case that all animals have capacities, that they, as individuals, may never be required to demonstrate, that allows them to survive the rarely occurring environmental change that would otherwise lead to their extinction.

It is interesting to speculate as to how such genes, which are responsible for rarely occurring but life-maintaining (or gene-pool-maintaining) behavior, have survived in the absence of the immediate selective pressures of daily survival. Perhaps they have become linked to genes that are strongly selected for survival on a day-to-day basis. In any case, if one accepts the notion of adaptive capacities that are rarely used but are critical to survival, we should not be surprised to find cognitive capacities in an animal that far exceed those that are observed in the animal's "natural" daily existence.

There are some researchers who argue that if one is interested in examining the properties of a particular capacity in animals (e.g., long-term memory), one should study species that show evidence for that capacity in their natural environment (see, e.g., arguments for studying food caching and retrieval behavior in nutcrackers, Kamil, 1989; and in marsh tits, Shettleworth, 1985). But according to the view presented here, one might expect to see evidence for high levels of memory performance in species that do not typically show it in nature (see, e.g., long-term memory for spatial choice in pigeons; Zentall, Steirn, & Jackson-Smith, 1990).

It may be that animals come equipped with (or develop) hierarchies of behavioral "strategies" that are ordered from least to most complex (or perhaps least to most time or energy consuming). Under most natural or even laboratory conditions the simplest of these strategies (e.g., reflexes and simple conditioning effects) are sufficient to ensure the animals' survival. Furthermore, it is likely that only when these simpler strategies are ineffective that evidence for higher level capacities will be found. Thus, it may be that the only way to bring out these latent cognitive strategies in animals is by placing them in unnatural environments in which use of these strategies results in higher levels of, or more efficient, performance. Therefore, it may be necessary to expose an animal to artifical procedures both to rule out explanations of performance in terms of simple learning principles, as well as to induce the animal to use hypothesized cognitive capacities.

ANIMAL COGNITION: WHERE DO WE STAND?

The roots of animal cognition go back at least as far as Romanes (1889), but because his conclusions were based largely on anecdote and analogical inference, rather than on objective observation and controlled experiment, his impact on psychology was relatively minor (see Wasserman, 1984).

The notion that animals might have complex internal representations of relations among external events was resurrected by Tolman (1932) whose views, were not widely accepted in spite of the fact that he was careful to base his theory on the data from carefully controlled experiments. But it was not until more recently that animal cognition research was accepted as one of the currents in the mainstream of research in animal psychology. In retrospect, one might identify Mackintosh's (1965) article on attention in animals or the more detailed book that followed (Sutherland & Mackintosh, 1971; see also Gilbert & Sutherland, 1969; Honig & James, 1971; Jarrard, 1971; Riley, 1968) as the beginning of the shift toward acceptance of animal cognition as a legitimate area of research. Clearly, however, it was Hulse, Fowler, and Honig's (1978) book and the even more ambitious Roitblat, Bever, and Terrace (1984) volume that signaled the acceptability of cognitive research with animals.

ORGANIZATION OF THE BOOK

The chapters that follow are intended to identify some of the recent trends in animal cognition research and bring us up to date in other areas of this rapidly expanding field. They also, not coincidentally, reflect some of the areas of research with which Al Riley has been associated. Al's eclectic (or functionalist) approach can be seen in the wide variety of problems he has studied in diverse organisms including humans (both adults and children), monkeys, rats, pigeons, Japanese quail, chickens, and octopuses.

The chapters presented here are not meant to be a comprehensive presentation of the breadth or depth of either the field of animal cognition research or of the scope of Al Riley's research, but I hope that they will serve to stimulate others to answer some of the many questions that remain. The remainder of the book has been somewhat arbitrarily divided into three sections: stimulus representation, memory processes, and perceptual processes. Many of the chapters deal with more than one of these processes and some deal with all three. The difficulty that I had in assigning the chapters to sections is an indication that the study of stimulus representation, memory, and perception may all be interrelated components of underlying cognitive processes.

Stimulus Representation. Attempts to identify the nature of the stimulus representation in humans and other animals is a theme that runs through much of Al's published research. His analysis of the effective stimulus in the transposition phenomenon (Riley, 1958), the role of generalization gradients in the easy-to-hard effect (Singer, Zentall, & Riley, 1969), and the unlearned characteristics of decremental generalization gradients (Riley & Leuin, 1971) represent some of his work. But perhaps the most complete presentation of Al's interest in stimulus representation can be seen in his book, *Discrimination Learning* (Riley, 1968).

The stimulus representation section of this book thus focuses on the complex ways in which stimuli can be represented, depending on their various relations to other stimuli (chapters 2 & 3), and the context (chapter 4) or conditions of reinforcement in which the stimuli are represented (chapter 5). In fact, the notion of context specificity is beginning to replace some traditional notions of interference, counterconditioning, and retention deficits in animals (chapter 6).

Memory Processes.   Al's involvement in memory research began with his interest in human memory (see, e.g., Riley, 1952, 1953, 1954, 1957). But his chapter investigating memory for form in humans (Riley, 1962) demonstrates the breadth of that interest. More recently, Al applied those interests to the study of prospective and retrospective memory strategies in rats (Cook, Brown, & Riley, 1985).

The memory section of this book starts with a chapter that deals with some persistent problems in how (and where) memories are stored and transferred in the brain (chapter 7). Parallels are drawn in the chapter between the increased persistence of memories learned in different external versus internal contexts. The role that natural foraging strategies play in animal memory is examined in both monkeys (chapter 8) and rats (chapter 9). In both chapters, one can see that animals will sometimes engage in patterns of behavior that are less cognitive or less efficient than recent literature would lead us to believe. Data presented in these chapters reinforce the view presented earlier that the use of cognitive capacity may be hierarchical and more complex strategies may be used only when simpler learning processes either fail or require considerably more effort.

The three chapters that conclude the memory section are all concerned with identifying the nature of the memory code. In the first two of these chapters (chapters 10 & 11), conditions are explored under which stimuli are coded in terms of prospective response intentions, rather than retrospective representations of initial stimuli. In chapter 12, evidence is presented that distinctive stimuli that are associated with the same outcome (or with a response to the same test stimulus) can be represented by an animal in terms of a common memory code.

Perceptual Processes.   Al Riley's early interest in perceptual processes can be seen in his chapter on memory for form in humans (Riley, 1962) as well as his work on echolocation in rats (Riley & Rosenzweig, 1957; Rosenzweig & Riley, 1955). But the area in which Al has had his greatest impact is clearly that of attentional mechanisms in pigeons (see e.g., Riley & Leith, 1976; Riley &

Roitblat, 1976). In these and related articles, attentional processes in animals are explored using principles and methods derived from research on attention in humans.

If there is a higher order thread that runs through Al Riley's approach to research, I believe it is the notion that evolutionary convergence (i.e., the principle that similar demands, such as finding food, a mate, and safety, placed on different species result in the evolution of similar behavioral mechanisms) can serve as an appropriate reference point for the examination of behavior across species (see Riley et al., 1986).

The perceptual processes section of this book deals with the examination of a number of perceptual processes. In chapter 13, it is argued that paradigms that have been used to help understand the brain mechanisms involved in memory processes may also be helpful in analyzing those involved in attentional processes.

The next three chapters of this section are concerned with models of selective attention. In the first two of these chapters (chapters 14 & 15), new methods for examining selective attention are examined. These methods, derived from research with humans, bring us closer to the goal of being able to make direct comparisons between humans and other animals. In chapter 16, connectionist models are applied to attentional learning data with the goal of developing a general model that not only can account for data from a variety of discrimination learning tasks but also can be compared with similar models that have been proposed to account for human learning processes.

The last two chapters of the third section, which also are closely related to perceptual research with humans, deal with other types of perceptual processes. Chapter 17 is concerned with two questions about pigeons' perceptual processing. First, what are the critical units in the perception of an object (e.g., a two-dimensional representation of a three-dimensional cube), edges or corners? The second question has to do with the pigeon's ability to anticipate the destination of an object moving through space. The final chapter in this section (chapter 18) is concerned with the absolute and relational properties of tone patterns in acoustical perception by songbirds. Here one can see novel approaches to answering questions that traditionally have been addressed using the visual modality.

It is my hope that this volume will not only serve as a *Festschrift* for Donald A. Riley, but will also provide the reader with a a an overview of the current status of the field of animal cognition.

ACKNOWLEDGMENTS

Preparation of this chapter was supported by National Institute of Mental Health Grant MH 45979 and National Science Foundation Grant BNS 9019080. I thank Janice Steirn and Lou Sherburne for their helpful comments on an earlier version of this chapter.

REFERENCES

Boice, R. (1973). Domestication. *Psychological Bulletin, 80*, 215–230.

Cook, R. G., Brown, M. F., & Riley, D. A. (1985). Flexible memory processing by rats: Use of prospective and retrospective information in the radial maze. *Journal of Experimental Psychology: Animal Behavior Processes, 11*, 453–469.

D'Amato, M. R., Salmon, D. P., Loukas, E., & Tomie, A. (1985). Symmetry and transitivity of conditional relations in monkeys (*Cebus apella*) and pigeons (*Columba livia*). *Journal of the Experimental Analysis of Behavior, 44*, 35–48.

Gilbert, R. M., & Sutherland, N. S. (Eds.). (1969). *Animal discrimination learning.* New York: Academic.

Harlow, H. F. (1949). The formation of learning sets. *Psychological Review, 56*, 51–65.

Honig, W. K., & James, P. H. R. (Eds.). (1971). *Animal memory.* New York: Academic.

Hulse, S. H., Fowler, H., & Honig, W. K. (Eds.). (1978). *Cognitive processes in animal behavior.* Hillsdale, NJ: Lawrence Erlbaum Associates.

Jarrard, L. E. (Ed.). (1971). *Cognitive processes of nonhuman primates.* New York: Academic.

Kamil, A. C. (1989). Studies of learning and memory in natural contexts: Integrating functional and mechanistic approaches to behavior. In R. J. Blanchard, P. Brain, D. C. Blanchard, & S. Parmigiani (Eds.), *Ethoexperimental approaches to the study of behavior* (pp. 30–50). Dordrecht, Germany: Kluwer.

Lombardi, C. M., Fachinelli, C. C., & Delius, J. D. (1984). Oddity of visual patterns conceptualized by pigeons. *Animal Learning and Behavior, 12*, 2–6.

Mackintosh, N. J. (1965). Selective attention in animal discrimination learning. *Psychological Bulletin, 64*, 124–150.

Remy, M., & Emmerton, J. (1991). Directional dependence of intraocular transfer of stimulus detection in pigeons. *Behavioral Neuroscience, 105*, 647–652.

Riley, D. A. (1952). Rote learning as a function of distribution of practice and the complexity of the situation. *Journal of Experimental Psychology, 43*, 88–95.

Riley, D. A. (1953). Reminiscence effects in paired-associate learning. *Journal of Experimental Psychology, 45*, 232–238.

Riley, D. A. (1954). Further studies of reminiscence effects with variations in stimulus-response relationships. *Journal of Experimental Psychology, 48*, 101–105.

Riley, D. A. (1957). The influence of the amount of pretest learning on reminiscence effects in paired-associate learning. *Journal of Experimental Psychology, 54*, 8–14.

Riley, D. A. (1958). The nature of the effective stimulus in animal discrimination learning: Transposition reconsidered. *Psychological Review, 65*, 1–7.

Riley, D. A. (1962). Memory for form. In L. Postman (Ed.), *Psychology in the making* (pp. 402–465). New York: Knopf.

Riley, D. A. (1968). *Discrimination learning.* Boston: Allyn & Bacon.

Riley, D. A., Brown, M. F., & Yoerg, S. I. (1986). Understanding animal cognition. In T. J. Knapp & L. C. Robertson (Eds.), *Approaches to cognition* (pp. 111–136). Hillsdale, NJ: Lawrence Erlbaum Associates.

Riley, D. A., & Leith, C. R. (1976). Multidimensional psychophysics and selective attention in animals. *Psychological Bulletin, 83*, 138–160.

Riley, D. A., & Leuin, T. C. (1971). Stimulus generalization gradients in chickens reared in monochromatic light and tested with a single wavelength value. *Journal of Comparative and Physiological Psychology, 75*, 389–402.

Riley, D. A., & Roitblat, H. L. (1976). Selective attention and related cognitive processes. In S. Hulse, W. K. Honig, & H. Fowler (Eds.), *Cognitive processes in animal behavior* (pp. 249–276). Hillsdale, NJ: Lawrence Erlbaum Associates.

Riley, D. A., & Rosenzweig, M. R. (1957). Echolocation in rats. *Journal of Comparative and Physiological Psychology, 50*, 323- 328.

Roitblat, H. L., Bever, T., & Terrace, H. S. (Eds.). (1984). *Animal cognition*. Hillsdale, NJ: Lawrence Erlbaum Associates.

Romanes, G. J. (1884). *Animal intelligence*. New York: Appleton.

Rosenzweig, M. R., & Riley, D. A. (1955). Evidence for echolocation in the rat. *Science, 121*, 600.

Shettleworth, S. J. (1985). Food storing by birds: Implications for comparative studies of memory. In M. Weinberger, J. McGaugh, & G. Lynch (Eds.), *Memory systems of the brain* (pp. 231–250). New York: Guilford.

Singer, B., Zentall T., & Riley, D. A. (1969). Stimulus generalization and the easy-to-hard effect. *Journal of Comparative and Physiological Psychology, 69*, 528–535.

Skinner, B. F. (1950). Are theories of learning necessary? *Psychological Review, 57*, 193–216.

Slotnick, B. M., & Katz, H. M. (1974). Olfactory learning-set formation in rats. *Science, 185*, 796–798.

Sutherland, N. S., & Mackintosh, N. J. (1971). *Mechanisms of animal discrimination learning*. New York: Academic.

Tolman, E. C. (1932). *Purposive behavior in animals and men*. New York: Century.

Warren, J. M. (1965). Primate learning in comparative perspective. In A. M. Schrier, H. F. Harlow, & F. Stollnitz (Eds.), *Behavior of nonhuman primates* (Vol. 1, pp. 249–281). New York: Academic.

Wasserman, E. A. (1984). Animal intelligence: Understanding the minds of animals through their behavioral "ambassadors." In H. L. Roitblat, T. G. Bever, & H. S. Terrace (Eds.), *Animal cognition* (pp. 45–60). Hillsdale NJ: Lawrence Erlbaum Associates.

Zentall, T. R., Edwards, C. A., Moore, B. S., & Hogan, D. E. (1981). Identity: The basis for both matching and oddity learning in pigeons. *Journal of Experimental Psychology: Animal Behavior Processes, 7*, 70–86.

Zentall, T. R., & Hogan, D. E. (1974). Abstract concept learning in the pigeon. *Journal of Experimental Psychology, 102*, 393–398.

Zentall, T. R., & Hogan, D. E. (1975). Concept learning in the pigeon: Transfer of matching and nonmatching to new stimuli. *American Journal of Psychology, 88*, 233–244.

Zentall, T. R., & Hogan, D. E. (1976). Pigeons can learn identity, difference, or both. *Science, 191*, 408–409.

Zentall, T. R., & Hogan, D. E. (1978). Same/different concept learning in the pigeon: The effect of negative instances and prior adaptation to the transfer stimuli. *Journal of the Experimental Analysis of Behavior, 30*, 177–186.

Zentall, T. R., Steirn, J. N., & Jackson-Smith, P. (1990). Memory strategies in pigeons' performance of a radial-arm-maze analog task. *Journal of Experimental Psychology: Animal Behavior Processes, 16*, 358–371.

II STIMULUS REPRESENTATION

2 The Stimulus Revisited: My, How You've Grown!

W. K. Honig
Dalhousie University

A traditional and productive strategy in many sciences involves the discovery and analysis of the elements that provide the empirical and conceptual basis for the science, and contribute to the explanation of the findings. We have all stared at the periodic table of elements that adorns every chemistry classroom. These are the "building blocks" of that science; a finite number of elements combine in various ways to produce a large number of compounds. In biology, the cell theory has been at the center of the conceptual stage for many years. The cells constitute the tissues, which in turn comprise organs and other parts of organisms, and these in turn comprise the organism itself. At other levels of biology there are other building blocks that are central to particular areas; I am thinking particularly of genetics, with its genes and chromosomes. At the opposite end of the scale in size and extent, astronomy is in a sense also built upon a set of elements, namely the stars, together with the few planets and moons that can be distinguished near the earth. These comprise planetary systems, clusters of stars, and galaxies, but none of these collections deprives the individual elements of their identity, although their interactions determine the course of their travel in space.

The initial field of experimental psychology was perception—*Sinnesphysiologie,* or the physiology of the senses. In looking to other sciences as models in the early years of our science, psychologists searched for a set of basic perceptual elements that would comprise the body of data. The research was carried out by the structuralists, who undertook an experimental analysis of those perceptual elements. The Gestalt psychologists claimed that the description of experience in terms of disparate elements was unsatisfactory, and they offered a different approach, which is often summed up in the hackneyed phrase that the whole array is more than, or at least different from, the sum of its parts. However, these parts still provided the building blocks for the science of perception, even if collections or arrays of elements possessed attributes that could not be ascribed to the elements themselves.

Research on stimulus control in animals has been carried out in a behavioristic tradition. Most of the work in this area has in recent years been based on operant principles and methods. Simple, uniform stimuli were brought into the laboratory. The classic example is the illuminated key or disk used for the study of stimulus control by pigeons (Guttman & Kalish, 1956). Uniform colors, shades of gray, and patterns such as line orientations have been used for the experimental analysis of stimulus control over the last 35 years (see Honig & Urcuioli, 1981, for a review). The pigeon, which tends to peck at small, distinguishable stimuli, was a convenient subject, and stimulus control could be studied even with extended stimuli like pure tones, while the pigeon pecked at the small target. Many aspects of stimulus control have been studied in detail with this general method. For example, reviews of attentional factors (Mackintosh, 1977) and of inhibitory processes (Rilling, 1977) each take up a chapter in the *Handbook of Operant Behavior* (Honig & Staddon, 1977). Simple stimuli that differ on a clearly defined dimension provide the basis for the study of empirical and conceptual questions in the area of stimulus control.

Although punctate stimuli have these advantages, they do not represent the real world in which the pigeon or any other animal has to determine its location, find its way about, and assess various aspects of its environment. Experimentally, and perhaps inadvertently, we have concentrated on the pigeon's foveas, which are at the back of the pigeon's retinas and are important for finding food on the ground and ingesting small particles in front of its head. To find food in the first place, the pigeon requires a wider view of the world. This view is wide indeed, because the eyes are set laterally rather than frontally, and binocular vision is pretty well limited to the foveas. In this chapter, I want to discuss some aspects of this wider view, and in my own wider view, I want to relate that to our conceptualization of the stimulus, and the process of stimulus control.

CONCEPT ATTAINMENT

In 1964, Herrnstein and Loveland published a classic article that initiated the study of visual concept attainment. Pigeons could readily learn to discriminate between pictures in which all or part of a person was visible and pictures that did not exemplify this concept. This experiment fostered many other interesting studies, and the range of categories was extensive; for example, one study had pigeons discriminate the presence of "man-made" objects in the picture (Lubow, 1974; see also Edwards & Honig, 1987, for reviews of other work), and we even showed that the pigeon could attain a concept of "pigeon" (Siegel & Honig, 1970). There have been many other convincing demonstrations of concept attainment, but an understanding of the discriminative process is harder to come by. Does the pigeon learn to "focus" upon the distinctive features of the categorical instances, or does it survey the entire scene? If the former, then the discrimination should be facilitated by training pigeons with matched pairs of positive and negative slides, that differ only, for example, in the presence of a person in the picture. According to some versions of learning theory, the irrelevant backgrounds, which would be very similar, should be "tuned out."

Edwards and I carried out an experiment that addressed this issue (Edwards & Honig, 1987). Matched pairs of positive and negative instances were presented to three groups of pigeons in training. The background in each pair was the same, but one or more people appeared in one member of each pair, whereas there was no person in the other. The slides were shown one at a time, and in a randomized order. For one group of birds, the person-present slides were positive; for a second group, the person-absent slides were positive. A third group of pigeons was trained with a *memorization control* procedure. The same slides were shown, but for half of the pairs, the presence of the person was positive, whereas for the other half, the person-absent slides were positive. Three other groups of pigeons were trained in a similar way with the same person-present slides, but with a different set of person-absent slides that did not match the backgrounds.

The rates of acquisition of these discriminations are compared in Fig. 2.1. These data provide an interesting, and, for humans, an unexpected outcome. The pigeons trained with the nonmatching backgrounds acquired the discrimination reasonably quickly, and to a high level. The corresponding groups with the matching backgrounds acquired the discrimination much less well. In fact, the memorization group trained with nonmatched pictures performed at a higher level than did the true-discrimination birds with matched slides. We asked people informally to identify the concept to be discriminated just from looking at sets of

FIG. 2.1. Acquisition of conceptual discriminations in pigeons. The distinguishing feature was the presence of one or more persons in pictures of natural settings. For independent groups of pigeons, this feature identified the positive category, or the negative category, or it was irrelevant. For one set of groups, the backgrounds of the feature-present and feature-absent slides were matched (solid lines); for another set of groups, the backgrounds were not matched (dashed lines).

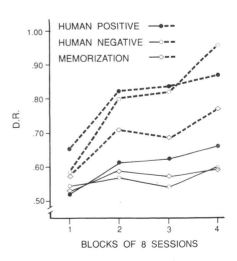

training slides, without any further information. They readily abstracted the feature when the backgrounds were similar, but not when the backgrounds were different. With pigeons, the opposite result was obtained. The pigeon learns such a successive discrimination by reducing responding to the negative instances, and this process is much slower when the irrelevant features are common to the training stimuli.

This finding implies that in the course of learning, the pigeons attended to the entire view, or array, of the presented slides. The irrelevant parts of the arrays gained control over responding. So the functional stimulus for the pigeon was extended and in this case also maladaptive, because it wasted a lot of effort pecking at the negative slides that shared many features with the positive slides.

DISCRIMINATION OF SPATIAL LOCATIONS

If the pigeon incorporates information about entire arrays in the process of discriminating among them, it also should be able to learn discriminations in which no single feature defines the concept. This can be accomplished with a discrimination between spatial locations. Several experiments on this topic have been carried out. Karen Stewart and I (Honig & Stewart, 1988) taught pigeons to discriminate between pictures of two different locations on the Dalhousie campus. One was an older part of the campus, a quadrangle flanked by handsome stone buildings rather typical of university architecture from the first half of this century. The other was a grassy area flanked by several units of the modern Life Sciences Building. These buildings are concrete, modern, and arid. Pictures were taken from several standpoints in each location, and the sum of the individual views would provide, if taken together, a panoramic view at each standpoint.

One group of four birds was trained on a true discrimination between these two areas. For two subjects, pictures of the Life Sciences area were positive, whereas for the other two, the quadrangle was the positive location. All of the subjects readily acquired the discrimination when they were trained with pictures taken at four different standpoints at each location, and they transferred the discrimination in a test session to a fifth standpoint not used in training. Four birds were trained on a pseudo-discrimination procedure, in which half of the pictures from each location were arbitrarily, but consistently positive, and the remaining views were negative. Two pigeons acquired this pseudodiscrimination only to a minimum extent. The other two did meet a criterion of 90% correct responses, but it took them much longer than the true-discrimination subjects. Similar research was done by Wilkie and his associates at the University of British Columbia (UBC; Wilkie, Willson, & Kardal, 1989). Pigeons successfully discriminated pictures of the general location of the UBC campus from other parts of Canada. Kendrick (1992) carried out similar discriminations between pictures of outdoor scenes visible from the birds' colony room at Middle Tennessee State University, and pictures of other parts of the university campus.

These discriminations can be acquired only if the pigeons attend to substantial portions of the visual arrays, and associate them with presence and absence of reward. Clearly, there was an overlap of features among the instances within each class, but there were surely some features in common even between classes, such as green grass and sky. The pigeons had to distinguish between the many elements that differed between the two settings, and yet classify the common elements as belonging to the same class. The process that enables such a discrimination to be acquired is not evident, at least to me, but it is clear that much of the information in the extended arrays has to be integrated from the different views of each location for the process to be effective.

Pictured Locations and Real Locations

The wide views of the world presented in these pictures were presented on a rather small screen, which would occupy a limited portion of the pigeon's visual fields. We do not know whether these spatial scenes represented the "real world" for the pigeon, or whether they were little more than visual concepts. If they do represent the real world, then the pigeon ought to transfer the discrimination from the pictures to the locations that they represent. Gene Ouellette and I (Ouellette, 1989) carried out a study to test this hypothesis. Pigeons were trained to discriminate between slides of two ends of a rectangular room. These contained some distinctive items and decorations, but there were many general features common to both ends of the room. After the pigeons learned the discrimination, they were trained to find food in one of two feeders. The feeders were placed in symmetrical positions in the opposite ends of the room. Four birds were run in a positive transfer condition. The baited feeder was in the end that was positive in the slide discrimination. Four birds in a negative transfer condition were trained to approach the end of the room that was negative in the slide discrimination.

These training conditions produced a marked difference in performance: The pigeons in the positive transfer group made very few entries into the negative side of the room and took much less time to reach the baited feeder, even from the first trial. The choice data are shown in Fig. 2.2. At the beginning of training, the negative transfer group actually favored the side of the room that was positive during slide training. We concluded that the discrimination of locations in pictures is not only a viable conceptual task, but that the information can be applied to real-world, three-dimensional settings. The training with the limited views provided by two-dimensional pictures transferred to a much wider set of views in the three-dimensional situation.

CONGRUENT TRANSFER

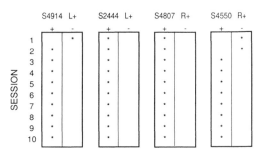

INCONGRUENT TRANSFER

FIG. 2.2. Acquisition by pigeons of a discrimination of the location of the baited feeder in a rectangular room. Four pigeons found food in the "negative" end of the room, as previously identified in pictures. Four pigeons found food in the "positive" end of the room. Errors and correct choices are indicated for the sequence of daily trials.

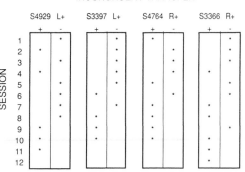

DIMENSIONAL ASPECTS OF STIMULUS ARRAYS

Discriminations of concepts and spatial locations support the notion that complex patterns and arrays, rather than individual elements, are the functional aspects of the environment that control behavior. However, these patterns do not lend themselves to the sort of dimensional analysis that can be carried out with uniform stimuli. To support the case that stimulus control is exercised by the several features that comprise the arrays, we require a method that provides a more systematic and quantitative analysis. Natural features of the environment contain elements in different proportions, such as the numbers of coniferous and deciduous trees in a forest, or the density of conspecific birds in different areas of a city square. For an animal to determine its location, or to select a location for foraging, it would be important to identify the relevant instances of these features, and to assess their numbers and spatial distributions. Even when the individual instances in an array are complex, they have to be integrated into a broader perceptual scheme.

Can we imitate nature and represent such aspects of the environment for study in the laboratory? We have made some initial attempts. In one experiment, pigeons discriminated between the numbers of persons in pictures of natural scenes. Slides of various scenes containing one person were discriminated from slides that showed many persons, when "many" was defined as seven or more. Two sets of eight slides each, all of them different, provided the training stimuli. The pigeons readily acquired this discrimination. We then tested them with other numbers of persons, ranging from two to seven. There were four different instances in each category. The pigeons provided orderly discrimination gradients (Honig, 1993) between the training values; the data are shown in Fig. 2.3. The birds were therefore sensitive to a dimension that could be identified only through the presentation of several stimulus arrays; different values on the dimension of numerosity are not identified by any particular instance.

However, absolute numerosity discriminations involve several problems. One is the inevitable confound between the number of items and other aspects of the array. The total extent of the array increases with the number of items if the average distance between items is constant. If that is compensated by varying this distance, a different confound is generated.

We therefore have concentrated on **relative** numerosity discriminations, which may in any case be more relevant to the real world. We present square arrays of small elements in different proportions. Pigeons are trained to discrimi-

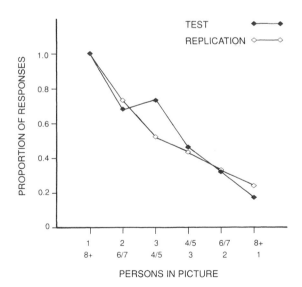

FIG. 2.3.   Gradients obtained in testing from pigeons that were trained to discriminate between one and "many" persons in pictures of natural scenes. The tests involved several new categories of numerosity: two persons, three persons, four or five persons, and six or seven persons.

nate between uniform arrays in which the elements differ in color, form, or size (Honig & Stewart, 1989). The pigeons are then tested with different proportions of elements on the relevant dimension. This method has yielded very orderly gradients on the dimension of relative numerosity. In line with the emphasis in this chapter on naturalistic stimuli, I have selected the discriminations that involve conceptual categories.

Pigeons readily learned to peck at arrays that contain little colored pictures of different kinds of birds or flowers. These were projected onto a response screen, as described earlier, and the general training procedures were also the same. Sixteen varieties of birds and flowers were represented by little stickers sold in school supply stores. A sample matrix is shown in Fig. 2.4. This array is comprised of eight different birds and eight different flowers.

The pigeons discriminated readily between the "uniform" arrays with 16 birds or 16 flowers; they reached a high criterion of discrimination as readily as they did with simple elements, such as blue and red dots, or large and small circles. In several tests, we presented the elements in different proportions, namely 4 birds and 12 flowers, 8 instances of each kind, and 12 birds and 4 flowers. Different subsets of the elements were presented on different trials. This test procedure provided orderly gradients of stimulus control, as shown in Fig. 2.5. Clearly, the pigeons were controlled by the proportions of each kind of element as they

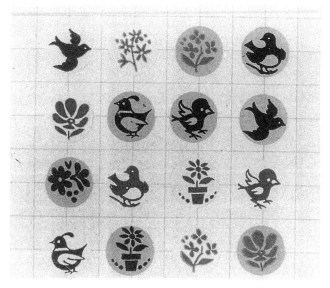

FIG. 2.4. A sample array of elements used for testing in a relative numerosity discrimination. This array, as displayed to the subject, contains colored sketches of eight different birds and eight different flowers.

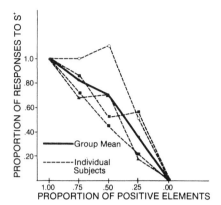

FIG. 2.5. Gradients obtained from individual pigeons when tested with different proportions of birds and flowers in the kinds of arrays shown in Fig. 2.4.

viewed the test patterns, and to do that, they would have to recognize the members of each category as similar.

It is possible that the pigeons memorized the individual instances of each category during training. Although this would not diminish the achievement of judgment the relative numbers in the test patterns, it would reduce the relevance of this research to arrays in real-world settings—for example, the discrimination of the relative numbers of different kinds of birds in a particular location. In order to test the possibility of memorization, we replicated this study, again with different classes of elements, but testing for transfer between specific instances of these classes.

We prepared a set of training slides and a set of test slides containing different members of the same categories. Eighteen species of flowers were available on the stickers, but not as many different birds, so we switched to unicorns, of which we could find 18 varieties. (Presumably this exhausts all known varieties of unicorns.) The pigeons were trained to discriminate between arrays of nine elements of each kind. When they learned this, they were tested with other proportions of these training elements in different arrays, and with transfer elements shown in the test slides.

The results from this test show that the birds again discriminated among the different proportions of elements. Furthermore, they transferred the discrimination very well from the training instances of flowers and unicorns to the test arrays. The data are shown in Fig. 2.6. Thus, the pigeon could assess new instances in a pattern of elements to provide a discrimination of their relative numerosities. The process of the classification and integration of the elements into distinct discriminative categories is not understood. Clearly, complex elements could not readily have been "recoded" in terms of some other dimension, such as the total areas of two kinds of items.

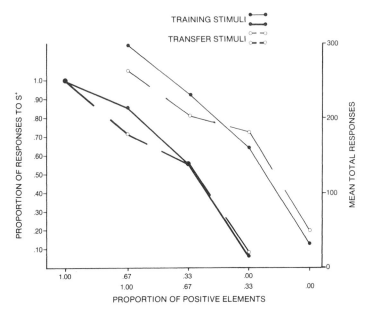

FIG. 2.6. Gradients obtained following discriminations between arrays of nine sketches of flowers and nine sketches of unicorns. Different sets of sketches were used for training and for transfer tests.

DISCRIMINATION AND THE PEAK SHIFT

The peak shift is a robust phenomenon in the area of stimulus control. It was first studied by Hanson (1959) with stimuli that differed in spectral value, and it has been replicated on other dimensions as well (see Honig & Urcuioli, 1981; Purtle, 1973; Rilling, 1977, for reviews). In fact, the peak shift supports the dimensional nature of any ordered set of stimuli from which it can be obtained. We have obtained the peak shift on the dimension of relative numerosity (Honig & Stewart, in press). The following procedures provided the most convincing results.

Pigeons were trained to respond to arrays of 18 blue and 18 red dots displayed in a square array on a response screen. The elements were randomly located in several different patterns. The birds were then trained in a maintained-discrimination procedure, with these patterns as the positive arrays. For two birds, the negative patterns were all of the arrays that contained more red than blue dots. These appeared in the following proportions: 21 red dots and 15 blue dots, 25 red and 11 blue, 30 red and 6 blue, and 36 red dots. Two other birds were trained in the same way, but with blue rather than red as the negative elements. Each of the equal (and positive) proportions of elements appeared twice in each block of training trials, whereas each of the other, negative proportions appeared once.

The pigeons acquired this discrimination within 16 training sessions. Orderly, declining gradients of responding reflected the proportions of positive and negative elements. The birds then were tested with the entire range of proportions of elements, including all of the patterns not used in training, in which there were more positive than negative elements (e.g., more blue than red dots in the example given previously).

This test yielded an orderly postdiscrimination gradient, with a peak shift of the kind obtained originally by Hanson (1959). On the average, responding to the test patterns, which contained more positive than negative elements, exceeded responding to the positive training arrays with equal numbers of elements. After a few further training sessions, the birds were tested again, this time with patterns of 64 elements that contained red and blue elements in the same proportions as those presented in the first test. These elements were somewhat smaller and more closely spaced than those used for training and the first test. Nonetheless, the gradient was very similar to that obtained with the patterns of 36 elements. The data from both tests are shown in Fig. 2.7.

We therefore obtained a result with complex arrays of stimuli that had previously been obtained only from simple, individual stimuli. The dimension of relative numerosity emerges from the complex patterns of elements. The findings suggest that for the pigeon, the entire array functioned in the same way as do

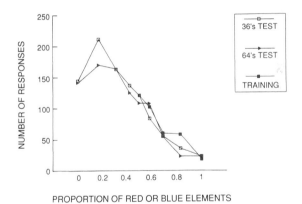

PROPORTION OF RED OR BLUE ELEMENTS

FIG. 2.7. Gradients obtained from different proportions of red and blue elements, following discrimination training in which arrays of equal numbers of red and blue elements were positive, whereas all proportions with more red than blue elements (or the reverse) were negative. A marked peak shift was obtained from test patterns containing more positive than negative elements. The "36's test" gradient was obtained from patterns of 36 elements. The "64's test" was obtained from transfer patterns containing 64 elements. The training data were taken from the last block of four training sessions. The values have been adjusted so that the value for the 50% mixture (S+) coincides with the data obtained from the same patterns during testing.

individual stimuli that differ on a given dimension. Although the different proportions established a stimulus dimension that is independent of the "carrier dimension" (in this case, a difference between the colors of the elements), it might be argued that arrays containing more red than blue elements are "redder" than arrays with the opposite proportions (and vice versa). Thus, the numerosity differences might be reduced to differences in color. The peak shift experiment was therefore replicated with arrays of little vertical and horizontal elements. Again, equal proportions comprised the positive arrays, whereas more vertical than horizontal elements (or the reverse) comprised the negative arrays. A reliable peak shift again was obtained with this procedure. It makes little sense to argue that vertical and horizontal can be applied to arrays of elements in the way that blue or red might be attributed to them. We conclude that the emergent dimension of relative numerosity controlled behavior in the same way as a dimension based on "simple" stimuli, even to the extent that we could observe a peak shift among the proportions of elements.

GENERAL DISCUSSION

The concept of the stimulus originated in the area of classical conditioning, in which brief, localized, and "neutral" cues preceded the reinforcer. The concept could sensibly be applied in instrumental learning to identify cues that indicate the proper location of the instrumental responses, the occasions or circumstances governing reinforcement, and the location of reinforcement. Such cues are generally neutral, so that they do not exercise any control over the subject at the start of training, and they are simple, so that they can be described precisely and varied on a given dimension for the study of generalization and discrimination. Following the landmark experiment by Guttman and Kalish (1956) on stimulus control with spectral values, this was a traditional approach, as I point out in the introduction to this chapter.

Although such "elemental" stimuli are useful for the study of generalization and discrimination, it was never assumed or intended that they would represent aspects of the real world. They provide convenient abstractions for the study for stimulus control. In most cases they represent different values on a perceptual dimension. However, animals are well adapted for coping with complex environments, and we could well expect that mechanisms for the discrimination of complex stimuli, and of aggregations of stimuli, would have been favored by evolution. Two aspects of the environment are particularly important. One is that many objects, including other animals, belong to natural classes or categories, and must be identified appropriately, in spite of differences in their appearance and location. The other is that the environment consists of locations that have to be recognized from different standpoints, and under various conditions of illumination, weather, and changes in local features. The identification and discrimination of the objects of interest, and of their environmental settings, are necessary for action, particularly for the direction of travel and the location of the action. But this could hardly be accomplished without a broader view—a spatial "window" on the world. Within that window, the elements of the view have to be incorporated into an integrated scene.

This perceptual task is facilitated by the identification of categorical features, such as trees or stones, that vary in their numbers, distances, and relative positions. However, this will be of little avail if irrelevant changes in the scene affect this process. The most common changes are probably produced by the animal's own movements. The changes in the scene provide the cues by which the animal (including the human animal) can assess changes in its position, but it still has to recognize that it remains in the same general location. Although this is no doubt a complex process, and not well understood, it must involve the integration of information from a large number of cues, including their relative positions. Furthermore, the animal has to discriminate between its own movements and the movements of features in the environment, particularly those of other animals.

The necessary cues for spatial location and spatial memory are often distal rather than local, and complex rather than simple. The data and the issues are reviewed elsewhere (Honig, 1987) but a couple of examples support this point. Suzuki, Augerinos, and Black (1980) trained rats to find food on all of the arms of an eight-arm radial maze. The arms could be identified from distinctive decorations hung from a curtain surrounding the maze. If a trial was interrupted, and the positions of these cues were changed, the rat's performance deteriorated, which indicates that the rats were using not only individual cues, but the relationships among them. Another example is provided by Morris' (1981) work on the so-called "water maze," which is not a maze but a tank containing opaque water—a "milk bath" for rats. The rat in the milk bath has to determine the location of a hidden platform on the basis of external cues through a process of triangulation. When such cues are present, the rat swims directly to the platform, even if it is placed into the pool at different locations. Without such cues, the rat tends to find the platform by swimming in a circle at an appropriate distance from the edge of the pool.

Such discriminations require a rapid perceptual analysis of complex landmarks. The locations of these landmarks would appear to change as the animal moves about, if it were to rely only on proximal (and generally retinal) cues. However, animals, including humans, do seem to have acquired, or evolved, a *location constancy.* They can integrate the information provided by their own movements with the apparent changes in the distal arrays. If such skills can be brought into the laboratory, we can study them. Data presented in this chapter represent an initial approach to this line of work.

A proper analysis of the necessary perceptual processes will require time and ingenuity. Our work on the transfer between pictured and real locations is only a beginning. We would like to know, for example, whether the subject can identify specific locations from a limited number of "local" cues, or whether this process is facilitated by a more extended scene. We know very little about the way that the elements are scanned so that the important features of the arrays are abstracted. The research described in the earlier part of this review shows that one such feature—the relative numerosity of different items—can be discriminated readily, and this is based on the relative, not the absolute numbers of the elements. Furthermore, the elements need not be simple; arrays of instances of perceptual categories control behavior in the same way as do the "simple" elements.

This review began with a study of concept attainment by Edwards and Honig (1987). The results indicated that the pigeons took a wide view of the slides, and associated many aspects of this view with reinforcement, so that, when these aspects were common to positive and negative instances, the discrimination was very difficult. For the experimenters, this was surprising, because humans have an analytic bias that sorts out the significant local features that define the discrimination. Pigeons seem to have an opposite bias—a tendency to include all of the features until they are found to be irrelevant. If this characterization is accurate, then the concept of the stimulus as an item that serves as focus for behavior would not be adequate for the analysis of at least some kinds of discriminative behaviors.

In the interest of tidy research, and in line with our original suppositions, the stimulus has usually been minified and isolated under experimental conditions. In more recent research, these conditions have changed to the extent that "stimulus" in its original sense is no longer an adequate term to describe the discriminanda. For the subject, the stimulus may never have been an isolated feature or event in the real world. Laboratory procedures and theoretical presuppositions imposed this isolation. If we abandon the stimulus in favor of the wider view, we will enhance our understanding of the processes under the heading of stimulus control. However, the emphasis will have shifted from control of behavior by the stimulus, to a perceptual assessment of the environment, that leads to movement and action.

ACKNOWLEDGMENTS

The research described here was supported by operating grant number AO-102 from the Natural Sciences and Engineering Research Council of Canada. The author is greatly indebted to Karen Stewart for collecting much of the data, and for advice and assistance in the preparation of the manuscript. Gene Ouelette collected the data in the experiment on the discrimination of locations of food, as part of an honors project carried out at Dalhousie University.

REFERENCES

Edwards, C. A., & Honig, W. K. (1987). Memorization and "feature selection" in the acquisition of natural concepts in pigeons. *Learning and Motivation, 18,* 235–260.

Guttman, N., & Kalish, H. I. (1956). Discriminability and stimulus generalization. *Journal of Experimental Psychology, 51,* 79–88.

Hanson, H. M. (1959). Effects of discrimination training on stimulus generalization. *Journal of Experimental Psychology, 58,* 321–344.

Herrnstein, R. J., & Loveland, D. H. (1964). Complex visual concept in the pigeon. *Science, 146,* 549–551.

Honig, W. K. (1987). Local cues and distal arrays in the control of spatial behavior. In P. Ellen & C. Thinus-Blanc (Eds.), *Cognitive processes and spatial orientation in animals and man* (pp. 73–88). Amsterdam, Netherlands: Martinus Nijhoff.

Honig, W. K. (1993). Numerosity as a dimension of stimulus control. In E. J. Capaldi & S. Boysen (Eds.), *Counting in animals* (pp. 61–86). Hillsdale, NJ: Lawrence Erlbaum Associates.

Honig, W. K., & Staddon, J. E. R. (1977). *Handbook of operant behavior.* Englewood Cliffs, NJ: Prentice-Hall.

Honig, W. K., & Stewart, K. E. (1988). Pigeons can discriminate locations presented in pictures. *Journal of the Experimental Analysis of Behavior, 50,* 541–551.

Honig, W. K., & Stewart, K. E. (1989). Discrimination of relative numerosity by pigeons. *Animal Learning and Behavior, 17,* 134–146.

Honig, W. K., & Stewart, K. E. (in press). Relative numerosity as a dimension of stimulus control: The peak shift. *Animal Learning and Behavior.*

Honig, W. K., & Urcuioli, P. J. (1981). The legacy of Guttman and Kalish (1956): Twenty-five years of stimulus generalization research. *Journal of the Experimental Analysis of Behavior, 36,* 405–445.

Kendrick, D. F. (1992). Pigeon's concept of experienced and nonexperienced real-world locations: Discrimination and generalization across seasonal variation. In W. K. Honig & J. G. Fetterman (Eds.), *Cognitive aspects of stimulus control* (pp. 113–134). Hillsdale, NJ: Lawrence Erlbaum Associates.

Lubow, R. E. (1974). High-order concept formation in the pigeon. *Journal of the Experimental Analysis of Behavior, 21,* 475–483.

Mackintosh, N. J. (1977). Stimulus control: Attentional factors. In W. K. Honig & J. E. R. Staddon (Eds.), *Handbook of operant behavior* (pp. 481–513). Englewood Cliffs, NJ: Prentice-Hall.

Morris, R. G. M. (1981). Spatial localization does not require the presence of local cues. *Learning and Motivation, 12,* 239–260.

Ouellette, G. (1989). *Transfer of a spatial discrimination between pictured and real locations in pigeons.* Unpublished honors thesis, Dalhousie University, Halifax.

Purtle, R. B. (1973). Peak shift: A review. *Psychological Bulletin, 80,* 408–421.

Rilling, M. (1977). Stimulus control and inhibitory processes. In W. K. Honig & J. E. R. Staddon (Eds.), *Handbook of operant behavior* (pp. 432–480). Englewood Cliffs, NJ: Prentice-Hall.

Siegel, R. K., & Honig, W. K. (1970). Pigeon concept formation: Successive and simultaneous acquisition. *Journal of the Experimental Analysis of Behavior, 13,* 385–390.

Suzuki, S., Augerinos, G., & Black, A. H. (1980). Stimulus control of behavior in the eight-arm maze in rats. *Learning and Motivation, 11,* 1–18.

Terrace, H. S. (1966). Stimulus control. In W. K. Honig (Ed.), *Operant behavior* (pp. 271–344.) New York: Appleton–Century Crofts.

Wilkie, D. M., Willson, R. J., & Kardal, S. (1989). Pigeons discriminate pictures of a geographic location. *Animal Learning and Behavior, 17,* 163–171.

3
When Is a Stimulus a Pattern?

Anthony A. Wright
University of Texas Health Science Center at Houston

The title "When Is a Stimulus a Pattern?" was used by Garner in his influential 1974 book to describe a theme central to much of his work; it can be equally used to describe a theme central to much of Al Riley's work. In a 1991 symposium honoring Tex Garner, Riley said that Garner's volume established a "theoretical framework within which intrinsic stimulus structure can be studied [and] stimulated much of [Riley's] research and theory." Actually, one can see Riley's interest in the integral and configural properties of stimuli in his earlier work as well, for example, his work on transposition effects (e.g., Riley, 1958; Riley, Ring, & Thomas, 1950; for a review see Riley, 1968). Transposition depends on the integral relationship among stimuli because when subjects are tested with stimuli shifted along a stimulus scale, they continue to choose, for example, the brighter of two stimuli even though the dimmer one may have been the original positive stimulus. Then in the 1960s Riley pursued similar transposition studies with children and auditory stimuli (McKee & Riley, 1962; Riley & McKee, 1963; Riley, McKee, Bell, & Schwartz, 1967; Riley, McKee, & Hadley, 1964).

More recently, Al Riley's research on configural pattern learning has been concerned with the configural properties of compound sample stimuli in matching-to-sample tasks, and how the spatial configurations of elements of the compound samples can affect analytical processing of these stimuli. Riley and his students, most notably Brown, Cook, and Lamb, have shown, among other things, that pigeons trained with spatially separated compounds cannot accurately perform with element samples, presumably because they process compound stimuli as integral patterns instead of analyzing them into elements. Subjects trained with element samples, however, can perform well with compound samples because they do process the (same) compound samples analytically. In

Riley's and Brown's (1991) words, "the results . . . suggest that these compound stimuli are represented in a unitary fashion unless the pigeons have been exposed to the elements in isolation from the compounds." (p. 242)

In this chapter I would like to describe an experiment that, like those of Riley, is concerned with the integral and configural properties of stimuli. This experiment is still in its preliminary stages, but the results, preliminary as they are, seem intriguing and its implications provocative. To put this experiment in the context of Al Riley's work, it is concerned with the integral nature of the whole matching-to-sample display—the sample plus both comparison stimuli—as opposed to the integral/separable nature of individual elements that make up the display.

I began these experiments through a question that has puzzled me for years. Pigeons trained in matching to sample with only two stimuli generally do not learn a concept of matching. While there is still some debate as to whether or not they ever learn such a concept with only two (or three) stimuli, it is clear that they do not in most situations. But they do learn. So, what do they learn? They cannot have learned any absolute stimulus strategies of choosing one stimulus and avoiding the other because the two stimuli are counterbalanced in the various possible roles of sample and comparisons and right/left positions. The focus of this chapter is on what they do learn.

The experiment to be described in this chapter employed a special training and testing procedure designed to determine what pigeons learn in such matching situations. To anticipate the conclusions, the results indicated that the pigeons learned each unique three stimulus array—the left comparison, the sample, and the right comparison. That is, they did not learn that the sample was an instructional stimulus, as we have usually assumed in such experiments, but instead the sample was just an element of a larger pattern that was learned.

These special training and test procedures involved dividing the set of possible stimulus arrays into two subsets and training with only one of the subsets. When three or more stimuli are used in matching it is possible to counterbalance all of the stimuli with only half the number of available stimulus arrays. That is, each stimulus appears equally often as the sample, as the correct comparison on the right and the left, and as the incorrect comparison on the right and the left. Thus, half the array patterns can be used to train subjects, without any possibility of training bias toward individual stimuli. The other patterns can be reserved for later testing. Built into this procedure is familiarization with individual stimuli. There cannot be a possibility of any novelty-averse reactions to the test stimuli because the test patterns are made up of the same stimuli as training patterns.

The task was a basic matching-to-sample task. A sample appeared. The pigeon pecked it. Comparison stimuli appeared to either side of the sample. A peck to one of these two comparison stimuli terminated the trial. If the chosen comparison matched the sample then grain reward was given. If the chosen comparison did not match the sample, then no grain reward was given. Trials were repeated following incorrect choices (correction procedure). Intertrial intervals of 15 s separated all trials.

The stimuli were cartoons presented from the floor of the experimental chamber on a computer CRT screen, responses were monitored with a computer "touch" screen, grain reward was placed on top of the correct comparison cartoon, and the pigeon ate the grain reward off of this comparison stimulus (see Wright, Cook, Rivera, Sands, & Delius, 1988).

The three training stimuli were three full-color cartoons: duck, apple, and grapes. Figure 3.1 shows the six different training patterns. Notice that each stimulus appears twice as the sample, twice as the left comparison, twice as the

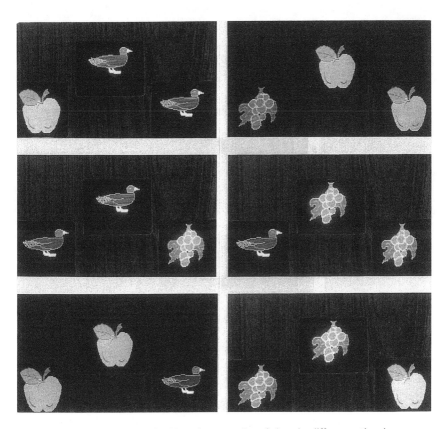

FIG. 3.1. Black and white photographs of the six different stimulus display patterns (left comparison, sample, right comparison) used to train the pigeons in each group. In the actual experiment, the three stimuli were full-color cartoons stored and displayed on a video monitor: duck—green head, yellow bill and feet, brown body, and blue and white tail; grapes— purple grapes with green leaves and brown stem; apple—red apple with green leaf and brown stem.

FIG. 3.2. Black and white photographs of the six different stimulus
display patterns used to test the pigeons in each group. The elements
of each pattern were identical to those used in training.

right comparison, and is the correct comparison once on the left and once on the
right.

Four groups of three pigeons each have been trained and tested in this experiment. The different groups are distinguished by the number of pecks (0, 1, 10,
20) required to the sample stimulus in order for the comparison stimuli to be
presented. The rationale here was that more pecks should encourage the pigeons
to analyze the display into its elements: the sample and the two comparisons.

After each pigeon learned the task to 80% correct or better (single session
criterion), they then were tested with the six test patterns shown in Fig. 3.2.
There were only six test trials, one of each test pattern, embedded in each 76-trial
daily session. Eight trials at the beginning and at the end of the session contained
no test trials, and each 10-trial block in between contained one test trial. Within
these restrictions the order and distribution of test trials was random. Correct

responses on all trials including test trials were rewarded. No correction procedure was used during testing. Although the experiment is incomplete, and is being reconducted due to some technical flaws, the results are encouraging.

Performance accuracy on test trials varied directly with the fixed-ratio (FR) pecking requirement. When no pecks were required to the sample stimulus and all three stimuli were presented at once, all subjects performed slightly below chance performance, a mean accuracy of 44% correct. When one peck to the sample was required, then mean test accuracy was 62%. When 10 pecks were required (FR10), mean test accuracy was 77%, and when 20 pecks were required (FR20), mean test accuracy was 75%. Thus, test performance generally increased with the FR requirement. This better test performance cannot be attributed to more training because the FR10 and FR20 groups reached the performance criterion in fewer training sessions than did the FR0 and FR1 groups. Apparently, pecking at the sample promotes analysis of the pattern into its stimulus elements.

Analysis into elements should be a prerequisite to rule learning. That is, if the rule "choose the comparison that matches the sample" is to be learned, subjects would have to analyze patterns into elements of sample and comparison stimuli in order to use the rule to perform accurately. In order to determine whether or not any subjects of these groups had learned the rule—the matching concept— they were tested with novel stimuli.

Immediately following the 10 test sessions with the six test patterns, subjects in each group were tested with novel stimuli to determine the degree to which they learned a matching-to-sample concept. Ten test trials were intermixed within 66 regular training trials, and five daily test sessions were conducted. One hundred different novel cartoons were used to make up these 50 test trials. Both stimuli were novel on each test trial. Reward was given for correct responses. Each novel stimulus was seen only once by each subject.

Performance accuracy on novel test trials increased with increasing FR requirement. With a FR0 it was only 48% correct. For the FR1 group, it was 50% correct. For the FR10 group, it was 59%, and for the FR20 group it was 69% correct. Although the novel-test performance level for the FR20 group is still 10% less than its training trial performance, it is 19% above chance performance.

Though still preliminary, the results from these two transfer tests indicate a possible explanation of what pigeons may be learning in the matching-to-sample setting. Pigeons tend to learn the correct response (right/left) to the particular pattern of the three stimuli, instead of the rule, "choose the comparison that matches the sample." It is my experience that pigeons will employ such absolute-stimulus strategies whenever possible. It is an absolute-stimulus strategy similar to learning the critical feature(s) in a go/no-go natural category study, because each configural pattern of three stimuli is uniquely associated with its correct response (right/left). By contrast, learning the matching rule is a relational stimulus strategy, and such relational strategies can transcend individual stimulus features allowing subjects to apply the rule and accurately perform with any (novel) stimuli.

The functional significance of the absolute-stimulus strategy of configural learning may be that this type of learning is easier (at least for pigeons). Relational rule/strategy learning apparently is more difficult for pigeons to learn than absolute strategy learning (Wright, Cook, & Kendrick, 1989), but it is not beyond their capability (Wright et al., 1988). Furthermore, relational rule learning is more vulnerable to the effects of proactive interference (PI), and PI is very high with a small number of repeating stimuli (Wright, Urcuioli, & Sands, 1986). That is, with a relational rule, elements are put together in a serial manner according to the conditional rule. With an absolute-stimulus strategy, on the other hand, each configural array is associated with its own response (right/left), and such strategies should not be subject to PI effects. Maybe pigeons prefer absolute-stimulus strategies to relational ones in order to avoid PI effects.

Although the tendency is to try and dichotomize learning into one or the another type of learning (absolute strategy or relational strategy), it seems more likely that both types of learning may proceed simultaneously with more emphasis on one or another depending on the specific task requirements. The intermediate results on some of transfer tests would support such an argument; otherwise transfer performance should have been either chance or equal to training performance.

If we accept the possibility that these two strategies can be learned simultaneously, then there also may be the possibility that the relative dominance of one or the other learning strategies might change during learning. For example, subjects who learned by a nonpreferred strategy of element analysis (due possibly from a larger number of stimuli) might during a period of overlearning be able to memorize all the different configural patterns and make a transition to their preferred configural-pattern learning. This might explain why, as I have noticed, some pigeons get "locked in" when overtrained on the same stimuli and become even less receptive than they once were to changes in the stimuli. Another possibility would be to try and capitalize on these (hypothesized) strategy shifts and use one as a transition to the other. For example, maybe the most efficient way for subjects to learn an abstract concept would be to initially learn by their preferred strategy—configural-pattern learning—and then gradually change to element analysis and rule learning by adding more and more elements and patterns to be learned. All techniques that encourage subjects to analyze stimuli rather than rely on configural-pattern learning should increase novel-stimulus performance and increase evidence for abstract concept learning. Although these possibilities are, at this point, speculative, this technique of divided-set training and testing along with novel stimulus testing may provide a way of determining whether or not there are learning strategy shifts during acquisition and continued training.

ACKNOWLEDGMENTS

Preparation of this article was supported in part by grants MH 35202 and MH 42881 to the author.

REFERENCES

Garner, W. R. (1974). *The processing of information and structure.* Potomac, MD: Lawrence Erlbaum Associates.

McKee, J. P., & Riley, D. A. (1962). Auditory transposition in six-year-old children. *Child Development, 33,* 469–476.

Riley, D. A. (1958). The nature of the effective stimulus in animal discrimination learning: Transposition reconsidered. *Psychological Review, 65,* 1–7.

Riley, D. A. (1968). *Discrimination learning.* Boston: Allyn & Bacon.

Riley, D. A., & Brown, M. F. (1991). Representation of multidimensional stimuli in pigeons. In G. Lockhead & J. R. Pomerantz (Eds.), *The perception of structure* (pp. 227–295). Washington, DC: American Psychological Association.

Riley, D. A., & McKee, J. P. (1963). Pitch and loudness transposition in children and adults. *Child Development, 34,* 471–482.

Riley, D. A., McKee, J. P., Bell, D., & Schwartz, C. (1967). Auditory discrimination in children: The effect of relative and absolute instructions on retention and transfer. *Journal of Experimental Psychology, 73,* 581–588.

Riley, D. A., McKee, J. P., & Hadley, R. W. (1964). Prediction of auditory discrimination learning and transposition from children's auditory ordering ability. *Journal of Experimental Psychology, 67,* 324–329.

Riley, D. A., Ring, K., & Thomas, J. (1950). The effect of stimulus comparison on discrimination learning and transposition. *Journal of Comparative and Physiological Psychology, 53,* 415–421.

Wright, A. A., Cook, R. G., & Kendrick, D. F. (1989). Relational and absolute stimulus learning by monkeys in a memory task. *Journal of the Experimental Analysis of Behavior, 52,* 237–248.

Wright, A. A., Cook, R. G., Rivera, J. J., Sands, S. F, & Delius, J. D. (1988). Concept learning by pigeons: Matching to sample with trial-unique video picture stimuli. *Animal Learning & Behavior, 16,* 436–444.

Wright, A. A., Urcuioli, P. J., & Sands, S. F. (1986). Proactive interference in animal memory research. In D. F. Kendrick, M. Rilling & R. Denny (Eds.) *Theories of animal memory* (pp. 101–125). Englewood Cliffs, NJ: Lawrence Erlbaum Associates.

4 Discriminative Stimulus Control: What You See is Not Necessarily What You Get

David R. Thomas
University of Colorado

In this chapter I am concerned with the discriminative stimulus control of operant behavior. The questions I ask are basically these three: How should we define discriminative stimulus control, how should we measure it, and once we have it, how can we get rid of it (if that is our intention)? As can be seen, these questions are interrelated, and the answers are neither obvious nor definitive.

DEFINITIONS

According to Skinner (1938) a discriminative stimulus is one that "sets the occasion" for a response. Skinner used this phrase to distinguish between a conditioned stimulus (CS) in Pavlovian conditioning, which elicits a response (i.e. a "respondent") from the organism, and a discriminative stimulus, which affects operant (i.e. "emitted") behavior in a less direct way. A (positive) discriminative stimulus is commonly defined as one in the presence of which a target operant behavior is more likely than in its absence. In another article (Thomas, 1985), I called this a "functional" definition of a discriminative stimulus, that is, one that defines the concept in terms of the way in which it can be measured.

It has become common practice to measure stimulus control by testing the subject for stimulus generalization by varying the training stimulus along a dimension such as color or tone frequency and noting the slope of the resulting gradient. The sharper the gradient, the greater the control exercised by the training stimulus. This measure has the advantage of specifying an attribute of the stimulus that actually controls the subjects' behavior. With a colored light, for example, it could be the color, the intensity, both of these, and a host of other attributes including size, spatial location, and so forth, that control responding. Generalization testing is particularly useful following discrimination training because subjects could master the task by learning to respond to a particular stimulus value, to withhold responding to a different particular stimulus value, or to do both. Generalization testing can help us to decide between these alternatives. It also can provide a more sensitive index of stimulus control than the difference between responding in the presence versus the absence of a particular training stimulus. An extensive discussion of alternative ways of defining and measuring stimulus control may be found in an important article by Hearst, Besley, and Farthing (1970). These investigators distinguished between excitatory stimulus control, which refers to the tendency for responding to be greater in the presence than in the absence of a particular stimulus, and excitatory dimensional control, which refers to the tendency for responding to be greatest in the presence of that value, along the dimension tested for generalization, that had accompanied reinforced training. These two definitions are orthogonal. It is easy to imagine a training stimulus that enhances responding yet that yields a flat generalization gradient when varied along a particular dimension (say, tone frequency) whereas responding may have been controlled by a different stimulus attribute (say, loudness).

It is useful to distinguish between a functional definition of stimulus control (i.e., differential responding in the presence of different stimuli) and a "procedural" definition, specifying the necessary training conditions to bring the behavior under stimulus control. Skinner (1938) proposed that a discriminative stimulus gains its power by virtue of the information that it provides regarding the availability of reinforcement. Skinner, of course, did not use the term information. Thus, a discriminative stimulus may be defined (procedurally) as a cue that provides information regarding the consequences of performing the operant response. If reinforcement is more likely in the presence of a particular stimulus, then responding should also be more likely (assuming, of course, that the subject's sensory systems have the capacity to detect that presence). Thus, for example, Skinner trained rats to bar press only when a light was on in the operant chamber by making reinforcement only available at such time. If the subjects had been blind, the procedure would not have been successful.

I propose in this chapter that the common functional definition of stimulus control is inadequate because differential responding can result from other sources. Consider, for example, the phenomenon called by Pavlov (1927) "external inhibition." When, during the course of conditioning, an extraneous stimulus is presented concurrently with the CS, the conditioned response (CR) is likely to be reduced in magnitude on that trial. This difference in responding surely reflects a discrimination between the CS by itself and the CS plus the added stimulus, but the process underlying this discrimination is likely to be different from that involved in discriminating between two tone frequencies or two light intensities. The terms *generalization decrement* and *stimulus control* seem appropriate in these cases but not when the effect is due to the added stimulus. Here an explanation in terms of "distraction" or a "shift in attention" seems called for. The procedural definition of stimulus control is also inappropriate because it presumes a conclusion not yet established. It remains to be determined what the necessary and sufficient conditions are for producing stimulus control, and despite extensive research activity over many years, the answer is, as yet, not determined.

THE DEMONSTRATION OF STIMULUS CONTROL

The issue of how to define discriminative stimulus control is inextricably tied to that of how to demonstrate such control, and it is to this issue that we now turn. There would seem to be no controversy here. Stimulus control is evidenced by differential responding in the presence of different stimuli. Clearly this is a necessary condition, but is it sufficient? I suggested earlier that it is not.

To say that a cue exercises discriminative stimulus control over behavior implies that the cue is associated with the behavior in question and through this association it has gained the capacity to elicit or facilitate the performance of that behavior. The situation is analogous to the way in which we define Pavlovian conditioning as a change in responding elicited by the CS as a result of its pairing with the US. Changes in responding that are due to sensitization or pseudoconditioning are ruled out by the inclusion of appropriate (i.e., CS alone; US, unconditioned stimulus, alone; or random) control groups.

In the case of discriminative stimulus control it is equally appropriate to demonstrate that the decrement in responding that accompanies a stimulus change depends on an association having been formed between the training stimulus and reinforced responding. In the study of retrieval cues in memory, this requirement is acknowledged (at least by some), but in the the study of stimulus control more generally, it has not heretofore been recognized.

Balsam (1985) described the *context change* procedure as employing a particular context during training and then testing subjects for the target response in that context or in a distinctively different one. The typical finding, reduced responding in the different context, may reflect retrieval cue value of the training context (i.e., that the context serves as a discriminative stimulus), but it also may reflect interfering behaviors induced by the novelty of the nontraining context (see Bindra, 1959). Spear (1978) warned that a decrement in performance in a changed context could be attributed only to the absence of retrieval cues if the disruptive effect of novelty could be ruled out. Unfortunately, he gave no instructions as to how this should be accomplished. Furthermore his admonition has not always been heeded.

Consider a study reported by Gordon, McCracken, Dess-Beech, and Mowrer

(1981). Rats were trained to perform an avoidance response in an apparatus that required them to run from a white to a black chamber. For some rats, training occurred in one experimental room (Context A) whereas for others it occurred in a room (Context B) that differed from the first room in terms of size, odor, lighting, background noise, and other features. Twenty-four hours later all rats were tested for retention of the avoidance response, either in the room in which they had been trained or in the alternative room. They performed better when tested in the room in which they had been trained and the difference was attributed to the features of the training room serving as effective retrieval cues. In a further experiment, these investigators demonstrated that the difference in performance in the two rooms could be eliminated if the subjects were familiarized with the alternative nontraining room before being tested in it. They interpreted this effect as indicating that as a result of placement in the nontraining room, the features of that room were incorporated into the memory of the training experience. It should be apparent, however, that the original finding of better performance in Context A could have resulted from a novelty effect elicited by Context B. Furthermore, the elimination of the difference between the two contexts that resulted from "placement" in Context B may have been due to the elimination of a novelty effect otherwise elicited by that context.

It is obviously a simple matter to eliminate the possibility that a novelty effect will be mistaken for evidence of stimulus control by contextual stimuli. One may abandon the context change method in favor of explicit discrimination training in which subjects have equal familiarity with two environments but are trained to make the target response in only one of them. This has been done successfully innumerable times but it begs the question. Research that demonstrates that contextual stimuli can be made to control behavior is irrelevant to the claim that they normally do so as a function of their "mere presence" during acquisition. To test this claim, the context change method must be employed.

How, then, can we separate a possible effect of novelty from a true instance of stimulus control? It first must be acknowledged that a decrement in performance may reflect both processes; they are not mutually exclusive. Stimulus control may exist but its magnitude may be overestimated due to its being confounded with novelty. Assuming these two factors do not interact with each other, we can provide an unconfounded measure of stimulus control by first obtaining a measure of the novelty effect so that the novelty component of the measured decrement in performance can be subtracted out. This is analogous to subtracting sensitization and pseudoconditioning effects from an acquisition function to get a truer picture of the Pavlovian conditioning process.

We have recently reported on several attempts to separate cue value from a novelty effect using odor as a potential discriminative cue for pigeons. There were several reasons behind the use of odor cues in these experiments. One was the virtual certainty that the odors used in these experiments were initially truly novel to the subjects. The novelty of the cues should facilitate the subjects' learning about them. The second reason is relevant to the interpretation of an effect, should one be obtained. Visual contextual stimuli may interact perceptually with visual discriminative cues (e.g., a green illuminated key in an illuminated operant chamber will look different from the same key in a darkened chamber). No such interaction should occur between olfactory and visual cues. Yet another reason for using odors as potential retrieval cues is the suggestion in the human memory literature that such cues may be especially powerful (see Schab, 1990). Finally, Honey, Willis, and Hall (1990) recently reported that pigeons autoshaped to respond to a particular key stimulus in the presence of a certain odor (either eucalyptus oil or isoamyl acetate) subsequently responded more to that stimulus in the presence of that odor than to a stimulus trained in the presence of the alternative odor. Thus we know that pigeons are sensitive to odor cues and can learn about them.

How does one demonstrate a novelty effect independent of any possible retrieval cue function? By producing familiarity with an odor in such a way that the odor cannot be associated with the target behavior. Suppose that pigeons are trained to key peck for food reinforcement in an odor-free environment and at another time, when key pecking is impossible and food is not present, the birds are exposed to a particular odor in the training chamber. That odor will now be familiar to the subjects but it should not function as a retrieval cue for key pecking because that behavior never occurred in its presence. Then, in a test in which subjects may key peck (under extinction conditions) in the presence of the familiar odor or a novel one, if the birds were to respond more in the familiar odor condition, that difference would necessarily reflect a novelty effect (i.e., interference with key pecking by the unfamiliar odor).

Two such experiments with just such a result were recently reported by Thomas and Empedocles (1992). The comparison of responding in familiar versus novel odors was done on a within-subject basis with the subjects switched back and forth between two chambers having the two different odors. The reason for doing two experiments was to vary whether the odor familiarization occurred subsequent to, or prior to, operant key peck training. If experienced subsequent to operant training, the odor had the potential to become a Pavlovian conditioned inhibitor; if experienced prior to operant training, this would necessitate a longer delay between odor exposure and testing, allowing for more forgetting to occur. It turned out that it did not make any difference. The result of the experiment involving a posttraining odor exposure is shown in Fig. 4.1. The group average result is typical of individual performance. Virtually every bird showed a decrease in key pecking when switched from the familiar to the novel odor condition and an increase in responding when switched in the opposite direction.

Having shown an effect of novelty per se on key peck responding, the next question was whether we could find evidence for a retrieval cue effect in addition to a novelty effect. The logic of the experimental question is straightforward. Suppose the subjects became familiar with a particular odor because it was

FIG. 4.1. Mean responses per 5-min test block under familiar (F) and novel (N) test conditions. Testing began with the novel odor condition for half of the subjects. Thus the data are pooled such that the first data point is based on the first test block under the familiar odor condition, whether or not a block under the novel odor condition had preceded it (from Thomas & Empedocles, 1992).

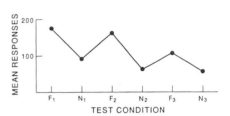

present during food-reinforced key pecking. Then they would have two reasons to peck more in the presence of this odor: (a) The odor should retrieve the memory of the key peck response and facilitate its performance, and (b) the alternative odor would not retrieve that memory and in addition might elicit interfering behavior. If both retrieval and novelty-elicited interference processes were at work, the result should be a greater "preference" for the training odor than occurs when that odor was merely familiar. Thomas and Empedocles (1992) reported two experiments to test this hypothesis; in one the odor exposure occurred early in operant training, in the other it occurred late. Again, the timing of the exposure made no difference. Again the subjects responded reliably more in the presence of the familiar odor. As shown in Fig. 4.2 (based on the experiment in which odor exposure followed substantial prior operant training without odor), the degree of preference was no greater than when the odor was merely familiar. Thus there was no evidence for a separate retrieval process.

In further experiments reported in the same article by Thomas and Empedocles (1992), it was exceedingly difficult to find evidence that the birds could form an association between a particular odor and the food-reinforced key pecking that had occurred in its presence. In one experiment, for example, the birds received equal exposure to each of two odors but key peck training in only one of them. In a subsequent test they responded comparably in the presence of both odors, the one in which they had been trained to peck and the alternative, with which they were equally familiar. Even training the birds to peck blue and not red in the presence of one odor followed by a single reversal, that is, training to peck red and not blue in the alternative odor, failed to establish the odor as a conditional cue. When tested for color generalization in each of the two odors the birds showed recency; that is, they pecked most at red, the most recently reinforced stimulus, regardless of the odor that was present. (See Thomas, Moye, & Kimose, 1984.)

The only procedure used by Thomas and Empedocles (1992) that was successful in making the odor an effective cue was to give the birds extensive reversal

FIG. 4.2. Mean responses per 5-min test block under familiar (F) and novel (N) test conditions. Testing began with the novel odor condition for half of the subjects. Thus the data are pooled such that the first data point is based on the first test block under the familiar odor condition, whether or not a block under the novel odor condition had preceded it (from Thomas & Empedocles, 1992).

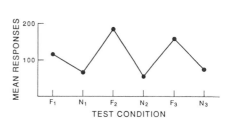

training with the red S+/blue S− and blue S+/red S− problems, each with its accompanying odor, alternated on a daily basis. In a generalization test in extinction these subjects tended to respond primarily to the S+ that had been reinforced when each of the training odors had been present. Of course, one would not call the odors retrieval cues in the sense of being irrelevant to the target task. The experiment does demonstrate that operant behavior in pigeons can be brought under stimulus control by odors but only after extensive and explicit discrimination training. The only prior study of operant behavior in pigeons controlled by odor cues was a demonstration by Michelson (1959) that pigeons could be trained to respond to one key when an odor was present and to another key when the odor was absent. No effort was made in that study to determine whether the birds could distinguish between two different odors. This was demonstrated in the Thomas and Empedocles study, but only when the two odors were used in an explicit discrimination training task to which they were relevant.

We can only speculate on why pigeons have such difficulty in learning to use odors as retrieval cues or even as explicit conditional cues in these experiments. Pigeons are highly visual organisms that discriminate grain from nonfood items (such as pebbles) based on their visual characteristics. There is no evidence to suggest that olfaction plays any role in pigeons' feeding behavior, thus the birds may be contraprepared to associate odors with food-reinforced behavior (see Seligman, 1970). Thomas, McKelvie, and Mah (1985) found similar difficulty in the use of auditory conditional cues with this species.

Recall that Hull (1943) and Spence (1936) believed that stimuli present when a response was reinforced would automatically gain in the capacity to control subsequent responding, presuming that the subjects were sensitive to the stimulus in question. The finding of a strong novelty effect in the Thomas and Empedocles (1992) experiments demonstrates that the pigeons are indeed sensitive to the odor cues and can learn about them. The requirements for associative learning clearly exceed those for habituation and stimulus recognition, however. We can state definitively that even when subjects "notice" environmental stimuli present during reinforced learning they do not "automatically" associate those stimuli with that learning. What else may be required is a matter for conjecture but probably depends on the species, the stimulus in question, and the amount and type of past experience that the subjects have had with that stimulus. It seems likely that if the Thomas and Empedocles experiments had used rats as subjects rather than pigeons, their results and conclusions would have been very different. Rats' feeding behavior is strongly influenced by olfactory cues.

The important lesson suggested by the Thomas and Empedocles (1922) experiments is that we cannot accept evidence of retrieval cue value based on the context change method without independent evidence that the effect cannot be attributed to novelty alone. It is rare that experiments employ such a control condition. The same caution about retrieval cues applies to the analysis of discriminative stimulus control by contextual cues more generally. Consider an experiment by Riccio, Urda, and Thomas (1966). In their experiment, two groups of pigeons received variable interval (VI) training to peck an illuminated key. For one group the floor was in the normal horizontal (0°) position; for the other it was tilted 30° counterclockwise. After the completion of training, a generalization test in extinction was performed with the floor tilt adjusted to 0°, 10°, 20°, and 30°. The result was a decremental gradient (in both groups) with maximal responding occurring under the training floor tilt condition. Riccio et al. interpreted this result as indicating discriminative stimulus control by the floor tilt used in training. This conclusion seems justified because it is based on the results of both groups. If only the 0° training group had been used, the obtained gradient might have merely reflected the novelty of the tilted floor test conditions. Because groups trained under the 30° floor tilt showed decreasing responding as the tilt was decreased and lowest responding when the floor was flat (as in their home cage), a novelty interpretation is inapplicable here.

Most claims of stimulus control by contextual stimuli lack such controls and thus the claim that the stimulus controls responding may be unjustified. Take the anecdotal observation that if the white noise normally experienced in the operant chamber is disrupted due to a blown fuse, a broken wire, or whatever, key pecking stops. Does this mean that the pecking is under the control of the white noise or rather that the change in stimulation elicits exploratory behaviors that compete with the key pecking? In the literature, when stimulus change results in depressed performance the effect is often attributed to "generalization decrement." The evidence suggests a more parsimonious alternative interpretation; that is, that it is due to a novelty effect. When McGeoch (1932) proposed that a change in contextual cues accounts for much of ordinary forgetting, he had in mind the retrieval function of stimuli that had accompanied original learning. He may have been right for the wrong reason. Forgetting may be due more to interference by novel environmental cues rather to failure to retrieve relevant information because of the absence of appropriate retrieval cues.

On the other hand, it is possible that the failure of Thomas and Empedocles

(1992) to find evidence for stimulus control in the sense of a retrieval function for odor cues was the exception rather than the rule. As pointed out previously, odors have no known role in pigeons' feeding behavior. Furthermore, novel odors elicit exploratory behaviors such as stretching of the neck and back-and-forth sidewise movements of the head, that could easily interfere with key pecking. Other novel contextual cues might not elicit exploratory (or fearful) behaviors or they might elicit behaviors that have no impact on the operant response being measured. It would be inappropriate to conclude from the Thomas and Empedocles results that global contextual cues do not exert stimulus control, but only that the novelty interpretation needs to be considered so that it can be ruled out.

We have recently completed an experiment in which we accomplished this, using houselight color as the contextual cue. Pilot work had indicated that training pigeons to key peck in the presence of a (red or blue) colored houselight that was on throughout the training session failed to produce control by houselight color unless substantial training was administered. In the experiment in its final form, 24 half-hour sessions of VI key peck training were given with a red or blue houselight present throughout the training sessions. The birds were tested for response rate in extinction with red and blue houselight colors randomly alternated in 5-min blocks. The degree of control by houselight color was extraordinary, as shown in Fig. 4.3.

All subjects responded substantially in the presence of the training color and very little in the presence of the novel color. But was the difference attributable to novelty? Two groups were run to test for this possibility. Both groups were key peck trained in the presence of a white houselight. For one of these groups, 24 sessions of placement in the training chamber with the key covered and the colored houselight on followed the VI training; for the other group, the placement sessions preceded the VI training. Thus both groups were familiar with one houselight color. These groups were tested like the group trained with the color present.

The group exposed to the houselight color subsequent to their operant training

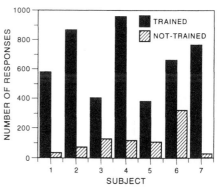

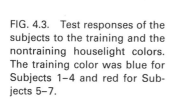

FIG. 4.3. Test responses of the subjects to the training and the nontraining houselight colors. The training color was blue for Subjects 1–4 and red for Subjects 5–7.

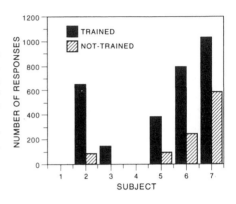

FIG. 4.4. Test responses of the subjects to the training and the nontraining houselight colors when training was followed by 24 sessions of placement under a white houselight condition. The training color was blue for Subjects 1–4 and red for Subjects 5–9. Note that Subjects 1 and 4 failed to respond during testing.

produced no usable data. Most subjects failed to respond at all during testing, suggesting that the placement had made the chamber cues (probably including but not restricted to the houselight color) Pavlovian inhibitors. The group that received placement prior to VI training responded substantially during testing but showed no reliable preference for the familiar color. This suggested that the strong "preference" seen in the group trained with the houselight color present is indeed due to the association formed during training. One alternative interpretation needed to be considered, however, and it was that the subjects may have forgotten the color of the houselight during the 24 subsequent sessions of training with the white houselight present. To eliminate this possibility, an additional control group was run, in which 24 sessions of placement in the chamber with the white houselight on followed VI training in the presence of the colored houselight. The results obtained with this group are shown in Fig. 4.4. Two of the subjects failed to respond during testing, suggesting that for these subjects the chamber cues had gained inhibitory properties during the placement sessions. Nevertheless, all subjects that responded showed a strong preference for the training houselight color, thus the failure to find this preference in the group for which houselight exposure preceded VI training cannot be attributed to forgetting of the color. This study demonstrates that unlike the odors used by Thomas and Empedocles (1992), houselight color merely present during reinforced training acquires discriminative control over responding by virtue of the association formed during that training. The procedural problems illustrated by this experiment may indicate why researchers have not typically empirically evaluated novelty as an alternative to true stimulus control. It is not easy to accomplish this, but it can and should be done.

CONTEXT SPECIFICITY

There is considerable interest in the contemporary animal learning literature in determining when the learning that occurs in various training paradigms is *con-*

text specific. To say that it is context specific means that the target behavior is under discriminative (or conditional) stimulus control by (generally global and often unspecified) environmental stimuli. The theoretical dispute underlying much of this research is whether such specificity, when it occurs, lies beyond the scope of associative theory because it requires us to postulate hierarchical relationships between events, that is, that the context signals the relationship between a CS and a US in Pavlovian preparations or between discriminative stimuli and the consequences of responding to them in operant procedures. According to this view, context serves as a retrieval cue, facilitating access to the memory of the events that transpired in its presence.

In the simple case of operant or Pavlovian conditioning, a retrieval interpretation is superfluous. As pointed out earlier, subjects might respond more in the training environment because of interference by the novelty of a test environment. If the test environment is associated with reward, subjects might respond more in the training environment because of a summation of associative strengths of that environment and of the explicit CS or discriminative cue. That is why the study of conditional stimulus control is so useful. Suppose that a pigeon is trained to peck red and not blue in a particular environmental context and then it is tested in a different context. If the test context is novel it may disrupt keypecking, but there is no reason to expect it to eliminate or reverse the subject's preference for red over blue. When this occurs it indicates that the environment is indeed exercising stimulus control over the subject's behavior.

There is no question but that environmental cues may come to exert conditional control over the subject's behavior. Pigeons, for example, can be trained to peck red and not blue when the houselight is on and blue rather than red when the houselight is off. Typically, such explicit conditional discrimination training is done with the four combinations of key color and houselight conditions experienced repeatedly within training sessions and the task is mastered after extensive training (see Boneau & Honig, 1964; Richards, 1979). With this procedure one can observe the gradual improvement of performance as training progresses.

An alternative procedure was used with surprising success by Thomas, McKelvie, Ranney, and Moye (1981) and again by Thomas et al. (1985). The pigeons learned a successive discrimination (green S+/yellow S−) in one context (houselight off plus white noise) and subsequently learned the reversal (yellow S+/green S−) in a different context (houselight on plus tone). With this single reversal procedure the conditional discrimination is implicit rather than explicit, because it is not revealed by performance during training. In posttraining generalization testing, however, the subjects respond appropriately to each context; that is, the gradient peaks at green in the houselight off—white noise context—and at yellow in the houselight on—tone context. That the subjects so readily acquire the conditional discrimination involving global environmental stimuli suggests that they may be predisposed to associate what is learned with the environment in which that learning took place.

In the aforementioned studies, as in typical experiments on context specificity, the two contexts were made very distinctively different. In some studies in the literature the contexts differ in visual, auditory, olfactory, and tactile characteristics. The intent is to make certain that the subjects distinguish between the different contexts without the necessity of explicitly training them to do so. Recent evidence suggests that this may be quite unnecessary, and that indeed contexts may acquire control over behavior despite explicit attempts to prevent this from happening.

Recall the experiment by Thomas and Empedocles (1992) that used a single reversal procedure in an unsuccessful attempt to establish conditional control over a key color discrimination by the odor that was present at the time. The birds learned to peck red and not blue in one odor (say, isoamyl acetate) and then they learned to peck blue and not red in the alternative odor (eucalyptus oil). After this the birds were tested for key color generalization while switching them between the two different odors.

Several practical problems had to be overcome to make this experiment feasible. Elaborate and expensive instrumentation would be required to present two different odors in the same chamber without an extensive delay in between them. The alternative that we chose was to switch the birds between two chambers, each of which would contain a different odor during testing. During initial training the birds were repeatedly switched between two chambers having the same odor in order to eliminate any reaction to the handling that might confound the test results. A second purpose of the frequent switching was to familiarize the subjects with both chambers that would contain different odors during eventual testing, so that the only basis for responding differently in each would be the difference in the odors present. With regard to the two chambers experienced by each bird this is explicit *irrelevance training* (see Mackintosh, 1973).

As mentioned previously, this experiment failed to reveal conditional control by the odors, with subjects showing only recency (i.e., they responded most to blue regardless of the test odor). The most surprising result, however, was the demonstration of conditional control by the different chambers in which each subject had been trained. This came about in the following manner. The switching between chambers took place during Phase 1 while the subjects were learning to peck red and not blue in a particular odor, which was present in both chambers. In Phase 2 when reversal training was started in the presence of the alternative odor in one of the chambers, switching between chambers was discontinued for a while so as to preclude interfering with the learning of the reversal. When reversal performance was well established, midway through the fourth session of training on the reversal problem, we switched the birds to the alternative chamber having the same odor. The results, as reported by Thomas and Empedocles (1991), were most unexpected. Despite the fact that the odor present was the one appropriate to the reversal and the reversal reinforcement contingency (blue

S+/red S−) remained in effect, performance appropriate to the original discrimination (red S+/blue S−) recurred.

Two interpretations of this unexpected finding were considered. One was that the handling and switching of birds between chambers during Phase 1 of training had come to serve as a cue signaling the red S+/blue S− contingency. This hypothesis was rejected based on the results of two tests: (a) Switching the birds back to the chambers in which they had learned the reversal immediately reinstated reversal performance, (b) a pseudoswitch, in which the birds were removed from and then returned to the same chambers, had no effect on their performance.

The alternative interpretation was that features of the different operant chambers had gained conditional control over the birds' discriminative performance despite the fact that the chambers were designed to be as similar to each other as possible and the frequent switching of the birds between chambers in initial training was expected to invalidate any cue value that detectable differences between the chambers might have had. In addition, the presence of highly salient odor cues during original discrimination and subsequent reversal training would be expected to overshadow any control that apparatus features might otherwise acquire, yet they clearly had no such effect.

Subsequent research has been directed at determining which aspects of the training procedures we had used were necessary and/or sufficient to produce the obtained result. The frequent switching of the pigeons between two chambers in initial discrimination training in the experiment was designed to invalidate apparatus cues but it may have had the opposite effect (or none at all). In the original experiment the subjects received half of their Phase 1 training in each of two chambers and were switched between the two every 5 min. To test for the effect of the frequent switching, a new group of pigeons was trained such that they experienced the entire initial phase of training in one box and reversal training in a different box. A control group was trained as in the original Thomas and Empedocles (1991) study for comparison purposes. The results of this control group are presented in column 1 of Table 4.1.

TABLE 4.1
Mean Discrimination Ratios at Various Stages of Training
for Four Training Groups

	Control Group	Group One-Switch	Group No-Odor
Last Day R+ B−	96.2	95.4	97.7
Last Full Day B+ R−	87.2	78.4	76.5
1st 10-Min Test Day (B+ R−)	81.9	85.7	84.1
2nd 10-Min Test Day (B+ R−)	38.6	36.1	55.6
Decrement	43.3	49.6	28.5

In the top row of column 1 is presented the mean discrimination ratio (i.e., percentage of responses to S+) obtained from the control group on the last (seventh) session of Phase 1 discrimination training (red S+/blue S−). The subjects had mastered the problem by this time.

In the second row the mean discrimination ratio is presented for the last (i.e., third) full session of reversal (blue S+/red S−) training. The subjects had reversed their preference between red and blue by this time. The next row in the table presents the discrimination ratio calculated over the first 10 min of the next training session while the birds were still in the box in which all reversal training had been carried out. The birds continued to perform well on the reversal problem. The next row is the critical one. It shows the result of the switch to the alternative chamber. In the control group, for every bird ($n = 8$) performance appropriate to the original red S+/blue S− discrimination contingency recurred despite the fact that the odor present and the reinforcement contingency were still the ones appropriate to the reversal. This is shown by the discrimination ratio dropping below 50%. Thus the results of the original Thomas and Empedocles (1991) study were replicated.

A convenient measure of the magnitude of the context specificity effect is the amount of change in the discrimination ratio from its preswitch to its postswitch value. This number, presented in the bottom row of the table, was a mean of 43.3% for the control group.

Column 2 in Table 4.1 presents the results for the group switched between boxes only once during training. It may be seen that the degree of context specificity in this group was at least as great as in the (multiswitch) control group. Thus the subjects learned about the apparatus cues without needing multiple opportunities to compare the two chambers. This suggests that those cues are extremely salient for the subjects. This is very surprising for several reasons. First of all the apparatus differences are so subtle that it is not obvious what the functional differences are. Furthermore, most studies of context specificity purposely make the different contexts as distinctively different as possible, by varying visual, auditory, and olfactory cues. Clearly under some circumstances this is "overkill" and quite unnecessary.

Is there anything unique about the procedure used in these experiments that would make the context manipulation so powerful? Certainly the use of different odors in the original and reversal phases of the experiment is unique. The acquisition of control by those odors was the intended focus of the original experiment, and manipulations like the (frequent) switching between boxes were included to prevent confounding of the measure of the odors' effectiveness. One would have expected the odors to overshadow any learning about apparatus differences, yet the opposite may have been true. For this reason, the original experimental procedure was replicated with an additional group of subjects, with the exception that no odors were present in the experimental chambers at any time. Column 3 in Table 4.1 presents the results from this group ("no odor"). It is clear that switching the birds to the chamber in which they had learned only the red S+/blue S− task disrupted reversal performance, reducing it to a chance level. Although the cues from that chamber were insufficient to reinstate Phase 1 performance, they were sufficient to prevent the transfer of reversal performance. The reduction in the discrimination ratio that resulted from the switch between boxes averaged 28.5%, which is significantly less than the 43.3% observed in the control group. Thus we may conclude that the use of odor cues in the control group potentiated learning about other contextual cues rather than overshadowing such learning.

Potentiation of learning is most frequently studied in the context of classical conditioning of aversions to stimuli paired with subsequent illness. The strongest evidence of such an effect comes from experiments with rats in which an odor that would not otherwise be an effective CS becomes one when paired with a taste cue. This effect has been interpreted in terms of a neuroanatomical link between olfactory and gustatory systems (see Garcia, 1989). Taste also can potentiate conditioning of auditory or visual CSs but the effect is much less strong (see, e.g., Galef & Osborne, 1978; Holder, Bermudez-Rattoni, & Garcia, 1988). Note, however, that the potentiating stimulus taste is a powerful CS in these situations.

In the present experiment the learning that is potentiated is of a higher order; it is the conditional relationship between apparatus cues and the key color discriminations learned in their presence. Furthermore, those odor cues are not, by themselves, effective as conditional cues. We may speculate that the odors elicit or promote exploratory behavior and it is through this means that visual cues that might not otherwise be noticed are attended to and learned about. Clearly, potentiation is not a unitary phenomenon and no doubt different mechanisms will be found to underlie its different forms.

Perhaps the use of a reversal paradigm is a necessary and sufficient condition for producing the context specificity effect. Kamin (1969) proposed that it is "surprise" or the violation of an expectancy that is essential for learning. What could be more surprising than the reversal of a previously learned discrimination? To test for the role of reversal training we performed an experiment in which subjects received single stimulus training to peck red (in Phase 1) and to peck blue (in Phase 2). There was no S- in either phase of training. To accommodate this difference in training procedure, there also had to be a difference in test procedure. It was decided to test for context specificity (i.e., conditional control) in the traditional way for this laboratory, that is, generalization testing in extinction with the two contexts (boxes) alternating in blocks of test trials (see Thomas, et al., 1981). In past research in this laboratory using a single reversal design, it has been shown that subjects tend to respond in accordance with the reversal contingency; that is, they show recency (see, e.g., Thomas et al., 1981; Thomas et al., 1984). Only when testing in the presence of contextual cues that had accompanied initial training and not reversal training is this recency effect overcome and responding appropriate to the original training contingency exhibited. This finding demonstrates the function of the original training context as a retrieval cue because appropriate responding in the reversal context would be expected on the basis of recency alone.

No odor was present in the chambers at any time during this experiment. In Phase 1, subjects were reinforced for responding to red, the only color present during training trials. Blue was the only color present during trials in Phase 2. In order to be as comparable as possible to past experiments, Phase 1 lasted for seven sessions whereas Phase 2 lasted for three and one-half sessions. As in previous experiments (with one exception) subjects were switched between two boxes every 5 min in Phase 1 but were trained in only one of the boxes in Phase 2.

Generalization testing was done in extinction with five colors presented in random order within blocks, and each subject experienced two test blocks in one chamber followed by two in the alternative chamber, and so on, until four blocks were experienced in each chamber. Separate generalization gradients were computed for responding in each of the chambers and the group mean generalization gradients are presented in Fig. 4.5.

The group average results are representative of the performance of the individual subjects. Even when tested in the chamber in which they had learned to peck blue in Phase 2, the subjects showed no recency effect. Rather they responded substantially to both red and blue and, on average, they responded slightly more to red. No doubt this reflects the fact that the birds had received substantially more training with the red stimulus than with the blue. The critical data come from the test results obtained from the other chamber in which the subjects had only learned to peck red. The gradients obtained in that chamber are similar to those described previously with the exception that they indicate a (marginally significantly) lower level of responding, on average. It is not surprising that the subjects would respond less in the chamber in which they had received less training and in which they had not been trained during the immediately preceding three sessions. The obtained difference is important because it suggests that the subjects discriminated between the different chambers. That discrimination is a precondition for the demonstration of context specificity but it does not guarantee

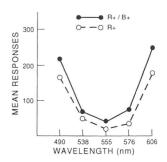

FIG. 4.5.   Group average wavelength generalization gradients obtained in two chambers, the one used in both Phase 1 and Phase 2 of training (R+/B+) and the one used only in Phase 1 (R+). Red = 606 nm, Blue = 490 nm.

FIG. 4.6. Group average wavelength generalization gradients obtained in two chambers, the one used in both original (Phase 1) discrimination training and subsequent (Phase 2) reversal training (R+ B−/B+ R−) and the one used only in original (Phase 1) discrimination training (R+ B−).

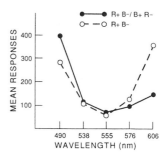

it. The critical finding is that the subjects responded virtually as much to blue as to red in a chamber in which they had never experienced blue prior to testing. It is this finding that indicates the absence of context specificity and suggests that reversal training (or something equally surprising to the subjects) may be necessary to produce it.

The aforementioned conclusion presumes that the generalization test procedure would have been adequate to demonstrate context specificity if the standard discrimination training procedure had been used. To test this hypothesis a group was trained as in the original Thomas and Empedocles (1991) experiment, but they were tested with the generalization test method with the odor appropriate to the reversal (blue S+/red S−) problem in both boxes. As shown in Fig. 4.6 the results demonstrated context specificity by the different chambers. Although the subjects responded substantially to both red and blue in both chambers, they responded substantially more to blue in the chamber used in reversal training and more to red in the chamber used only in Phase 1.

What may we conclude about context specificity based on the results of these experiments? First of all it must be acknowledged that context specificity is not absolute. The magnitude of the disruption of performance following the context shift varies with whether or not odor cues are available to the subject. Even under the condition that produced the strongest effect, there was still some transfer of learning between the chambers. When returned to the chamber used only in Phase 1, the subjects invariably performed more poorly in that chamber than they did prior to reversal training. Part of the difference is doubtlessly due to the fact that the reversal reinforcement contingency was still in effect, but poorer performance was evident immediately prior to any reinforcement. Thus, not surprisingly, the degree of control over behavior acquired through incidental learning about the contexts is much less than would be obtained if those contexts were employed as explicit conditional cues. The surprise is that the contexts gain control at all given (a) the opportunity for overshadowing by the more informative odor cues, and (b) the explicit irrelevance training performed with the box cues in Phase 1 of the experiments.

In 1963, Prokasy and Hall noted that "what represents an important dimension of the physical event for the experimenter may not even exist as part of the effective stimulus for the subject. Similarly, the subject may perceive aspects of an experimenter event which have been ignored by, or unknown to, the experimenter" (p. 315). Indeed, when behavior is not predictable based on experimenter-defined and manipulated stimulus-response relationships, it is often possible to demonstrate unintended and unexpected but simultaneously operating stimulus-response relationships. Bickel and Etzel (1985) reviewed several different studies in which unintended stimulus control was shown to exist. A convenient example is an experiment by Ray and Sidman (1970), in which two monkeys were trained with an eight-key simultaneous discrete-trial discrimination procedure. The monkeys learned to select a vertical line from an array that contained seven alternative orientations. The position of the different lines varied from trial to trial. Both subjects showed a position bias (i.e., control by key location) that, in the case of one of the animals, substantially conflicted with control by the line orientation dimension. The present results are similar in that the source of control was unintended and unexpected. In addition, in both studies two sources of control were in conflict; if only the box cues were relevant in the present experiments, the subjects would not have simply responded more to red than to blue when first switched between chambers in Phase 2. Instead, they would have resumed the near-perfect performance they had achieved at the end of Phase 1 training. Our results differ from those of Ray and Sidman in that the unexpected controlling stimuli served a conditional function, indicating the stimulus-response-reinforcer contingency that had been in effect in previous training. As was true of key location in the Ray and Sidman study, in our study the training procedure was expressly designed to eliminate differences between the chambers as potential controlling stimuli. It must be acknowledged that we, as yet, have no understanding of the conditions necessary and sufficient to invalidate or reduce control by particular stimuli.

SUMMARY

It is time to summarize the points made in this chapter. We have distinguished between functional and procedural definitions of discriminative stimulus control and found them both inadequate, although for different reasons. The functional definition (i.e., a decrement in behavior when a stimulus is changed) fails to distinguish between a novelty effect based on interfering responses in the presence of the nontraining value and a true case of control by the training value that results from an association having been formed between that value and the target behavior. It is probably the case that many claims of stimulus control by global contextual stimuli, in the absence of discrimination training or relevant past experience, are unjustified and are better interpreted as instances of a novelty effect. The use of the term generalization decrement to describe all decrements in performance that result from stimulus change is misleading and should be discontinued.

Just as stimulus control may be inferred when it does not exist, it may be overlooked when it exists in unexpected places. A failure to observe stimulus control when a stimulus is explicitly manipulated may mean that the behavior is under the control of stimuli (or stimulus attributes) that the experimenter is unaware of. Sometimes the controlling stimuli are ones that the experimenter has attempted to invalidate. Perhaps because we cannot specify the necessary and sufficient conditions for producing discriminative stimulus control, it should come as no surprise that we are no closer to specifying how to get rid of it.

REFERENCES

Balsam, P. D. (1985). The functions of context in learning and performance. In P. D. Balsam & A. Tomie (Eds.), *Context and learning* (pp. 1–21). Hillsdale, NJ: Lawrence Erlbaum Associates.

Bickel, W. K., & Etzel, B. C. (1985). The quantal nature of controlling stimulus-response relations as measured in tests of stimulus generalization. *Journal of the Experimental Analysis of Behavior, 44,* 245–270.

Bindra, D. (1959). Stimulus change, reactions to novelty, and response decrement. *Psychological Review, 66,* 96–103.

Boneau, C. A., & Honig, W. K. (1964). Opposed generalization gradients based upon conditional discrimination training. *Journal of Experimental Psychology, 66,* 89–93.

Galef, B. G., Jr., & Osborne, B. (1978). Novel taste facilitation of the association of visual cues with toxicosis in rats. *Journal of Comparative and Physiological Psychology, 92,* 907–916.

Garcia, J. (1989). Food for Tolman: Cognition and cathexis in concert. In T. Archer & L. G. Nilsson (Eds.), *Aversion, avoidance, and anxiety. Perspectives on aversively motivated behavior* (pp. 45–85). Hillsdale, NJ: Lawrence Erlbaum Associates.

Gordon, W. C., McCracken, K. M., Dess-Beech, N., & Mowrer, R. R. (1981). Mechanisms for the cueing phenomenon: The addition of cueing context to the training memory. *Learning and Motivation, 12,* 196–211.

Hearst, E., Besley, S., & Farthing, G. W. (1970). Inhibition and the stimulus control of operant behavior. *Journal of the Experimental Analysis of Behavior, 14,* 373–409.

Holder, M. D., Bermudez-Rattoni, F., & Garcia, J. (1988). Taste-potentiated noise-illness associations. *Behavioral Neuroscience, 102,* 363–370.

Honey, R. C., Willis, A., & Hall, G. (1990). Context specificity in pigeon autoshaping. *Learning and Motivation, 21,* 137–152.

Hull, C. L. (1943). *Principles of behavior.* New York: Appleton–Century–Crofts.

Kamin, L. J. (1969). Predictability, surprise, attention, and conditioning. In B. A. Campbell & R. M. Church (Eds.), *Punishment and aversive behavior* (pp. 279–296). New York: Appleton–Century–Crofts.

Mackintosh, N. J. (1973). Stimulus selection: Learning to ignore stimuli that predict no change in reinforcement. In R. A. Hinde & J. Stevenson-Hinde (Eds.), *Constraints on learning* (pp. 75–96). London: Academic Press.

McGeoch, J. A. (1932). Forgetting and the law of disuse. *Psychological Review, 39,* 352–370.

Michelson, W. J. (1959). Procedure for studying olfactory discrimination in pigeons. *Science, 130,* 630–631.

Pavlov, I. P. (1927). *Conditioned reflexes.* London: Oxford University Press.

Prokasy, W. F., & Hall, J. F. (1963). Primary stimulus generalization. *Psychological Review*, *70*, 310–322.

Ray, B. A., & Sidman, M. (1970). Reinforcement schedules and stimulus control. In W. N. Schoenfeld (Ed.), *The theory of reinforcement schedules* (pp. 187–214). New York: Appleton–Century–Crofts.

Riccio, D. C., Urda, M., & Thomas, D. R. (1966). Stimulus control in pigeons based on proprioceptive stimuli from floor inclination. *Science*, *153*, 434–436.

Richards, R. W. (1979). Stimulus control following training on a conditional discrimination. *Animal Learning & Behavior*, *17*, 309–312.

Schab, F. R. (1990). Odors and the remembrance of things past. *Journal of Experimental Psychology: Learning, Memory and Cognition*, *16*, 638–655.

Seligman, M. E. P. (1970). On the generality of the laws of learning. *Psychological Review*, *77*, 406–418.

Skinner, B. F. (1938). *The behavior of organisms*. New York: Appleton–Century–Crofts.

Spear, N. E. (1978). *The processing of memories: Forgetting and retention*. Hillsdale, NJ: Lawrence Erlbaum Associates.

Spence, K. W. (1936). The nature of discrimination learning in animals. *Psychological Review*, *43*, 427–449.

Thomas, D. R. (1985). Contextual stimulus control of operant responding in pigeons. In P. D. Balsam & A. Tomie (Eds.), *Context and learning* (pp. 295–321). Hillsdale, NJ: Lawrence Erlbaum Associates.

Thomas, D. R., & Empedocles, S. (1991). Context specificity of operant discrimination performance in pigeons. *Journal of the Experimental Analysis of Behavior*, *55*, 267–274.

Thomas, D. R., & Empedocles, S. (1992). Novelty vs retrieval cue value in the study of long-term memory in pigeons. *Journal of Experimental Psychology: Animal Behavior Processes*, *18*, 22–33.

Thomas, D. R., McKelvie, A. R., & Mah, W. L. (1985). Context as a conditional cue in operant discrimination reversal learning. *Journal of Experimental Psychology: Animal Behavior Processes*, *11*, 317–330.

Thomas, D. R., McKelvie, A. R., Ranney, M., & Moye, T. B. (1981). Interference in pigeons' long-term memory viewed as a retrieval problem. *Animal Learning & Behavior*, *9*, 581–586.

Thomas, D. R., Moye, T. B., & Kimose, E. (1984). The recency effect in pigeons' long-term memory. *Animal Learning & Behavior*, *12*, 21–28.

5
Generalization Gradients of Excitation and Inhibition: Long-Term Memory for Dimensional Control and Curious Inversions During Repeated Tests With Reinforcement

Eliot Hearst
Indiana University

Serena Sutton
University of Missouri

The analysis of operant generalization gradients along dimensions of reinforcement-associated (excitatory, S^+) and extinction-associated (inhibitory, S^-) stimuli was a major research area for students of animal learning during the 1960s and early 1970s (see Hearst, Besley, & Farthing, 1970; Honig & Urcuioli, 1981; Mackintosh, 1974, 1977; Mostofsky, 1965; Rilling, 1977; Terrace, 1966). Historically, interest in this topic was fostered by the results and theories of pioneers like Hull, Kohler, Lashley, Miller, Pavlov, Skinner, and Spence, who explored aspects of the relationships between stimulus generalization and such phenomena as discrimination learning, selective attention, induction-contrast, conflict, and transposition. Riley's (1968) book on *Discrimination Learning* provided an important summary and evaluation of classic work and enduring controversies in the field and suggested useful directions for future experimentation and theorizing, many of which were later examined.

Unfortunately, however, the topic and corresponding techniques have been neglected for more than a decade—even though virtually every contemporary textbook on animal learning and behavior includes a major section devoted to stimulus generalization and its presumed importance. These books also describe and discuss various factors that have been demonstrated to affect the shape and slope of response gradients produced by testing a subject with new stimulus values that lie at varying distances from the original training stimulus along some dimension like visual wavelength or auditory frequency.

This neglect is apparently not due to any consensus that all the principal questions about stimulus generalization have been resolved or that technical problems (e.g., difficulties associated with the use of response rate as a measure during tests in extinction) have seriously hampered interpretation of the data.

Rather, the research interests of workers who examine stimulus control in animals have shifted toward the analysis of more complex, compound training stimuli (e.g., combinations of lights and tones), and toward the assessment of various theories designed to explain why certain elements of a compound come to block, overshadow, or potentiate other elements. Furthermore, by manipulating stimuli along several dimensions, many cognitive psychologists using human and nonhuman subjects attempt to compare different mathematical models concerned with how such stimuli are represented, coded, or categorized in memory. Thus the study of generalization in its "pure" sense of simple transfer of a response from some training stimulus to various new stimuli, usually differing from the original stimulus along only one dimension, would hardly be appropriate for assessing popular models of, say, identification and categorization.

Viewed from another standpoint, however, the phenomenon of simple stimulus generalization remains a truly basic process that deserves continued empirical and theoretical analysis in the field of animal cognition and behavior. Virtually all attempts to explain associative learning, transfer, and memory imply the fundamental nature of stimulus generalization. These attempts extend from Aristotle's discussion of the role of resemblance in evoking certain memories, through the British empiricist philosophers' views about similarity as a factor in the association of ideas, up to more modern discussions of the probability of a new stimulus activating neuronal populations or containing "identical elements" that correspond to those possessed by previously experienced stimuli. Furthermore, to assess the nature of the "representation" that an animal or human has formed of some stimulus, a test involving other stimulus items must be instituted. This type of procedure is necessary both in the study of stimulus generalization and in work on human categorization and concept learning. Studies of stimulus generalization may continue to reveal basic facts as important to contemporary cognitive psychology as they have previously been to more behavioristically oriented approaches.

In this connection Shepard's (1965, 1987) views are noteworthy. He discussed his attempts to establish a "psychological space" enabling calculation of metric distances between members of any set of stimuli; his proposed solution makes the probability of a response learned to a stimulus an invariant monotonic function of the distance between that stimulus and any test stimulus. Moreover, he nominated stimulus generalization as psychology's most fundamental, universal law (eclipsing even contiguity or reinforcement) and argued that the regularities he derived from animal and human data are favored by natural selection "in sentient organisms wherever they evolve" (1987, p. 1323). Thus one of the most respected of today's cognitive psychologists assigns a basic role to stimulus generalization.

This chapter does not present specific tests of any popular mathematical models of learning, memory, or categorization, nor does it attempt to extensively relate our work on stimulus generalization to various sophisticated analyses of complex cognitive processes in animals. Instead, we describe some basic, previously unpublished data, including several curious and surprising results, that have emerged from our research over the years. We believe that empirical findings of this sort should receive consideration in any full theoretical account of animal behavior, neural functioning, or cognition.

The first section of the chapter is rather straightforward and deals with work on simple memory for prior excitatory or inhibitory learning. Different groups of pigeons learned either to respond or not to respond to a specific stimulus, and then they were divided into subgroups and tested in extinction for generalization to various stimulus values after the passage of varying amounts of time (from 30 min to 21 days) since the completion of training. Shades of Ebbinghaus! Previous experiments concerning memory for features or dimensions of excitatory stimuli suggest a progressive loss of precise stimulus control as the retention interval increases. But there are reasons for expecting that losses of this kind would be even greater in the case of inhibitory stimuli. That is not what happened.

The second part of the chapter describes results obtained through application of a somewhat different procedure for determining excitatory and inhibitory gradients after original training. Instead of the usual technique of delivering no reinforcement during tests for stimulus generalization, we provided equivalent opportunities for food at all the different test values. (Some early data using this procedure to obtain inhibitory gradients were reported in Hearst et al., 1970). Not only did this method yield highly reliable excitatory and inhibitory gradients immediately after completion of training, but it also permitted generalization tests to be continued for many sessions; of course, responding would not extinguish on this procedure, as it does on the typical test involving nonreinforcement. Some unexpected changes took place for a good number of subjects in the shapes of their excitatory and inhibitory gradients as daily "reinforcement tests" progressed. And yet the form of their long-lost, original gradients often could be retrieved by means of changes in the generalization test procedure.

LONG-TERM MEMORY FOR EXCITATORY AND INHIBITORY DIMENSIONAL CONTROL

Pavlov (1927) believed that two opponent processes, excitation and inhibition—which he conceived of as involving actual physiological changes in the cerebral cortex—were established by reinforcement or nonreinforcement of conditioned stimuli (CSs), respectively. He further claimed that such phenomena as the spontaneous recovery and disinhibition of conditioned responses (CRs) to a CS could not be explained without the use of such a dual-process approach. Basic to this claim were his assumptions that, as time passes, inhibition dissipates more rapidly than excitation and that inhibition is inherently more labile, fragile, or unstable (see Boakes & Halliday, 1972; Bottjer, 1979, 1982; Hearst, 1972).

Although there are certainly alternatives to Pavlov's interpretations, few data are available to assess his statement about relative forgetting or decay rates for excitation versus inhibition. Unsurprisingly, studies of the long-term retention of excitatory stimuli have generally revealed overall decrements in response output and in control by features or dimensions of the original training stimulus (see Honig & James, 1971; Spear & Miller, 1981; Thomas, 1981). However, there have been virtually no studies that directly compare the forgetting of excitatory and inhibitory stimuli. Only Hendersen (1978) made a real attempt to accomplish this goal, through the use of fear-conditioning procedures in rats. He found that a conditioned excitor of fear showed little or no change in its effects after the passage of 25–35 days, whereas a conditioned inhibitor of fear was greatly reduced in effectiveness—a conclusion supported in a somewhat less direct way by Thomas (1979).

Even though there are unquestionable problems in "equating" original levels of excitation and inhibition for a fair comparison of the forgetting of associated stimulus and response functions, the general topic has for years seemed to us worthy of special experimental attention. Relevant results would not only be of considerable empirical value but would also have implications for a variety of memory theories and models, especially those that do invoke the opponent processes of excitation and inhibition. Since 1969 we and our colleagues have performed several studies in which S^+s or S^-s involving a particular line orientation on the response key have been established as signals for the presence or absence of food, respectively, in pigeon operant discriminations; and then, after original training is complete, different time intervals have been allowed to elapse before subgroups are returned to the experimental chamber and tested for generalization to a variety of different line orientations. The interpretations and scanty data mentioned previously would suggest that, as time passes, gradients following inhibitory training should become relatively flatter than excitatory gradients.

In an unpublished master's thesis sponsored by D. A. Riley, B. Ritchie, and the first author of this chapter, Richard Desman (1969) trained 12 birds to peck an illuminated white key bisected by a thin black vertical line (S^+) for variable-interval (VI 1-min) food reinforcement. Trials with this stimulus were randomly interspersed with presentations of S^-, when reinforcement was not possible and no black line appeared on the illuminated white key. Brief blackouts (when the key was not lit and the chamber was completely dark) separated each trial from the next. Twelve other birds learned the opposite discrimination; S^+ was the blank illuminated key and S^- the key illuminated by the vertical line. All birds were maintained at 75% of their ad lib weights throughout the experiment.

Thus the two main groups in the experiment were trained either to respond to (excitatory) or not to respond to (inhibitory) a vertical line, on procedures that were otherwise symmetrical. After a bird reached the preset discrimination criterion—at least 90% of its total pecks during a session had to occur to S^+—it was assigned to either a 30-min or 24-hr delay group for tests of line-orientation generalization (Thomas [1981] has typically used a 24-hr retention interval to examine delay-induced changes in responding along the visual wavelength dimension after various types of pretraining and reactivation treatments). During the initial discrimination training Desman found no statistically significant differences between the excitatory and inhibitory groups in mean number of days (about 4–6) needed to meet the 90% discrimination criterion.

Generalization tests took place in the total absence of reinforcement at any of the different test stimuli. There were seven blocks containing seven different stimuli (the original training values—blank and vertical line [0°]—plus lines tilted $+30°$ or $-30°$, $+60°$ or $-60°$, or 90° from the vertical). Thirty minutes before their generalization test, birds in the 24-hr delay subgroups were fed an amount of food equal to the average amount that birds in the 30-min delay subgroup had earned in their last discrimination session. These feedings were given in the birds' home cages and were intended to minimize possible motivational differences between the two delay groups. It is worth noting that the mere delivery of food shortly before generalization testing has been found ineffective as a simple reactivation or reminder treatment, by itself having no significant influence on gradient form and slope (Thomas, 1981).

Clear decremental line-tilt gradients (that is, a maximum at the vertical line, 0°) were obtained in all 12 individual excitatory birds, whereas all 12 inhibitory birds exhibited a minimum at the 0° value, with a much lower overall response output to the different line tilts than in the excitatory gradients. Figure 5.1 presents group absolute and relative gradients for the four experimental conditions. For purposes of symmetry, the same point is plotted at $+90°$ and $-90°$.

Differences between indices of gradient slope, as well as between total numbers of responses, in the 30-min versus 24-hr delay groups did not approach statistical significance for either the excitatory or inhibitory case—although Fig. 5.1 suggests a flattening of the excitatory but not the inhibitory gradients over a 24-hr delay. Thus there was no definite evidence of a retention loss for either group over the course of a 24-hr delay, whether one considers the shape and slope of generalization gradients or the strength of the key pecking response itself. Desman (1969) concluded that the mere passage of time leads to no greater "dissipation" of inhibition than excitation, at least over a 24-hr period following completion of training.

In continued pursuit of possible delay-induced differences between conditioned excitation and inhibition, with respect to both loss of appropriate responding and weakening of dimensional stimulus control, we later systematically replicated Desman's experiment, but this time included much longer retention intervals: 7 and 21 days, as well as 1 day. All birds were maintained at their 75% weights throughout the assigned delay and were never placed in the experimental

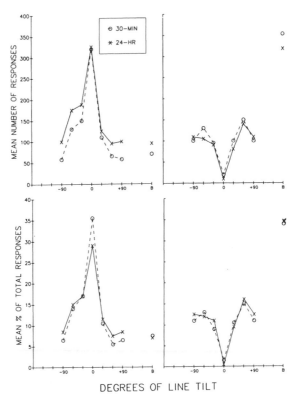

FIG. 5.1. Mean absolute (top) and relative (bottom) generalization gradients for two groups of pigeons, one (excitatory, on the left) previously trained with a vertical line (0°) on the key as S⁺ and a blank key (B) as S⁻, and the other (inhibitory, on the right) previously trained with the blank as S⁺ and the vertical line as S⁻. Different subgroups were tested for generalization after either a 30-min or 24-hr retention interval. The blank-stimulus value was included in the calculation of the relative gradients.

chambers. In this study no extra feedings were required for any groups before their generalization tests, because all birds had not received food for approximately 24 hr and were presumably equated for motivational level.

The absolute and relative gradients for the various groups (ns = 6) are displayed in Fig. 5.2. Although visual inspection of the graphed results indicates a relatively small but consistent flattening of excitatory gradients as the delay interval is lengthened (the expected general result, on the basis of prior work; see Thomas, 1981), application of conventional statistical tests to all the data in the excitatory absolute and relative curves, as well as to standard measures of gradient slope (e.g., percentage of total gradient responding that occurred to the S⁺)

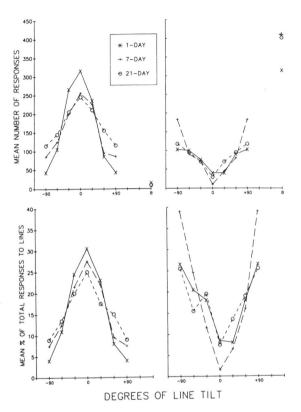

FIG. 5.2. Mean absolute (top) and relative (bottom) gradients for excitatory (left) and inhibitory (right) birds tested for generalization after either 1-, 7-, or 21-day retention intervals. The respective training stimuli corresponded to those in Fig. 5.1.

did not yield statistically reliable support for this conclusion or for any apparent differences in absolute number of responses to S^+ itself.

However, remarkably enough, and as in Fig. 5.1, inhibitory gradients exhibited no evidence whatsoever of flattening with increasing delay intervals. Neither visual inspection nor statistical analysis of the absolute and relative gradients revealed any consistent or reliable differences between the 1- and 21-day groups; but the 7-day group's relative gradient (and standard measures of gradient slope) did prove significantly *steeper* than gradients at the shorter and longer retention intervals.

Without further replication of the nonmonotonicity of the inhibitory delay function, we do not wish to draw any really strong conclusions or implications from this surprising result. However, a safe statement is that although excitatory gradients may become flatter with increased delays between training and testing (see Thomas, 1981, and the trends in our Figs. 5.1 and 5.2—an effect that Thomas et al., 1985, related to the difficulty of the original discrimination), there is no evidence from our work for parallel changes in the inhibitory functions. Inhibitory control certainly does not appear to dissipate faster than excitatory control, as Pavlov (1927) maintained and as was suggested by Moye and Thomas (1982); but cf. Hendersen's (1978) and Thomas' (1979) work with aversive stimuli and conditioned fear in rats, and the general points raised by Kraemer and Spear, 1992.

GENERALIZATION TESTS WITH REINFORCEMENT: INVERSIONS OF EXCITATORY AND INHIBITORY GRADIENTS AND LONG-TERM RETRIEVAL OF ORIGINAL GRADIENT FORM

Early Research with a New Method.   Beginning with the seminal studies of Guttman and Kalish (1956), a standard technique has typically been used for obtaining generalization gradients after operant conditioning (usually on a VI schedule) in individual subjects: A variety of test stimuli are presented during complete extinction of the response (in a direct test for stimulus generalization it is important not to treat any stimulus values differentially; the interest is in the degree of generalization from the original stimulus value to other values, rather than in the effects of differential reinforcement along a dimension). Because of the intermittent reinforcement schedule employed during training, the resistance to extinction of the response is relatively strong. Thus the behavior persists for a long enough time to obtain reliable gradients in individual subjects. However, because of the continued absence of reinforcement, responding eventually disappears and the gradient becomes indeterminate.

This was the general method used in the memory experiments described in the previous section of this chapter. The inhibitory counterpart of Guttman and Kalish's test procedure was exemplified there, too, and involved variation of a stimulus (S⁻) to which birds had previously been trained not to peck, while receiving VI reinforcement for pecks at some orthogonal stimulus. Although sufficient responding occurred along the S⁻ dimension to allow examination of inhibitory gradients, overall behavioral output for those gradients in Figs. 5.1 and 5.2 was considerably lower than for the excitatory gradients (see Hearst et al., 1970, and Jenkins, 1965, for a discussion of probable reasons for this difference). With repeated stimulus presentations, close-to-zero responding would soon have made determination of reliable inhibitory gradients impossible and the continued administration of generalization tests over several sessions would not have been worthwhile.

Mainly because of these practical problems, Hearst et al. (1970, especially pp. 392–398) proposed a different method for obtaining inhibitory gradients, which, as the present chapter documents, also proved quite applicable for the determination of excitatory gradients. Furthermore, use of this procedure unexpectedly yields some rather novel and curious kinds of effects that frequently develop during repeated generalization testing. On this new procedure we simply arranged for equal VI reinforcement at all the test values—rather than the standard method of extinguishing responses at every value. Compared to the use of extinction, this arrangement should produce considerably greater and more persistent responding during tests for generalization along some dimension of S−. If certain values are more inhibitory than others, they ought to resist the influence of reinforcement more strongly—paralleling the logic underlying the original Guttman–Kalish (1956) procedure, on which highly excitatory values should resist extinction more strongly than weakly excitatory ones. Readers also will notice the similarity of our resistance-to-reinforcement procedure to the type of Pavlovian inhibitory assessment technique labeled a *retardation test* by Rescorla (1969).

In our initial experiments with the new procedure, we first trained pigeons on discriminations that were very similar to those used for the inhibitory subjects described in the previous section of this chapter. Trials involved illumination of a blank white key as S+ (to which pecks produced food on a 30-s VI schedule), and illumination of the white key bisected by a particular orientation of a black line as S− (to which pecks were never reinforced). Training continued until the ratio of S− responses to S+ responses first reached .04 or lower during a session. The day after a bird met this criterion, it received a generalization test, during which pecks at all stimulus values (line orientations and the blank key) were reinforced on the same 30-s VI schedule previously in force for S+. The seven values were each presented a total of 12 times during a test session, according to a randomized blocks design.

The top panel of Fig. 5.3 displays, for the first day of generalization testing, a sample of six representative individual gradients as well as the group gradient for all birds that had been trained to discriminate S+ (blank) from S− (0°, a vertical line). The inhibitory gradients were steep, with minima at or very close to S−.

Gradients from a sample of six representative individuals as well as the group gradient for birds trained to discriminate S+ (blank) from S− (+30°), a line tilted 30° clockwise from vertical, are presented in the bottom panel of Fig. 5.3. Unlike those of the 0° group, these gradients are characteristically asymmetrical. A good number of the individual gradients showed minima both at S− and its mirror image, −30°; the group gradient illustrates this general outcome. We have almost never observed a definite mirror-image effect in previous determinations of inhibitory gradients that employed tilted-line S−s and extinction at all values during testing, when response output is usually much lower. (For some examples of mirror-image effects in generalization gradients obtained after excitatory training to a tilted line, see Thomas, Klipec, & Lyons, 1966.)

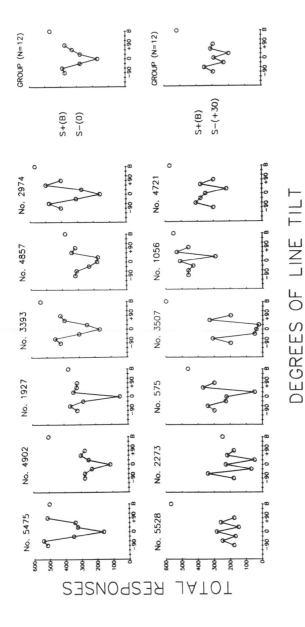

FIG. 5.3. Group (median) absolute gradients, as well as 12 representative individual gradients for inhibitory birds tested with equal VI reinforcement obtainable at all stimulus values. The top curves were secured from subjects trained with a blank key as S+ and a vertical line on the key as S−, whereas the bottom curves were from subjects trained with the same S+ but with a +30° line as S−. All the gradients come from the first day of resistance-to-reinforcement testing.

Because the gradients were still steep toward the end of the first generalization test, we decided to continue daily VI-reinforced tests, just to see what would happen to the gradients with repeated tests under the same conditions. We expected the gradients to become flat quickly, because the birds would be receiving approximately equal numbers of food deliveries at all test values. Even during the first test, the total number of reinforcers delivered at each value was about equal in almost every individual subject; responding at the former S⁻ value rapidly reached high enough levels for most subjects so that they earned virtually all the reinforcers available at that value, too. Nevertheless, rather surprisingly, the effects of prior discrimination training did not disappear quickly; gradients typically retained a minimum at S⁻ for at least a few sessions. The most interesting result, however, concerned changes in the gradients as birds were exposed to many repeated generalization tests with reinforcement.

The top panel of Fig. 5.4 shows the record of a single bird (not one of those in Fig. 5.3) over eight successive tests. No discrimination training or any other type of experimental experience was given between tests. Clear-cut inhibitory gradients with minima at S⁻ were obtained for the first 3 days. During the fourth session, however, the gradient began to change. By the fifth test the entire gradient was inverted and this effect persisted for several more sessions. Furthermore, the relative amounts of responding to the original S⁺ (blank stimulus) and S⁻ (0°) changed during the fourth test and remained consistently reversed (another kind of inversion: interdimensional).

A test sequence for a bird whose S⁻ was +30° is shown in the bottom panel of Fig. 5.4. During the fourth test there is an inversion involving both +30° and −30°, in relation to the other line-tilt values and to the blank.

Altogether, 17 birds underwent extended testing. At least 12 of them produced inverted gradients. Because inversions occurred on different days for different birds and were usually transitory, group gradients do not clearly reflect the phenomenon. So far as we know, the only similar result in the published generalization literature was obtained in an experiment of Lyons (1969), who trained pigeons with a solid green key as S⁺ and a vertical white line on an otherwise dark key as S⁻ (cf. Davis, 1971; Rilling, 1977, p. 446; Rilling, Caplan, Howard, & Brown, 1975; they present related research and commentary, although Rilling and his colleagues used an errorless-discrimination procedure). Testing in the standard way with nonreinforcement at all stimulus values, Lyons reported that when line orientation was varied on the dark background, minimum responding did occur at the vertical line, as expected. But when the tilt of the line was varied on a green background—that is, when S⁻ was superimposed on the S⁺ color—maximum responding occurred at the vertical line.

Lyons' (1969) and our procedure have in common the superimposition of equivalent "excitation" upon values along the S⁻ dimension. Lyons accomplished this by combining cues and we achieved it via nondifferential reinforcement all along the S⁻ dimension. Unfortunately, however, we have never been

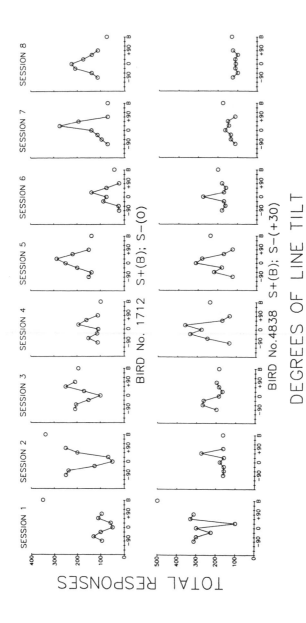

FIG. 5.4. Sample absolute gradients over eight successive sessions of generalization testing with equal VI reinforcement obtainable at all stimulus values. The top curve is from an individual bird that had received inhibitory training with an S⁻ of 0°, the bottom curve from a bird whose training had been with an S⁻ of +30°.

able to convincingly replicate Lyons' finding, despite several attempts with a variety of cue-combination procedures similar to his.

The Hearst et al. (1970) monograph on inhibition and operant stimulus control was written at about the time we first began to try out the resistance-to-reinforcement procedure. We briefly suggested several potential difficulties with the technique (see p. 397 ff of the monograph). None of those possible complications ever posed serious technical or interpretive problems in our further experiments of a similar kind, some of which were designed to allow assessment of such factors. This is not the place to burden the reader with details of specific control studies, but it is important to note that birds trained with S^+ (blank) and no S^- trials at all do not show a vertical-line preference when given repeated resistance-to-reinforcement tests with the various line tilts (see also Rilling et al., 1975). Furthermore, it seems implausible to argue that the inversions are traceable to a summation of excitation at the vertical line or nearby, because the line-tilt dimension is a circular one and has no "center." We also have observed inversions in birds trained with horizontal or tilted lines as S^- (see, e.g., Fig. 5.4). However, the most telling argument against interpretations of this kind is provided shortly, when data are presented showing the opposite kind of inversion following excitatory training to a vertical line.

More Inversions and Some Later Revelations.   Although we were impressed with the general technical benefits of the reinforcement procedure for research on the stimulus generalization of inhibition, outlined earlier, it was the frequent appearance of unexpected gradient inversions during repeated testing that especially intrigued us. We wanted to find out whether inversions would occur not only in the case of response inhibition, but also for excitation. In other words, would gradients initially displaying a definite peak at the S^+ line tilt, frequently come to exhibit a clear-cut minimum at that value, while repeated tests with reinforcement were administered? And, by some final change in context or procedure, not involving differential reinforcement, could we recapture or reinstate the type of gradient obtained long before, immediately following discrimination training?

In this section of the chapter we concentrate on the excitatory case, but also provide a few more examples of inversions occurring after inhibitory training. Furthermore, we show that, even after numerous tests with reinforcement, the institution of complete nonreinforcement at all stimulus values unexpectedly but often yields a retrieval of the gradient forms exhibited immediately after original training.

As part of her unpublished dissertation project (Besley, 1975), the second author of this chapter investigated a variety of phenomena and issues in the visual discrimination learning of pigeons: behavioral contrast, interresponse time distributions, and several aspects of the types of generalization effects described previously. Unfortunately, the detailed computer printouts for the generalization work have been lost and we hesitate to present data from the incomplete, hand-drawn figures that remain. However, except for the use of white lines on a red background, with the "blank" stimulus thus becoming a key illuminated by a completely red field, the apparatus and general procedures were similar to the arrangements employed in the aforementioned research. A variety of other comparison groups were included besides the standard excitatory and inhibitory treatments: for example, birds given (a) intradimensional training with two line tilts serving as S^+ and S^-, or (b) single stimulus ("control") training, with the S^+ line tilt or blank occurring on all trials, or (c) overtraining that involved initial VI reinforcement for 36 sessions to S^+ only and then exposure either to discrimination procedures (excitatory, inhibitory, intradimensional) or to continued S^+-only training for 36 more sessions. Then 2–4 weeks of daily resistance-to-reinforcement tests were given and, finally, subjects received several test sessions in which everything was the same as during the previous tests except that no reinforcement was ever delivered.

Except for the "control" birds, gradient inversions were observed in many subjects. For example, Besley's (1975) "excitatory" subjects exhibited steep line-tilt gradients during Sessions 1–3 of the reinforcement tests, with peaks at the S^+ value and very little responding at the S^- value. However, both intradimensional and interdimensional inversions typically occurred over the course of testing: The original peak at S^+ became the gradient minimum and responding to the old S^- value became higher than to the former S^+.

When extinction was introduced at all stimulus values after the reinforcement tests had been completed, responding slowly disappeared and birds in the different groups consistently showed a gradual reinstatement of their respective original gradients—despite 2–4 weeks of daily tests with nondifferential reinforcement at all values and no clear exhibition of gradients of that form since the early sessions of reinforcement testing. Thus, even though at the end of many reinforcement tests any differential responding at various stimulus values had more or less vanished, the introduction of tests with no reinforcement led to a retrieval of the original stimulus control expected and initially obtained immediately after training.

Besley's (1975) results conformed generally to our prior findings with respect to inversions and, additionally, demonstrated a remarkable extinction-induced retrieval of original dimensional control. However, one weakness of her set of experiments was the use of a solid (red) color as the blank stimulus value. This procedural detail complicated matters because of the great possibility of the color's overshadowing or masking control by the line-tilt dimension (see also Rilling, 1977; Tomie & Kruse, 1980). The line-tilt gradients, especially in the inhibitory groups, were not as steep as in our previous work with white light as the blank stimulus, perhaps also because of the greater possibility of the formation of within-compound associations. This outcome made it difficult to draw strong conclusions about when inversions occurred.

However, in later experiments at Indiana University we have frequently observed inversions during successive tests with reinforcement, as well as subsequent extinction-induced "revelations" of original stimulus control. For example, after the excitatory and inhibitory birds whose data are displayed in Fig. 5.2 had received that initial generalization test without any reinforcement, following either the 1-, 7-, or 21-day retention interval, they were given at least nine daily tests with a 1-min VI schedule in effect during all the stimulus values. About half the 36 subjects showed definite inversions during the course of such testing, the likelihood of which bore no clear relation to whether they were excitatory or inhibitory birds or to which of the retention intervals they had been exposed.

A few striking examples of such inversions are presented in Figs. 5.5 and 5.6. Unfortunately, no final tests without any reinforcement were included in that experiment, and therefore the opportunity to observe reinstatements of the "original" gradients (i.e., like those displayed in Fig. 5.2) was lost.

A later experiment concentrated on replication of both the novel phenomena revealed in the aforementioned work, that is, (a) the inversions of excitatory and inhibitory gradients occurring in individual subjects during repeated and prolonged generalization tests with equivalent reinforcement at all test values, and (b) the subsequent retrievals of original gradient form produced by removing all reinforcement during the tests. The training stimuli in this study were either a blank white key or a white key with a black vertical line (0°) bisecting it. For 20 sessions after initial VI training to S^+, eight "excitatory" birds received discrimination training with the vertical-line stimulus as S^+ (when food was available on a 1-min VI schedule) and the blank stimulus as S^-, and eight "inhibitory" birds had the S^+ and S^- stimuli reversed.

At the end of this phase all birds were given 20 consecutive daily sessions of generalization tests with VI 1-min reinforcement at all values; seven test stimuli were presented nine times during each session, according to the usual randomized blocks design. This 20-day phase was followed by a period lasting at least eight sessions, in which generalization tests were programmed as before except that food was never delivered. During original discrimination training and all subsequent generalization tests, stimulus presentations lasted 30 s and were separated by 10-s blackouts.

As the phase of reinforced generalization tests continued, about half the birds in both groups displayed the kinds of inversions we had obtained in previous work. Individual examples of such effects are depicted in Fig. 5.7 for two subjects in the excitatory group. When extinction was introduced at all values after 20 days of nondifferential reinforcement, it did not take long for recovery of the original gradient, with a peak at or near S^+, to occur in these two subjects.

For most subjects generalization gradients were essentially flat during the last 4 days of the 20 tests with reinforcement; there was little or no evidence of differential stimulus control along the line-tilt dimension. However, during the subsequent tests without any reinforcement, gradients resembling the form of

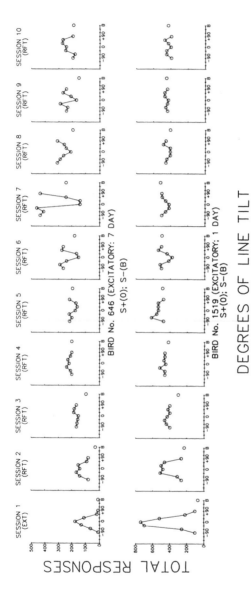

DEGREES OF LINE TILT

FIG. 5.5. Sample absolute gradients for two individual birds tested for generalization after training with a vertical line as S⁺ and a blank key as S⁻. Their first test was given with no reinforcement available at any value (extinction), whereas the next nine sessions involved equal VI reinforcement obtainable at all values. The subjects come from the experiment whose complete results are summarized for Session 1 in Fig. 5.2, and the birds were members of the 7-day (top) or 1-day (bottom) excitatory retention groups.

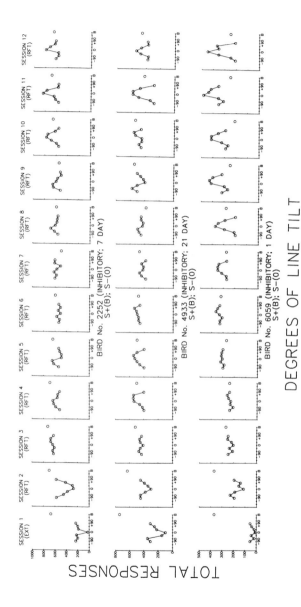

DEGREES OF LINE TILT

FIG. 5.6. Sample absolute gradients for three individual birds tested for generalization after training with a blank key as S+ and a vertical line as S-. Their first test was given with no reinforcement available at any value (extinction), whereas the next 11 sessions involved equal VI reinforcement obtainable at all values. The subjects come from the experiment whose complete results are summarized for Session 1 in Fig. 5.2, and the birds were members of the 1-, 7-, or 21-day inhibitory retention groups.

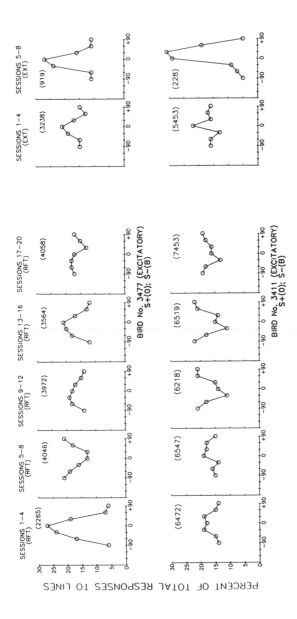

DEGREES OF LINE TILT

FIG. 5.7. Sample relative gradients for two excitatory birds first tested with equal VI reinforcement obtainable at all stimulus values for 20 sessions after original training. These 20 sessions are presented in separate blocks of 4 sessions; the number above each curve is the total number of responses made to all the stimulus values in that curve. The final eight sessions (on the right) were conducted during complete extinction; they are blocked and numbered as for the prior sessions.

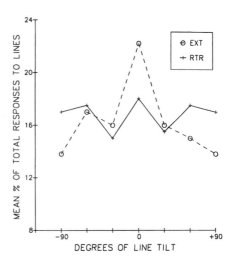

FIG. 5.8. Group mean relative gradients for eight excitatory birds (the S$^+$ had been 0°, S$^-$ the blank) during the final five sessions of 20 resistance-to-reinforcement tests (RTR) and the first five subsequent sessions of testing in complete extinction (EXT). A mean of 7,477 total responses were involved in calculation of the RTR curve, and 4,033 total responses for the EXT curve.

those obtained immediately after original discrimination training were displayed by seven of the eight excitatory birds; that is, they yielded gradients with a clear maximum at the vertical-line stimulus. Figure 5.8 provides a group summary illustrating this reinstatement effect, which was highly reliable statistically in terms of the Stimulus by Test Condition interaction. Unfortunately, the inhibitory birds did not yield a statistically significant reinstatement effect.

Thus this final experiment produced more examples of the curious inversions that we reported earlier for repeated reinforcement tests. Furthermore, after extended testing with reinforcement had more or less equalized responding at the different stimulus values, the removal of all food deliveries led to a dramatic retrieval of the excitatory-gradient shapes that had been obtained right after those subjects' initial discrimination training, almost 3 weeks before.

CONCLUDING COMMENTS

The main goal of this chapter has been to stimulate the reader's interest in some phenomena and findings about excitatory and inhibitory generalization that seem rather unexpected or curious to us. In the first section of the article we reported that line-tilt generalization gradients obtained from pigeons after mastery of an appetitive discrimination involving a particular line orientation as S$^-$ did not become flatter even after 21 days without any exposure to the original discrimination. In fact, gradients were significantly steeper around S$^-$ following a 7-day as compared to a 1- or 21-day retention interval. On the other hand, when for other birds the same line orientation had served as S$^+$ during discrimination training, gradients tended to become progressively flatter as the duration of the retention interval was extended. The results with the latter birds are in conformity with prior research on excitatory generalization. Thus, contrary to what Pavlov (1927) and others would presumably have predicted, control by a supposed conditioned inhibitor was not more fragile or transitory than control by a conditioned excitor. If anything, the opposite was true.

Additional, future corroboration of our results—or analysis of what conditions produce better or worse memory for aspects of and response strength to excitatory versus inhibitory stimuli—has obvious significance for the understanding of possible dynamic changes and interactions that occur as time passes since initial experience and discrimination training with positive and negative stimuli.

The second section of the chapter dealt mainly with changes in generalization gradients that take place over repeated sessions in which equivalent intermittent reinforcement is provided for pecking at every line-orientation test stimulus, after different birds have learned either to peck (excitatory) or not to peck (inhibitory) at a particular one of them. Normally, such generalization tests are conducted during complete extinction, rather than with uniform reinforcement. Surprisingly, a large number of subjects eventually displayed inversions or "flopovers" in their generalization gradients: After initially showing the expected minimum at their specific S^-, inhibitory subjects often would produce several gradients with a maximum at or near that value, whereas excitatory subjects would exhibit the opposite effect with respect to their S^+ and values along the line-orientation dimension.

A final interesting result was the extinction-induced recovery or reinstatement of gradient shapes that had occurred immediately after initial training, when the extended series of generalization tests with reinforcement first began. This unmasking or retrieval effect of removing reinforcement at all values was clear-cut for excitatory birds, but not convincing for inhibitory birds in the work reported here (although over the years we have obtained the effect often enough in individual inhibitory subjects for us to have reasonable confidence that such reinstatements could be achieved more frequently if we had better understanding of the factors that produce them).

Although we do not have any simple explanation to offer for the excitatory and inhibitory inversions observed during resistance-to-reinforcement tests, it may be valuable to mention some other, perhaps only superficially relevant findings or conclusions from work by others. A search for theoretical linkage among these disparate results and notions could yield insights into the mechanisms or processes responsible for our inversions and could help fit the pieces of this puzzle together. In discussing temporal and spatial induction (contrast), Pavlov (1927) mentioned procedures that produced "paradoxical" and "ultraparadoxical" effects—for example, positive CSs surprisingly came to yield weaker salivation than negative CSs. These findings did not occur reliably in all dogs, nor were they necessarily repeatable in a single dog. Pavlov talked about "rhythmic undulations" and speculated that cortical regions may show alternating periods during which excitation or inhibition predominate. Malone and his colleagues (see, e.g., Rowe & Malone, 1981) reported unexpectedly high responding to previously extinction-associated stimuli in studies of dimensional contrast, along with additional results that could not easily be explained by Blough's (1975; see also Blough, 1983) quantitative model of operant generalization and discrimination (an extension of the well-known Rescorla–Wagner, 1972, model). Readers familiar with the extensive literature on other types of contrast effects will certainly see some parallels between those phenomena and the gradient inversions reported here.

Bolles (1985) worried about the inability of researchers to predict which of the permanent representations presumably evoked by a stimulus with a complex, ambiguous history of reinforcement and extinction will be selected and acted upon at a particular time. In a masterful summary of the literature Durlach (1989) concluded that the same stimulus may well display excitatory properties in one context or according to one assay, and inhibitory properties in other contexts or with other assays.

Even social psychologists may have studied phenomena that relate to effects reported in this chapter. For example, Aronson and Lindner (1965) found that subjects liked best a confederate who first said negative and then positive things about them, as compared to other types of confederates—even those who always made positive comments.

The complete removal of reinforcement has been successful in revealing or reinstating hidden, silent, or old associative properties in several past experiments (see Hearst, 1987). However, exactly why extinction frequently produced a "regression" to the form of generalization gradient to be expected right after original training in our work is not easy to pinpoint. Perhaps relatively long periods of time without reinforcement partially reinstated the context of original training, when S^- trials constituted half the session. At any rate, Hearst (1987; see also Lindblom & Jenkins, 1981) discussed several studies in which extinction procedures unmasked or improved control by various stimulus elements or dimensions—latent learning that did not manifest itself under conditions or assays involving reinforcement. It seems likely that analysis of such outcomes will not only tell us more about the richness of what is learned by animals in various settings, but also help us to understand more about the mechanisms of extinction itself, a procedure or process whose effects have never, in our opinion, been adequately explained or integrated into behavior theory, especially with respect to their evolutionary and ecological significance.

This chapter has not been concerned with the details of definitional and measurement problems involving conditioned excitation and inhibition. Of course, there is a vast literature on these issues that has accumulated over the last 20 years (see, e.g., the chapters in Miller & Spear, 1985), and in the past we have ourselves contributed a few words to the various debates. Our own inhibitory behavior with respect to these topics in the present chapter was deliberate. The emphasis here was much more empirical than theoretical, because our curiosity has long been piqued by the major phenomena we reported—even though we have never formulated adequate explanations for them. However, any complete theory of generalization, discrimination, and contrast will have to take these unexpected findings into account.

ACKNOWLEDGMENTS

This chapter is dedicated to Al Riley as he retires from an academic life filled with diversity, curiosity, and enthusiasm. E. Hearst vividly recalls our sometimes breathless discussions about stimulus generalization, problems in psychology, and the meaning of life as we climbed the steep hill from Tolman Hall to Grizzley Peak Boulevard in Berkeley during long daily walks home together in 1968–1969. Now that Al is concentrating on some new activities—for example, learning how to play the piano and using this personal experience to better understand the psychology of music—his expertise about behavior and cognition will subtly but inevitably keep increasing.

The research reported here was supported by National Institute of Mental Health research grants to Eliot Hearst at the University of Missouri and Indiana University, and by a National Defense Education Act Predoctoral Fellowship to Serena (Besley) Sutton. The initial work revealing the generalization-gradient inversions and extinction-induced reinstatements that are described in this chapter was performed by Sutton at the University of Missouri more than 20 years ago, as part of her unpublished graduate-student research and doctoral project, but additional unpublished experiments to substantiate and extend her findings have been performed and analyzed by E. Hearst at Indiana University since 1970. The assistance and collaboration of John Duncanson, Roberta Ewing, Stanley Franklin, Dexter Gormley, Mary Janssen, and Sharron Taus at Indiana University are gratefully acknowledged. We think the major results are of particular interest now, in the context of much recent work in the fields of animal learning, memory, and cognition and therefore we do not terribly regret the lag between performance of the original experiments and publication of their and later findings.

REFERENCES

Aronson, E., & Linder, D. (1965). Gain and loss of esteem as determinants of interpersonal attractiveness. *Journal of Experimental Social Psychology, 1*, 156–171.

Besley, S. S. (1975). *Processes of visual discrimination learning in pigeons: Variable-interval responding, behavioral contrast, and stimulus generalization.* Unpublished manuscript, University of Missouri, Department of Psychology, Columbia.

Blough, D. S. (1975). Steady-state data and a quantitative model of operant generalization and discrimination. *Journal of Experimental Psychology: Animal Behavior Processes, 104*, 3–21.

Blough, D. S. (1983). Alternative accounts of dimensional stimulus control. In M. L. Commons, R. J. Herrnstein, & A. R. Wagner (Eds.), *Quantitative analyses of behavior*: Vol. 4. *Discrimination processes* (pp. 59–72). Cambridge, MA: Ballinger.

Boakes, R. A., & Halliday, M. S. (Eds.). (1972). *Inhibition and learning.* New York: Academic.

Bolles, R. C. (1985). A cognitive, nonassociative view of inhibition. In R. R. Miller & N. E. Spear (Eds.), *Information processing in animals: Conditioned inhibition* (pp. 355–367). Hillsdale, NJ: Lawrence Erlbaum Associates.

Bottjer, S. W. (1979). *Extinction and "disinhibition" of conditioned excitation and inhibition.* Unpublished doctoral dissertation, Indiana University, Bloomington.

Bottjer, S. W. (1982). Conditioned approach and withdrawal behavior in pigeons: Effects of a novel extraneous stimulus during acquisition and extinction. *Learning and Motivation, 13*, 44–67.

Davis, J. M. (1971). Testing for inhibitory stimulus control with S− superimposed on S+. *Journal of the Experimental Analysis of Behavior, 15*, 365–369.

Desman, R. P. (1969). *Generalization of excitation and inhibition as a function of time between training and testing.* Unpublished master's thesis, University of California, Berkeley.

Durlach, P. J. (1989). Learning and performance in Pavlovian conditioning: Are failures of contiguity failures of learning or performance? In S. B. Klein & R. R. Mowrer (Eds.), *Contemporary learning theories: Pavlovian conditioning and the status of traditional learning theory* (pp. 19–59). Hillsdale, NJ: Lawrence Erlbaum Associates.

Guttman, N., & Kalish, H. I. (1956). Discriminability and stimulus generalization. *Journal of Experimental Psychology, 51*, 79–88.

Hearst, E. (1972). Some persistent problems in the analysis of conditioned inhibition. In R. A. Boakes & M. S. Halliday (Eds.), *Inhibition and learning* (pp. 5–39). London: Academic.

Hearst, E. (1987). Extinction reveals stimulus control: Latent learning of feature-negative discriminations in pigeons. *Journal of Experimental Psychology: Animal Behavior Processes, 13*, 52–64.

Hearst, E., Besley, S., & Farthing, G. W. (1970). Inhibition and the stimulus control of operant behavior [Monograph]. *Journal of the Experimental Analysis of Behavior, 14*, 373–409.

Hendersen, R. W. (1978). Forgetting of conditioned fear inhibition. *Learning and Motivation, 9*, 16–30.

Honig, W. K., & James, P. H. R. (Eds.). (1971). *Animal memory.* New York: Academic.

Honig, W. K., & Urcuioli, P. J. (1981). The legacy of Guttman and Kalish (1956): 25 years of research on stimulus generalization. *Journal of the Experimental Analysis of Behavior, 36*, 405–445.

Jenkins, H. M. (1965). Generalization gradients and the concept of inhibition. In D. Mostofsky (Ed.), *Stimulus generalization* (pp. 55–61). Stanford, CA: Stanford University Press.

Kraemer, P. J., & Spear, N. E. (1992). The effect of nonreinforced stimulus exposure on the strength of a conditioned taste aversion as a function of retention interval: Do latent inhibition and extinction involve a shared process? *Animal Learning and Behavior, 20*, 1–7.

Lindblom, L. L., & Jenkins, H. M. (1981). Responses eliminated by noncontingent or negatively contingent reinforcement recover in extinction. *Journal of Experimental Psychology: Animal Behavior Processes, 7*, 175–190.

Lyons, J. (1969). Stimulus generalization as a function of discrimination learning with and without errors. *Science, 163*, 490–491.

Mackintosh, N. J. (1974). *The psychology of animal learning.* London: Academic.

Mackintosh, N. J. (1977). Stimulus control: Attentional factors. In W. K. Honig & J. E. R. Staddon (Eds.), *Handbook of operant behavior* (pp. 481–513). Englewood Cliffs, NJ: Prentice-Hall.

Miller, R. R., & Spear, N. E. (Eds.). (1985). *Information processing in animals: Conditioned inhibition.* Hillsdale, NJ: Lawrence Erlbaum Associates.

Mostofsky, D. (Ed.). (1965). *Stimulus generalization.* Stanford, CA: Stanford University Press.

Moye, T. B., & Thomas, D. R. (1982). Effects of memory reactivation treatments on post-discrimination generalization performance in pigeons. *Animal Learning and Behavior, 10*, 159–166.

Pavlov, I. P. (1927). *Conditioned reflexes.* London: Oxford.

Rescorla, R. A. (1969). Pavlovian conditioned inhibition. *Psychological Bulletin, 72*, 77–94.

Rescorla, R. A., & Wagner, A. R. (1972). A theory of Pavlovian conditioning: Variations in the effectiveness of reinforcement and nonreinforcement. In A. H. Black & W. F. Prokasy (Eds.), *Classical conditioning II: Current research and theory* (pp. 64–99). New York: Appleton–Century–Crofts.

Riley, D. A. (1968). *Discrimination learning.* Boston: Allyn & Bacon.

Rilling, M. (1977). Stimulus control and inhibitory processes. In W. K. Honig & J. E. R. Staddon (Eds.), *Handbook of operant behavior* (pp. 432–480). Englewood Cliffs, NJ: Prentice-Hall.

Rilling, M., Caplan, H. J., Howard, R. C., & Brown, C. H. (1975). Inhibitory stimulus control following errorless discrimination learning. *Journal of the Experimental Analysis of Behavior, 24*, 121–133.

Rowe, D. W., & Malone, J. C. (1981). Multiple-schedule interactions and discrimination. *Animal Learning and Behavior, 9*, 115–126.

Shepard, R. N. (1965). Approximation to uniform gradients of generalization by monotone transformations of scale. In D. Mostofsky (Ed.), *Stimulus generalization* (pp. 94–110). Stanford, CA: Stanford University Press.

Shepard, R. N. (1987). Toward a universal law of generalization for psychological science. *Science, 237*, 1317–1323.

Spear, N. E., & Miller, R. R. (Eds.). (1981). *Information processing in animals: Memory mechanisms.* Hillsdale, NJ: Lawrence Erlbaum Associates.

Terrace, H. S. (1966). Stimulus control. In W. K. Honig (Ed.), *Operant behavior: Areas of research and application* (pp. 271–344). New York: Appleton–Century–Crofts.

Thomas, D. A. (1979). Retention of conditioned inhibition in a bar-press suppression paradigm. *Learning and Motivation, 10*, 161–177.

Thomas, D. R. (1981). Studies of long-term memory in the pigeon. In N. E. Spear & R. R. Miller (Eds.), *Information processing in animals: Memory mechanisms* (pp. 257–290). Hillsdale, NJ: Lawrence Erlbaum Associates.

Thomas, D. R., Klipec, W., & Lyons, J. (1966). Investigations of a mirror-image transfer effect in pigeons. *Journal of the Experimental Analysis of Behavior, 9*, 567–571.

Thomas, D. R., Windell, B. T., Bakke, I., Kreye, J., Kimose, E., & Aposhyan, H. (1985). Long-term memory in pigeons: I. The role of discrimination problem difficulty assessed by reacquisition measures. II. The role of stimulus modality assessed by generalization slope. *Learning and Motivation, 16*, 464–477.

Tomie, A., & Kruse, J. (1980). Retardation tests of inhibition following discriminative autoshaping. *Animal Learning and Behavior, 8*, 401–408.

6 Retrieval Processes and Conditioning

Philipp J. Kraemer
University of Kentucky

Norman E. Spear
SUNY Binghamton

This chapter takes the view that the understanding of certain features of basic conditioning requires consideration of variables and processes usually dealt with under the heading, "memory." These features, often referred to as instances of stimulus selection, include the consequences of nonreinforced exposure to a conditioned stimulus before or after conditioning and of the occurrence of one stimulus in compound with another in such a way as to invoke stimulus selection (overshadowing).

This consideration of memory and stimulus selection is appropriate for a book honoring Dr. Donald A. Riley. About half of Dr. Riley's research and scholarly publications has been devoted to clarifying the issues of stimulus selection, and roughly the other half has dealt with the memory issues of retention and forgetting (or its inverse, reminiscence) and the ontogeny of behaviors relevant to cognition and memory. So it is significant more than coincidental that the present chapter—both the experiments and the theory—focuses on the influences of stimulus selection, retention and forgetting, and includes considerations of the ontogeny of learning and memory; the content of the present chapter is indebted to the research, theoretical formulations, and ideas of Dr. Riley.

The specific purpose of the present chapter is to present a model involving a limited number of simple assumptions that, when considered together, can help us account for and understand significant instances of variation in the consequences of Pavlovian conditioning. The variation is induced by particular postacquisition circumstances and is attributed generally to the interaction between conflicting memories.

BACKGROUND

After behavior has been modified by the relatively consistent occurrence of a reinforcer, the eventual consequences of entirely eliminating the reinforcer are to decrease or eliminate the behavior. Despite the unanimous agreement on this solid fact of behavioral extinction, there have been a variety of versions of how it is to be interpreted. Pavlov (1927) and later Hull (1943) and Amsel (1958) attributed these consequences of nonreward primarily to motivational factors such as inhibition or emotion. Extinction operations do not, however, yield consequences that formally qualify as inhibitory (e.g., they do not pass summation and retardation tests). Also, what some have viewed as motivational consequences of nonreward have successfully been interpreted differently—for example, in ecological terms (e.g., Staddon, 1970, 1974) or in terms of new learning (Capaldi, 1967). The view we favor is that a new, conflicting memory is acquired when consistent nonreinforcement is introduced before or after related experience with reinforcement (Capaldi, 1971; Estes, 1951, 1959; Spear, 1971). This orientation was applied recently, with good effect, by Bouton (1991).

We apply this orientation in terms of a model that might begin to account for the variance in retention and forgetting caused by having acquired a memory that conflicts with the target memory. The model is intended also to account for key phenomena of Pavlovian conditioning that seem to depend on circumstances that *follow* acquisition of the conditioning, such as the introduction of a long retention interval. For instance, it is well known that nonreinforced exposure to the conditioned stimulus (CS) prior to its pairing with an unconditioned stimulus (US) leads to less effective conditioning than would have occurred without the prior exposure. This effect, conventionally termed "latent inhibition" (LI) (Lubow, 1973), is reduced and even eliminated, however, if a long delay intervenes between the conditioning episode and a test for conditioning (Kraemer, Randall, & Carbary, 1991; Kraemer & Roberts, 1984).

Following a brief review of each of these general effects—the influence of a conflicting memory on retention and postacquisition effects on Pavlovian conditioning phenomena—we describe the features of the model.

INTERFERENCE EFFECTS ON RETENTION: THE CONSEQUENCES OF ACQUIRING A CONFLICTING MEMORY FOR RETENTION OF A TARGET MEMORY

The study of associative interference in animal memory now has a fairly long history. The first reviews of these effects were published over 20 years ago (Gleitman, 1971; Spear, 1971). There was little doubt then that retention of a target memory after several days or weeks could be impaired by a previously acquired (proactive interference) or a subsequently acquired (retroactive interference) conflicting memory.

Proactive interference in retention seemed at times surprisingly ineffective in animals, however, relative to its pervasiveness in human verbal memory (Underwood, 1957). This was especially so for retention of simple acquired discriminations between alternative locations. For several tests up to that time and later, having previously learned the reverse discrimination did not seem to increase forgetting rate (Spear, 1978).

Interference in Long-Term Retention

Proactive and retroactive interference on retention after long intervals is likely to be understood by taking advantage of the circumstances in which interference has its greatest effect. For the case of discrimination learning, proactive interference has seemed most clear for relatively complex discriminations such as a brightness discrimination in which the location dimension is irrelevant (e.g., Chiszar & Spear, 1968; Maier & Gleitman, 1967), or in terms of memory for multichoice discriminations required with relatively complex mazes (Spear, 1978). Another variable that is known to control interference in such tests of retention of discrimination learning is the similarity between the context in which the target memory and conflicting memories are acquired and that in which the target memory is tested (Zentall, 1970).

Although proactive interference and retention by animals has seemed especially clear after short intervals or with relatively complex discriminations (Thomas, 1981), a strong effect of proactive interference after long intervals and without explicit manipulation of context was reported by Kraemer (1984). Pigeons in this study learned a discrimination between an array of three green dots and another of two green dots; pecking the key when three dots appeared (S+) was followed by access to food whereas pecks to two dots (S−) was not, or vice versa. After the birds had achieved a criterion of about 90% accuracy, half were removed to their home cages to wait out a retention interval of 1, 10, or 20 days before testing; others learned the reverse discrimination, and thereby acquired a new memory in clear conflict with the previously acquired memory. Then they too were tested after a retention interval of 1, 10, or 20 days. The test was relearning of the most recent discrimination.

The results were in accord with the suggestions of the earlier experiments with rats. They were more informative, though, due to methodological advantages such as independent measurement of response to S− and S+. There was a good deal of forgetting, even without the prior conflicting learning, primarily in terms of increased responding to S−. There was significantly greater forgetting for the reversal group that had the conflicting memory as the source of proactive interference.

Independence of Previously Correct Responses and Proactive Interference

A notable feature of proactive interference in animals is illustrated in an experiment in which proactive interference was induced by a nonreversal shift rather than a reversal shift (Spear, 1971). These experiments tested rats in a single-choice maze. The correct (reinforced) alternative was one of two values of either of two uncorrelated dimensions, brightness and location.

As an example of the reversal shift case, if the animal learned that the brighter alternative was reinforced but the darker alternative was not, the source of proactive interference was previous learning that the darker alternative was reinforced but the brighter was not. For the nonreversal shift procedure, the conflicting memory that preceded learning of reinforcement in the brighter alternative was having learned that the left location was reinforced and the right one was not, regardless of which was the brighter.

With the nonreversal shift procedure, one can measure the number of formerly correct choices (e.g., correct location) independently of the number of correct choices relevant to the target memory (e.g., the correct brightness). Although this study was fairly extensive and included a variety of experimental conditions, in terms of accuracy of the target memory the results indicated only that rate of forgetting between 1 and 28 days was not increased by the source of proactive interference. This merely added to the substantial number of negative results of this kind that have appeared. More important, however, it was also found that there was no relationship between retention of the target memory and the number of incorrect test responses that would have been correct for the prior conflicting learning. A similar lack of relationship between the intrusion of previously correct responses at the test and retention of the target memory was reported by Kehoe (1963). These findings are consistent with the realization long ago in the field of human verbal memory, that associative interference in retention is not due simply to response competition (e.g., Melton & Irwin, 1940).

Associative Interference in Retention of Nondiscriminative Instrumental (GO/NO-GO) Learning

Simple appetitive conditioning to traverse a straight runway for a food reinforcer has been used to study proactive and retroactive interference of the memory for magnitude of the reinforcer. In these experiments, the rat might first learn to run to a large reinforcer followed by a series of similar experiences with only a small reinforcer, or vice versa; analogously, the rat might first learn to traverse the runway for a particular reinforcer followed by a series of experiences in which no reinforcer was present, or vice versa. For present purposes, the question is whether prior experience with a particular reinforcer magnitude influences retention of performance appropriate for a different, more recent reinforcer magnitude (a variety of examples of associative interference in retention of go/no-go learning also may be found in terms of avoidance learning or Pavlovian aversive conditioning: (Bouton, 1991; Spear et al., 1980).

There seems little doubt that proactive interference occurs in these circumstances, in the sense that performance in accord with the most recently learned reinforcer magnitude declines more rapidly over a subsequent, relatively long interval (i.e., is forgotten more rapidly) if the rat first had learned to traverse the runway for a quite different (larger or smaller) reinforcer magnitude (Spear, 1967; Spear & Spitzner, 1968). Similarly, rats that learned to run for a food reinforcer and then were given a series of running experiences in which no reinforcer was present (extinction) show a greater increase in their running speed after a subsequent retention interval than do animals that never received the first experience of running for a reinforcer (spontaneous recovery). A similar proactive interference effect on retention occurs if rats receive a mild shock along with food in the goal box of a runway after having learned to approach the food in the absence of shock (a sort of recovery from punishment over time; Bintz, Braud, & Brown, 1970).

A final case of the influence of such conflicting memories of reinforcer magnitude on retention is the effect of first allowing animals to run a runway without a food reinforcer present in the goal box, and then allowing them to run in the presence of food reward in the goal box (contrary to intuition, hungry rats will learn to traverse a straight runway even without food at the end; e.g., Spear, Hill, & O'Sullivan, 1965). This case is of special interest because although perhaps less familiar than the instance of spontaneous recovery, it provides a case of instrumental conditioning that is somewhat analogous to the phenomenon of LI in Pavlovian conditioning. The effect of initial nonrewarded trials has not seemed to impair learning as it does in Pavlovian conditioning, at least not in any simple sense. There are in fact indications that prior nonreinforced exposure to the circumstances of instrumental conditioning facilitates acquisition (Spear & Spitzner, 1967), an effect that also has appeared, however, with the LI paradigm (Hoffmann & Spear, 1989).

Two relatively clear effects of initial nonrewarded emissions of instrumental behavior are of interest here. The first is straightforward and is quite consistent with the aforementioned examples of proactive interference in retention induced by prior experience with a different reinforcer magnitude: For rats given initial nonrewarded trials (for which they run slowly) followed by a series of rewarded trials (for which they run rapidly), running speeds after a 24-hr interval initially are significantly slower that those for animals given the rewarded trials but not the prior experience with nonrewarded trials (Spear et al., 1965; Spear & Spitzner, 1967).

The second effect of interest here is related to the fact that initial nonrewarded experiences make the rat more resistant to extinction of subsequent instrumental running reinforced by food: after very long retention intervals, this influence of the prior nonreinforced experiences shows a surprising decrease (Spear & Spitzner, 1967). In several conditions prior nonrewarded experiences substantially increased resistance to extinction 1 day after the reinforced experiences, but if the extinction test was not given until after a retention interval of 8 days, the initial nonrewarded trials exerted relatively little or no influence on resistance to extinction. This result implicitly supports the notion that the occurrence of nonreward is forgotten more rapidly than the occurrence of reward, a point that is relevant to, and in fact constitutes the core of, an important assumption (Rule 4) in the model we present.

Although not often tested directly, the assumption that memory for reinforcement may outlast memory for nonreinforcement has at least one other precedent. To account for behavior patterns that accompany periodic schedules of reinforcement (e.g., fixed-interval schedules), Staddon (1974) explicitly assumed reward omission to be " . . . much less memorable than R (reward)" (p. 380).

Although this discussion serves to illustrate support for one of our theoretical assumptions, it also introduces our consideration of the effect of the retention interval on prior nonreinforced exposure to the CS in Pavlovian conditioning.

FORGETTING OF LATENT INHIBITION: AN EXAMPLE OF POSTACQUISITION MODIFICATION OF PAVLOVIAN STIMULUS SELECTION EFFECTS

It has been reported in a number of experiments that the effect of nonrewarded presentations of the CS prior to conditioning dissipates markedly if a long interval elapses between conditioning and the test (Ackil, Carman, Bakner, & Riccio, 1992; Kraemer et al., 1991; Kraemer & Roberts, 1984).

This effect is interesting for two reasons. The first reason pertains to the empirical definition of the effect: The effect involves an *increase* over time in the expression of conditioning that was preceded by CS exposure. The other reason the effect is interesting is that the effect cannot be accounted for by any existing theories of conditioning and learning: Theories that assume CS preexposure disrupts conditioning are unable to explain the increase in expression of conditioning over time (Lubow, 1989).

The first experiment to observe this "release from latent inhibition" did so unexpectedly in a study of the effect of flavor preexposure and retention interval on conditioned flavor aversion in rats (Kraemer & Roberts, 1984). Rats with or without prior flavor experience were exposed to a flavor paired with scopolamine. When ingestion of a flavor was tested 1 day after conditioning, rats given flavor preexposure had an attenuated magnitude of the learned aversion to the flavor, relative to the nonpreexposed control group. This provided a conventional demonstration of latent inhibition. When testing was delayed 21 days, however, preexposed and nonpreexposed subjects expressed equally strong aversions. The LI that was evident after a short retention interval seemed to dissipate over time.

Subsequent experiments tested the effect further for its generality and for conditions that might indicate limitations of the effect. Release from latent inhibition was found with the more conventional unconditioned stimulus used in such experiments, LiCl, in addition to scopolamine (Kraemer & Roberts, 1984). The effect occurred whether the test was a 1-bottle consumption test or a two-bottle preference test (Kraemer & Ossenkopp, 1986), and it occurred after primary CS exposure (i.e., prior experience with the CS itself) as well as generalized exposure (prior experience with a flavor different from the CS) (Ackil et al., 1992; Kraemer, Hoffmann, & Spear, 1988; Kraemer & Roberts, 1984; Strohen, Bakner, Nordeen, & Riccio, 1990). Although release from latent inhibition has not always occurred with primary preexposure, it has been observed in all experiments using the generalized preexposure procedure.

Release from LI over time suggests that preexposed flavor memories may themselves undergo forgetting. This is an important part of the theoretical analysis presented later in this chapter, which treats the memory of the CS preexposure and the memory of the CS conditioning as conflicting memories.

The generality of release from LI as been extended by its occurrence in conditioning tests other than conditioned flavor avoidance. One test was in terms of a conditioned emotional response developed by pairing the onset of a light with a mild footshock (Kraemer et al., 1991). Half the rats in this experiment were given only the 12 pairings of the light and footshock whereas the other half received in addition 20 light-alone presentations 1 hr before the pairings. Separate subgroups of preexposed and nonpreexposed subjects were tested after 1, 7, or 21 days following termination of conditioning. Conditioned suppression of activity to the light was quite clear in the conditioned groups at all retention intervals (forgetting in these circumstances is relatively slow). After a 1-day interval there was clear latent inhibition; the conditioned suppression exhibited by the preexposed animals was significantly weaker than that of the nonpreexposed animals. After the 7- and 21-day retention intervals, however, conditioned suppression was equivalent in preexposed subjects; conditioning by the preexposed subjects was more strongly expressed after these longer retention intervals. Subsequent control experiments verified that the time-dependent increase in conditioned suppression among animals given prior CS exposure was a true reflection of Pavlovian conditioning; these control experiments indicated that increases in neophobia, sensitization, or pseudoconditioning over the retention interval could not explain the release from LI.

Finally, an experiment was conducted to assess the functional relationship between the effects of CS exposure prior to conditioning (LI) and CS exposure subsequent to conditioning (extinction). The question was whether the introduction of a retention interval between treatment and test would have similar or different effects on these two phenomena. Conditioned flavor aversion for a maple-flavored solution was induced by pairing it with injection of LiCl. This treatment was preceded or followed by nonreinforced exposure to either the same maple-flavored solution or to a different novel-flavored solution (saccharin). A test for relative preference of the maple-flavored solution and water was given 1 or 21 days after treatment. The results indicated quite similar effects of retention interval and of generalized versus primary CS exposure on extinction and LI (Kraemer & Spear, 1992).

Postacquisition Modulation of Other Pavlovian Conditioning Effects

On the basis of most contemporary theories of conditioning and learning, one might tend to view release from LI as an anomaly. The effect of CS preexposure has been interpreted almost unanimously as an effect on acquisition. Due either to decreased attention or decreased accumulation of associative strength, less learning of the CS-US association is assumed to have occurred for animals given CS preexposure (e.g., Lubow, 1973, 1989; Pearce & Hall, 1980; Rescorla & Wagner, 1972). The release from LI indicates, however, that the effect of CS preexposure is in large part due to postacquisition processes, those associated with memory retrieval and expression.

Yet release from LI is unlikely to be an anomaly. A larger number of apparent acquisition effects on which conditioning theories are based are now known to be due at least in part, and perhaps primarily, to postacquisition effects associated with memory retrieval and expression (for reviews see Miller, Kasprow, & Schachtman, 1986; Spear, 1981; Spear, Miller, & Jagielo, 1990; Spear & Riccio, in press). In the case of LI, for instance, this conclusion has been verified by at least one other postacquisition manipulation in addition to the introduction of a retention interval. In this experiment pairings of a white noise and a mild foot-shock were preceded for some animals by nonreinforced presentations of the white noise (Kasprow, Catterson, Schachtman, & Miller, 1984). The test for conditioning was the extent to which licking for water was suppressed in the presence of the white noise. As expected, those animals given nonreinforced prior exposure to the white noise showed less conditioned suppression than those not given the prior exposure. Some of the conditioned animals, however, were given a special prior cueing treatment just before the conditioned suppression test. This consisted of presentations of only the US (footshock) in a context different from that of conditioning. The consequence of this prior cueing treatment was to greatly reduce the degree of LI.

To illustrate concretely that release from LI is not an anomaly among conditioning effects, we review the case of overshadowing, which is such a viable and well-known effect that it has been a prime consideration of conditioning theories.

Pavlov (1927) discovered that Stimuli A and B presented in compound as a CS paired with a US do not share equally in controlling conditioned behavior. If A is more "salient" than B, less conditioning to B will occur than if only B had been paired with the US. That this is not merely a matter of less acquisition of B in the presence of A (Rescorla & Wagner, 1972) has been shown by experiments in which nonreinforced exposure to the A component is given after acquisition and B is found to have greater associative strength; overshadowing is decreased (Kasprow, Cacheiro, Balaz, & Miller, 1982; Kaufman & Bolles, 1981; Matzel, Schachtman, & Miller, 1985). It is now known that the consequences of over-shadowing, like those of LI, also decrease as time passes following acquisition (Kraemer, Lariviere, & Spear, 1988; Miller, Jagielo, & Spear, 1990).

THE MODEL

The model is designed to provide a retrieval-oriented account of the kinds of interference effects described earlier. The cornerstone of the model is the con-struct of retrievability, which refers to the relative probability that a given memo-ry will be retrieved in response to presentation of appropriate cues. It is assumed that in addition to its content, each memory has a characteristic level of retriev-ability, which governs the extent to which a memory remains accessible. Retriev-ability provides a simple mechanism for the suppression and reinstatement of conflicting behavioral tendencies. One obvious advantage of this mechanism is that it allows the organism to benefit from the retention of knowledge that may become only temporarily invalid or irrelevant. If a suppressed memory is once again useful, the organism does not need to encode that information again; it need only make it more accessible. This type of process provides a more flexible alternative to the notion of memory substitution, whereby conflicting memories are continually replaced.

The main assumptions of the model are as follows: (a) retrievability of a memory can vary, (b) changes in retrievability can be induced by experience, and (c) changes in retrievability influence both acquisition and retention of learning episodes. Consistent with several contemporary analyses (e.g., Lewis, 1978), the model adopts an activation view of memory retrieval. A memory is defined as a list or set of attributes (Bower, 1967; Spear, 1973; Underwood, 1969). Each attribute corresponds to an encoded version of some perceived feature of a learning episode. These features include characteristics of external stimuli (e.g., CSs, USs, and contextual cues), internal stimuli associated with affective and motivational aspects of the event, and properties of the organism's behavioral reaction to the event. Memory storage consists of the establishment (encoding) of individual memory attributes and the organization of these attributes into attri-bute sets (memories). An organized set of attributes defines a memory, although each individual attribute can be contained in more than one memory. Memories are more or less similar to each other to the degree to which they share attributes. At the extreme, it is possible for two or more memories to share the same stimulus attributes and differ only with respect to attributes representing response consequences, such as reward or nonreward.

Retrieval consequences are governed by output lines. Output lines link specific memories to other memories or other processing units. Output lines are energized once a memory is retrieved. When an output line is energized, it can have several different consequences. It can activate a response generator, which then translates a retrieved memory into behavior; it can activate attributes within other memories through associative linkage; or it can do both. It is also assumed that higher level processing units can block or inhibit the actions of an energized output line. Thus, it is possible for a memory to be retrieved (i.e., by energizing its output line) without that memory being expressed in behavior (cf. Spear, 1981).

Memory attributes are aroused or activated when the corresponding stimulus features are perceived by the organism. The current model assumes that attribute activation is a discrete all or none process—attributes are active or inactive (Lewis, 1978). Once activated, however, attributes can vary in terms of their level of activation. This allows for the possibility for some stimulus features of a learning episode to exert more influence than other features over retrieval. Each attribute normally remains active for a brief interval, but an attribute can be reactivated persistently when the feature it represents continues to be perceived. At any point in time, the overall state of activation for a memory is determined by the cumulative level of activation of all of its aroused attributes.

There are two additional assumptions that are critical. First, each memory is assumed to have a baseline *activation threshold* (AT), which refers to the level of cumulative attribute activation that must be reached in order for the memory to be retrieved, that is, for its output line to be energized. The baseline AT is set initially at encoding, and it is a function of the nature of the constituent attributes (i.e., memorial content). Baseline ATs are usually set so that not all attributes included in the memory need to be activated in order for the memory to be retrieved; only a sufficient number, type, or combination of attributes must be aroused in order to surpass the AT (cf., Spear, 1973).

The second assumption is that ATs can change; they can be increased above or decreased below baseline by various factors, including experience. As an AT is increased above baseline, it will become increasingly more difficult to retrieve that memory. At the extreme, an AT may be so high that even with the activation of all of its attributes the memory still cannot be retrieved. When an AT is decreased below baseline, it becomes progressively easier to retrieve that memory.

Processing Rules

The model is characterized by a set of hypothetical processing rules that govern how ATs change, both as a result of experience and as a function of time. By describing the rules governing changes in ATs, the model attempts to explain a number of diverse acquisition and retention phenomena.

Rule 1.   The continued occurrence of events consistent with the content of a memory will lower its AT, and the continued occurrence of events inconsistent with its content will raise its AT. This rule addresses the impact of event consistency; for example, the continued occurrence or omission of a previously expected US.

Rule 2.   When two or more conflicting memories exist (i.e., memories that share attributes), the ATs for these memories will gradually converge over a retention interval, unless conditions exist that are stipulated by other rules. The reciprocal changes in ATs will tend to equalize the relative retrievability of the competing memories. This rule will be used to account for various interference effects.

Rule 3.   Subthreshold activation of a memory, due to the arousal of attributes that are insufficient to surpass its AT, can decrease the AT itself. This rule will be applied to phenomena such as reactivation and reinstatement, in which brief exposure to some aspects of an original learning episode can decrease forgetting. The important feature of this rule is that it allows for the future retrievability of a memory to be affected by events that by themselves are insufficient to actually retrieve the memory.

Rule 4.   Motivational and emotive feedback associated with a learning episode can modify the degree to which other events influence an AT. This rule has precedence over all other rules. Consequently, under conditions to be described later, ATs may not change as expected by other rules. Rule 4 is especially important in considering conflicting learning episodes that are asymmetrical with respect to their affective and motivational properties.

Application of the Model

Interference Effects and Retention.   The model is explicitly designed to deal with situations that involve processing of conflicting information. Critical to the model is a distinction between two types of memorial interference, which differ in terms of the nature of the interfering events themselves. Conflicting events can be either symmetrical or asymmetrical. Symmetrical events share the same affective and motivational properties, whereas asymmetrical events involve salient differences in these properties.

The vast research on the role of interference in forgetting, both in humans and nonhumans, has concentrated on situations that incorporate two or more conflicting episodes that are symmetrical. Reverse discrimination training involves this type of interference. For example in Kraemer's (1984) study described earlier,

the reverse discrimination procedure involved the same response-reinforcer relation as that of the original discrimination: Food followed responses to S+ and no
food followed responses to S−. The only difference between the two discriminations was the reversal of the stimuli that served as S+ and S−. As found by
Kraemer, the general result has been that reversal groups show more forgetting
over a long retention interval than do control groups, which are trained on only
the original discrimination (cf. Kehoe, 1963).

The important point is that the conflicting events involved in reverse discriminations are symmetrical with respect to their emotive and motivational attributes.
According to the model, the learning of the initial discrimination involves the
formation of the originally acquired memory (OM), which when retrieved results
in a correct choice response. In addition to the storage of OM, changes in its AT
will occur as training continues. According to Rule 1, continued rewarding of
correct choices will lower the AT from baseline, which will increase the retrievability of OM over training.

Reversal training will involve both the formation of the recent memory (RM),
as well as changes in the AT for OM. Based on Rule 1, the absence of anticipated
consequences associated with responding to the original S+ is expected to increase the AT for OM, and rewarding of choices consistent with the reversal
problem should lower the AT for RM. Thus, immediately after reversal training
RM will have a selective retrieval advantage over OM.

The model predicts that further changes will occur in ATs for the two memories over a long retention interval. Given the affective symmetry between OM
and RM, these changes will be governed solely by Rule 2. Accordingly, the ATs
for the two memories are expected to change in opposite directions as to approach a state of parity, relative to the status of the two memories immediately
after reversal training. The AT for OM, which had been increased during reversal
training, will now decline toward baseline, whereas the AT for RM will increase
above baseline. The retrieval consequence of these changes will be that OM will
become more accessible while RM becomes less accessible, relative to the state
of affairs that existed at the end of reversal training. The behavioral consequence
of these changes will be proactive interference: a decreased tendency for the
subject to respond according to the contingencies associated with RM as the
retention interval passes, relative to performance of a subject trained on only one
discrimination.

Consistent with traditional interference theory, the model postulates a recovery in the accessibility of OM, which is partly responsible for the proactive
interference effect. It should be emphasized, however, that the present model
does not regard the increased accessibility of OM as the cause of the decreased
accessibility of RM. Note that the model does not assume that proactive interference entails only response competition; there is also expected to be a decrease
in the accessibility of the more recent memory, which is independent of the
recovery of responding associated with OM. Thus, the model predicts that proac-

tive interference can obtain in situations in which there is no behavioral evidence of recovery of a competing memory, which more accurately reflects the evidence cited previously (Spear, 1971).

A different process is assumed to operate with conflicting memories that are asymmetrical. It is reasonable to expect that memory processing will be sensitive to the affective and motivational features of a learning episode, and the model explicitly allows for this possibility. The general idea is that emotive feedback associated with particular events will influence retrievability.

Extinction.   The case of Pavlovian extinction provides a good illustration of a situation that involves asymmetrical memories. In contrast to views that have emphasized or restricted interpretation to memorial content (e.g., Pearce & Hall, 1980; Rescorla & Wagner, 1972), the model adopts a retrieval processing view of extinction, which is quite similar to the framework offered by Bouton (1991) (see also Capaldi, 1966, 1967, 1971; Spear, 1971).

The model assumes that the extinction episode (EM) and the original conditioning episode (OM) are represented by two independent memories. Each memory has a different output line, and different behaviors are associated with the retrieval of each; a conditioned response appears when OM is retrieved, and the absence of a conditioned response (or perhaps some response tendency that is incompatible with the conditioned response) occurs when EM is retrieved. The process of extinction involves encoding of EM, as well as changes in the retrievability of OM. According to Rule 1, the withholding of an expected US, which defines extinction training, will increase the AT for OM. This change will decrease the probability that OM will be retrieved, which explains the dissipation of conditioned responding that occurs as extinction training progresses.

In addition to the mere occurrence of extinction, the model can successfully account for a number of other extinction phenomena. For instance, the model can adequately address the context specificity of extinction. If animals are conditioned in one context and given extinction training in a distinctly different context, conditioned responding often reappears when tested in the original learning context (Bouton, 1991; Bouton & Bolles, 1985).

According to the model, a change of context between original conditioning and extinction will diminish the attribute similarity between OM and EM, which will promote selective retrieval. Contextual cues available during testing will arouse attributes within the memory associated with the test context. Although activation of contextual attributes will contribute to the overall state of activation of a memory, arousal of these attributes alone usually will not be sufficient to reach an AT. Presentation of a CS, however, will provide the necessary remaining activation to surpass the AT, but only in the memory containing the already aroused contextual attributes; the level of activation in the other memory will remain below threshold. Thus, with different contextual cues associated with EM and CM, the CS will selectively retrieve the memory associated with the test context, and behavior consistent with the content of the retrieved memory will be expressed.

Another extinction phenomenon to which the model can be applied is reinstatement. Conditioned responding following extinction can be reestablished by merely presenting the US just prior to testing (Bouton, 1984; Rescorla & Heth, 1975). The model assumes that features of the US are represented by attributes contained in OM. Presentation of the US prior to testing will consequently activate attributes in OM, but the cumulative level of activation will be insufficient to surpass its AT. Rule 3 stipulates, however, that subthreshold activation of OM will decrease its AT, which in turn will increase the likelihood that OM will be retrieved when the CS is subsequently presented.

The previous discussion of context specificity can be combined with the analysis of reinstatement to account for the fact that reinstatement itself appears to be context specific. Presentation of a US prior to testing does not always evoke conditioned responding to a previously extinguished CS; the reinstatement experience must occur in the same context as extinction (Bouton, 1984; Bouton & Bolles, 1979; Bouton & King, 1983). According to the model, US presentation outside of the extinction context will fail to arouse attributes within OM, and consequently the AT for OM will be unaffected by Rule 3.

The final extinction phenomenon discussed is spontaneous recovery, which perhaps best illustrates the scope and value of the model. Spontaneous recovery has proven sufficiently troublesome for other models of conditioning as to be generally ignored (e.g., Mackintosh, 1975, 1983; Pearce & Hall, 1980; Rescorla & Wagner, 1972). The model approaches spontaneous recovery as a form of forgetting, which is consistent with some previous analyses (Capaldi, 1967; Spear, 1971). The existence of two conflicting memories (OM and EM) will precipitate changes in ATs for these memories over a retention interval. The two memories are presumed to be asymmetrical, in that the affective consequences of experiencing the US (OM) are expected to differ substantially from the consequences of not receiving the US (EM). Thus, changes in ATs will be governed by Rule 4, rather than Rule 2.

Rule 4 reflects the assumption that forgetting sometimes reflects an adaptive process whereby memories of greater biological importance (i.e., memories associated with more salient affective or motivational significance) gain a retrieval advantage over memories of less biological importance; in short, animals remember better that which is more important (e.g., memory for reward may be better than memory for nonreward; Staddon, 1974). Thus, rather than the ATs for OM and EM reaching an equilibrium as stipulated by Rule 2, the model regards spontaneous recovery as a case in which a more biologically significant memory (OM) gains a retrieval advantage over a relatively less important memory (EM); the AT for OM will become lower than the AT for EM over a retention interval, giving OM a retrieval advantage that is expressed as spontaneous recovery.

Another situation that would alter memorial processing through Rule 4 is counterconditioning. Rather than omitting the US, a subject might instead receive the CS paired with some event opposite in affective value to that of the original US. For example, a CS that had been paired with shock during conditioning might be subsequently paired with food. The model predicts less spontaneous recovery in this situation, relative to a US-omission group. The positive affect of the US in the second phase of training would counteract the affective imbalance between learning episodes that is assumed for the case of simple extinction. In a sense, with counterconditioning neither memory would be unimportant, relative to the other, and the situation would be more like that found with symmetrical interference.

Latent Inhibition.   Perhaps the most novel feature of this proposal is found in the application of the model to LI, which represents yet another instance where two conflicting memories are affectively imbalanced. Explanations of LI have almost universally emphasized the acquisition process (Lubow, 1989), yet as described earlier it has been shown that LI can be reduced or eliminated either through reinstatement (i.e., presentations of the US given after conditioning; Kasprow et al., 1984) or by delayed testing (Kraemer et al., 1991; Kraemer & Roberts, 1984). Both reinstatement and release from LI suggest that stimulus preexposure does not necessarily produce a storage deficit. If nonreinforced exposure to a stimulus disrupts what an animal can learn about that stimulus, then it is difficult to explain how this putative learning deficit could be eliminated by events that occur after the learning episode.

As an alternative to the storage failure view, the model regards LI as another instance of retrieval failure, one induced by changes in ATs for conflicting memories that are affectively asymmetrical. It is assumed that nonreinforced stimulus exposure is represented by the preexposure memory (PM) and subsequent exposure to CS-US presentations is represented by the conditioning memory (CM). These two independent memories are highly similar, and may differ only with respect to the event characteristics associated with each: nonreinforcement during preexposure and contingent reinforcement during conditioning. Retrieval of each memory generates behavior consistent with its content: conditioned responding when CM is retrieved and some other response tendency when PM is retrieved.

When testing occurs soon after conditioning, both of the competing memories are highly accessible, and selective retrieval failure may occur. Available retrieval cues (e.g., the CS and contextual cues) are able to activate attributes within both PM and CM, which means that CM will not be exclusively retrieved. Consequently, weaker conditioned responding to the CS will appear, relative to a nonpreexposed subject that is not given CS-alone exposure prior to conditioning.

When testing is delayed, however, ATs are expected to change in accord with Rule 4, given that CM and PM are affectively asymmetrical. Presentation of the US during conditioning should imbue CM with a higher affective value than that associated with PM. Thus, over a retention interval the AT for CM is expected to become lower than the AT for PM, which will give CM a retrieval advantage. The enhanced retrievability of CM will appear as a enhanced conditioned responding, relative to responding soon after conditioning; in other words, there will be a release from LI.

It is also possible, however, that despite changes in ATs the CS may still be able to activate sufficient attributes within PM to surpass its AT and release from LI may not always occur (Kraemer & Roberts, 1984). In addition, the retrieval process is assumed to be further influenced by the degree of similarity between the contents of PM and CM. Memorial similarity is expected to influence not only the initial magnitude of LI, but also the rate and extent of any release from LI. With generalized preexposure (nonreinforced exposure to some stimulus other than the CS), the degree of LI will decrease as a function of decreasing similarity between the preexposure stimulus and the CS; the tendency for the CS to retrieve PM will be greater when it shares more attributes with the preexposure stimulus. The available evidence supports this contention (Dawley, 1979; Siegel, 1969).

The attribute similarity between PM and CM will also affect the rate and magnitude of release from LI. In general, the greater the discrepancy between the CS and the preexposure stimulus, the more difficult it will be for the CS to retrieve PM. As the AT for PM increases over a retention interval, it will become even more difficult for the CS to retrieve PM. Consequently, it is expected that release from LI will be more rapid and more potent as the similarity between the CS and the preexposure stimulus decreases. The available evidence tends to support this prediction (Kraemer & Ossenkopp, 1986; Kraemer & Roberts, 1984; Kraemer & Spear, 1992).

Another reliable constraint on LI involves the role of context. When preexposed to the CS in one context and conditioned in a distinctly different context, subjects often fail to demonstrate LI, at least when tested in the conditioning context (Bouton, 1991; Hall & Channell, 1986; Hall & Minor, 1984; Kaye, Preston, Szabo, Druiff, & Mackintosh, 1987). As described previously, the model assumes that contextual cues play an important role in retrieval. It is the cumulative state of activation of all aroused attributes within a memory that determines whether or not the AT will be reached. When the contextual cues represented in a memory are not present at the time the retrieval process is initiated, the probability of retrieving the memory containing those attributes will consequently be diminished. Thus, similar to extinction, changing contexts between nonreinforced and reinforced exposure to the CS can eliminate LI, through enhanced selective retrieval resulting from a decrease in the attribute overlap between PM and CM.

The model predicts that LI will be a function of where testing occurs. The model predicts stronger LI when testing occurs in the preexposure context (PM will be retrieved) than when it occurs in the conditioning context (CM will be retrieved), and the degree to which conditioned responding will be attenuated should be greater than that expected by merely changing contexts between conditioning and testing. This effect has already been reported (Bouton, 1991; Bouton & Bolles, 1985; Wright, Skala, & Peuser, 1986).

The reinstatement effect reported by Kasprow et al. (1984) also can be accounted for by the model. Kasprow et al. found that exposure to the US prior to testing was sufficient to alleviate LI. According to Rule 3, activation of memorial attributes can lower an AT, even when the memory itself is not retrieved. Lowering of the AT for CM, by presentation of the US, will give CM a retrieval advantage over PM, resulting in less retrieval competition between the two memories and less LI.

An important implication of the present analysis is that it suggests that a common process may underlie LI and extinction, which is incongruent with many theories of conditioning. In Pavlovian conditioning, extinction and LI share an undeniable methodological symmetry; they both involve exposing the subject to the CS alone and to the CS paired with a US. The only difference between the two procedures is the order in which these two events occur. Despite this procedural symmetry, the two phenomena have not been given parallel theoretical treatment.

A notable exception is the analysis provided by Bouton (1991). He suggested that extinction and LI may depend on a common processes, one that entails retrieval rather than storage operations. The present analysis is in strong agreement with Bouton's orientation. Although the process commonality of the two phenomena may be difficult to establish, some of the initial research discussed previously indicates that there already exist some functional similarities between LI and EXT. This evidence includes the influence of contextual cues, the impact of delayed testing, the effect of reinstatement, and the interaction between stimulus similarity and retention interval.

Extension of the Model

The model represents an initial attempt to provide a detailed theoretical account of how retrieval failure can be used to account for an array of divergent phenomena, some of which are clearly beyond the scope of alternative theories. We want to emphasize, however, the preliminary status of the model. It may be necessary to alter some aspects of the model, perhaps by adding or changing some of the processing rules, in order to expand the scope of the basic idea of retrievability. For example, an additional rule that specifies that ATs can change merely through the passage of time could be used to explain a result reported by Kraemer and Roberts (1984). They found that a retention interval imposed between generalized flavor preexposure and conditioning eliminated generalized LI. By allowing ATs to change even in the absence of conflicting memories, the model could accommodate this result.

Another challenge for the model concerns the results of postacquisition changes in overshadowing. It has been shown that overshadowing can be eliminated by extinguishing the "overshadowing stimulus" (Kaufman & Bolles, 1981; Matzel et al., 1985) or by delayed testing (Kraemer et al., 1988; Miller et al., 1990). In addition, delayed testing also has been found to induce overshadowing (Miller et al., 1990). Although these effects do not readily follow from the current version of the model, it may be possible to include additional assumptions or processing rules that will enable the model to capture these kinds of phenomena. What remains nonetheless firm, however, is that there is a value in emphasizing the role of retrieval processes in conditioning, and as alternative models are developed and tested we can expect to discover important new insights about the conditioning process.

REFERENCES

Ackil, J. K., Carman, H. M., Bakner, L., & Riccio, D. C. (1992, May). *Reinstatement of latent inhibition through a reminder treatment.* Paper presented at the annual meeting of the Midwestern Psychological Association, Chicago.

Amsel, A. (1958). The role of frustrative nonreward in noncontinuous reward situations. *Psychological Bulletin, 55,* 102–119.

Bintz, J., Braud, W. G., & Brown, J. S. (1970). An analysis of the role of fear in the Kamin Effect. *Learning and Motivation, 1,* 170–176.

Bouton, M. E. (1984). Differential control by context in the inflation and reinstatement paradigms. *Journal of Experimental Psychology: Animal Behavior Processes, 10,* 56–74.

Bouton, M. E. (1991). Context and retrieval in extinction and in other examples of interference in simple associative learning. In L. W. Dachowski & C. F. Flaherty (Eds.), *Current topics in animal learning: Brain, emotion, and cognition.* (pp. 25–52). Hillsdale, N.J.: Lawrence Erlbaum Associates.

Bouton, M. E., & Bolles, R. C. (1979). Role of conditioned contextual stimuli in reinstatement of extinguished fear. *Journal of Experimental Psychology: Animal Behavior Processes, 5,* 368–378.

Bouton, M. E., & Bolles, R. C. (1985). Contexts, event-memories, and extinction. In P. D. Balsam & A. Tomie (Eds.), *Context and learning* (pp. 133–166). Hillsdale, N. J.: Lawrence Erlbaum Associates.

Bouton, M. E., & King, D. A. (1983). Contextual control of the extinction of conditioned fear: Tests for the associative value of the context. *Journal of Experimental Psychology: Animal Behavior Processes, 9,* 248–265.

Bower, G. (1967). A multicomponent theory of the memory trace. In G. Bower (Ed.), *The psychology of learning and motivation: Advances in theory and research* (Vol. 1, pp. 229–325). New York: Academic.

Capaldi, E. J. (1966). Partial reinforcement: A hypothesis of sequential effects. *Psychological Review, 73,* 459–477.

Capaldi, E. J. (1967). A sequential hypothesis of instrumental learning. In K. W. Spence & J. T. Spence (Eds.), *The psychology of learning and motivation* (Vol. 1, pp. 67–156). New York: Academic.

Capaldi, E. J. (1971). Memory and learning: A sequential viewpoint. In W. K. Honig & P. H. R. James (Eds.), *Animal memory* (pp. 115–154). New York: Academic.

Chiszar, D. A., & Spear, N. E. (1968). Proactive interference in retention of nondiscriminative learning. *Psychonomic Society, 12*, 87–88.

Dawley, J. M. (1979). Generalization of the CS-preexposure effect transfers to taste aversion learning. *Animal Learning & Behavior, 7*, 23–24.

Estes, W. K. (1951). Toward a statistical theory of learning. *Psychological Review, 57*, 94–107.

Estes, W. K. (1959). The statistical approach to learning theory. In S. Koch (Ed.), *Psychology: A study of a science* (Vol. 2, pp. 380–491). New York: McGraw-Hill.

Gleitman, H. (1971). Forgetting of long-term memories in animals. In W. K. Honig & P. H. R. James (Eds.), *Animal memory* (pp. 1–44). New York: Academic.

Hall, G., & Channell, S. (1986). Context specificity of latent inhibition in taste aversion learning. *Quarterly Journal of Experimental Psychology, 38B*, 121–139.

Hall, G., & Minor, H. (1984). A search for context-stimulus associations in latent inhibition. *Quarterly Journal of Experimental Psychology, 36B*, 145–169.

Hoffmann, H., & Spear, N. E. (1989). Facilitation and impairment of conditioning in the preweanling rat after prior exposure to the conditioned stimulus. *Animal Learning & Behavior, 17*, 63–69.

Hull, C. L. (1943). *Principles of Behavior*. New York: Appleton.

Kasprow, W. J., Cacheiro, H., Balaz, M., & Miller, R. R. (1982). Reminder-induced recovery of associations to an overshadowed stimulus. *Learning and Motivation, 13*, 155–166.

Kasprow, W. J., Catterson, D., Schachtman, T. R., & Miller, R. R. (1984). Attenuation of latent inhibition by post-acquisition reminder. *Quarterly Journal of Experimental Psychology, 36B*, 53–63.

Kaufman, M. A., & Bolles, R. C. (1981). A nonassociative aspect of overshadowing. *Bulletin of the Psychonomic Society, 18*, 318–320.

Kaye, H., Preston, G. C., Szabo, L., Druiff, H., & Mackintosh, N. J. (1987). Context specificity of conditioning and latent inhibition: Evidence for a dissociation of latent inhibition and associative interference. *The Quarterly Journal of Experimental Psychology, 39B*, 127–145.

Kehoe, J. (1963). Effects of prior and interpolated learning on retention in pigeons. *Journal of Experimental Psychology, 65*, 537–545.

Kraemer, P. J. (1984). Forgetting of visual discriminations by pigeons. *Journal of Experimental Psychology: Animal Behavior Processes, 10*, 530–542.

Kraemer, P. J., Hoffmann, H., & Spear, N. E. (1988). Attenuation of the CS-preexposure effect after a retention interval in preweanling rats. *Animal Learning & Behavior, 16*, 185–190.

Kraemer, P. J., Lariviere, N. A., & Spear, N. E. (1988). Expression of a taste aversion conditioned with an odor-taste compound: Overshadowing is relatively weak in weanlings and decreases over a retention interval in adults. *Animal Learning & Behavior, 16*, 164–168.

Kraemer, P. J., & Ossenkopp, K. P. (1986). The effects of flavor exposure and test interval on conditioned taste aversions in rats. *Bulletin of the Psychonomic Society, 24*, 219–221.

Kraemer, P. J., Randall, C., & Carbary, T. (1991). Release from latent inhibition with delayed testing. *Animal Learning & Behavior, 19*, 139–145.

Kraemer, P. J., & Roberts, W. A. (1984). The influence of flavor preexposure and test interval on conditioned taste aversions in the rat. *Learning and Motivation, 15*, 259–278.

Kraemer, P. J., & Spear, N. E. (1992). The effect of nonreinforced stimulus exposure on the strength of a conditioned taste aversion as a function of retention interval: Do latent inhibition and extinction involve a shared process? *Animal Learning & Behavior, 20*, 1–7.

Lewis, D. J. (1978). Psychobiology of active and inactive memory. *Psychological Bulletin, 86*, 1054–1083.

Lubow, R. E. (1973). Latent inhibition. *Psychological Review, 79*, 398–407.

Lubow, R. R. (1989). *Latent inhibition and conditioned attention theory*. New York: Cambridge University Press.

Mackintosh, N. J. (1975). A theory of attention: Variations in the associability of stimuli with reinforcement. *Psychological Review*, *82*, 276–298.

Mackintosh, N. J. (1983). *Conditioning and associative learning*. New York: Oxford University Press.

Maier, S. F., & Gleitman, H. (1967). Proactive interference in rats. *Psychonomic Society*, *7*, 25–26.

Matzel, L. D., Schachtman, T. R., & Miller, R. R. (1985). Recovery of an overshadowed association achieved by extinction of the overshadowing stimulus. *Learning and Motivation*, *16*, 398–412.

Melton, A. W., & Irwin, J. M. (1940). The influence of the degree of interpolated learning on retroactive inhibition and overt transfer of specific responses. *American Journal of Psychology*, *53*, 173–203.

Miller, J. S., Jagielo, J. A., & Spear, N. E. (1990). Age-related differences in short-term retention of separable elements of an odor aversion. *Journal of Experimental Psychology: Animal Behavior Processes*, *15*, 194–201.

Miller, R. R., Kasprow, W. J., & Schachtman, T. R. (1986). Retrieval variability: Sources and consequences. *American Journal of Psychology*, *99*, 145–218.

Pavlov, I. P. (1927). *Conditioned reflexes*. New York: Dover.

Pearce, J. M., & Hall, G. (1980). A model for Pavlovian learning: Variations in the effectiveness of conditioned but not of unconditioned stimuli. *Psychological Review*, *87*, 532–552.

Rescorla, R. A., & Heth, C. D. (1975). Reinstatement of fear to an extinguished conditioned stimulus. *Journal of Experimental Psychology: Animal Behavior Processes*, *1*, 88–96.

Rescorla, R. A., & Wagner, A. R. (1972). A theory of Pavlovian conditioning: Variations in the effectiveness of reinforcement and nonreinforcement. In A. H. Black & W. F. Prokasy (Eds.), *Classical conditioning II: Current research and theory* (pp. 64–99). New York: Appleton–Century–Crofts.

Siegel, S. (1969). Generalization of latent inhibition. *Journal of Comparative and Physiological Psychology*, *69*, 157–159.

Spear, N. E. (1967). Retention of reinforcer magnitude. *Psychological Review*, *64*, 216–234.

Spear, N. E. (1971). Forgetting as retrieval failure. In W. K. Honig & P. H. R. James (Eds.), *Animal memory* (pp. 45–109). New York: Academic.

Spear, N. E. (1973). Retrieval of memory in animals. *Psychological Review*, *80*, 163–194.

Spear, N. E. (1978). *The processing of memories: Forgetting and retention*. Hillsdale, NJ.: Lawrence Erlbaum Associates.

Spear, N. E. (1981). Extending the domain of memory retrieval. In N. E. Spear & R. R. Miller (Eds.), *Information processing in animals: Memory mechanisms* (pp. 341–378). Hillsdale, NJ.: Lawrence Erlbaum Associates.

Spear, N. E., Hill, W. F., & O'Sullivan, D. J. (1965). Acquisition and extinction after initial trials without reward. *Journal of Experimental Psychology*, *69*, 25–29.

Spear, N. E., Miller, J. S., & Jagielo, J. A. (1990). Animal memory and learning. *Annual Review of Psychology 41*, 169–211.

Spear, N. E., & Riccio, D. C. (in press). *Memory: Phenomena and Principles*. Boston: Allyn & Bacon.

Spear, N. E., Smith, G. J. Bryan, R. G., Gordon, W. C., Timmons, R., & Chiszar, D. (1980). Contextual influences on the interaction between conflicting memories in the rat. *Animal Learning & Behavior*, *8*, 273–281.

Spear, N. E., & Spitzner, J. H. (1967). Effect of initial nonrewarded trials: Factors responsible for increased resistance to extinction. *Journal of Experimental Psychology*, *74*, 525–537.

Spear, N. E., & Spitzner, J. H. (1968). Residual effects of reinforcer magnitude. *Journal of Experimental Psychology*, *77*, 135–149.

Staddon, J. E. R. (1970). Temporal effects of reinforcement: A negative "frustration" effect. *Learning and Motivation*, *1*, 227–247.

Staddon, J. E. R. (1974). Temporal control, attention, and memory. *Psychological Review*, *81*, 375–191.

Strohen, K., Bakner, L., Nordeen, M., & Riccio, D. C. (1990, May). *Post-conditioning recovery from the latent inhibition effect in conditioned taste aversion*. Paper presented at the annual meeting of the Midwestern Psychological Association, Chicago.

Thomas, D. R. (1981). Studies of long-term memory. In N. E. Spear & R. R. Miller (Eds.), *Information processing in animals: Memory mechanisms* (pp. 257–290). Hillsdale, NJ.: Lawrence Erlbaum Associates.

Underwood, B. J. (1957). Interference and forgetting. *Psychological Review*, *64*, 49–60.

Underwood, B. J. (1969). Attributes of memory. *Psychological Review*, *76*, 559–573.

Wright, D. C., Skala, K. D., & Peuser, K. A. (1986). Latent inhibition from context-dependent retrieval of conflicting information. *Bulletin of the Psychonomic Society*, *24*, 152–154.

Zentall, T. R. (1970). Effects of context change on forgetting in rats. *Journal of Experimental Psychology*, *86*, 440–448.

III MEMORY PROCESSES

7 When Memory Fails to Fail

Dennis C. Wright
University of Missouri

It is significant that this festschrift for Al Riley is titled "Animal Cognition." Not all that long ago the notion that we could be studying cognitive processes in the rat lab would have seemed bizarre to many. To be sure many did (and do) pay tribute to the notion that principles of operant and classical conditioning could be generalized from rats to humans, but few "rat-runners" and "pigeon-peckers" assumed that generalization was a two-way street. I had the great good fortune to study with a man who not only found nothing bizarre in the notion that, for example, monkeys make similarity judgments (Wright, French, & Riley, 1968), but also team-taught a seminar with a verbal-learner and actually did both human and animal research. Even if I had learned nothing else from Al Riley, I would owe him a debt of gratitude for showing me that it was possible to both under-stand the literature in animal learning and comprehend the arcane language of the human-learning literature. I do not know that I have managed to do either, but at least I have learned to tell my students that it may pay to read the human cognition (née human learning) literature on the off-chance that results and theory from that literature can be generalized to rats. I hope that students of human cognition are getting an equivalent message. Perhaps they will all come to realize what Al Riley knew all along, both rat-runners and "people-runners" have much to learn from each other.

IMPLICIT AND EXPLICIT MEMORY

In the recent human cognition literature there has been a staggering number of both demonstrations that indicate that success or failure in remembering is not independent of the way in which memory is tested, and suggestions that there is a seeming multitude of "types" of memory. For example, *direct tests* can indicate *explicit memory* failure at the same time that *indirect tests* indicate that *implicit memory* for prior events is influencing current behavior. In humans, distinctions can be made between tests that direct the subject to remember a prior event and indirect tests that make no reference to the need to remember the prior event, but measure the effects of that event on current behavior (for review, see Richardson-Klavehn & Bjork, 1988). I know of no way to make exactly the same distinction in nonhumans. However, there are parallel instances in humans and nonhumans in which memory success or failure can be dependent on the measure of memory that is used. I discuss two of them briefly. In the first of these, the affective consequences of learning, indirect tests of memory have been defined in exactly the same way for humans and nonhumans. As Wagner and Brandon (1989) noted of their *AESOP* model: "In particular line with Konorski (1967) and in general sympathy with numerous 'two-process' theorists . . . , we also assume that whereas the prominent behavioral consequence of conditioned (A2) activity of a sensory node is the elicitation of a discrete response, the prominent behavioral consequence of similar conditioned activity (A2) of an emotional node is the *diffuse modulation of behavior otherwise initiated*" (p. 170).

Affective and Autonomic Responding

Human patients can show affective responses that indicate memory for a training event even when other measures indicate amnesia for that event (Johnson, Kim, & Risse, 1985; Claparede, 1911, described in Weiskrantz & Warrington, 1979). There are a number of similar findings in nonhumans. Goldfish show no interocular transfer of a swimming avoidance response when the discriminative stimulus for that response is presented to the "untrained" eye, but do show a conditioned heart rate response to untrained-eye stimulus presentation (McCleary, 1960). Rats given posttraining electroconvulsive shock can show both autonomic indices of retention and skeletal-motor response indications of retrieval failure (e.g., Hine & Paolino, 1970; Springer, 1975). With doses of pentobarbital that can produce drug-state dependent retrieval deficits in one-way avoidance or black/white discrimination tasks, rats that are conditioned drugged and tested nondrugged show no loss of a conditioned heart rate response (Rickett, 1970). Theory generated to explain these types of "preserved" learning in either humans or nonhumans may well help us understand the striking robustness of affective responding in both humans and nonhumans.

Procedural Learning

Procedural knowledge for problem-solving rules and perceptual-motor skills involved in task performance also appears to be preserved in human amnesic patients who fail to show retention of declarative knowledge for the task (Richardson-Klavehn & Bjork, 1988). The rules and skills required for task performance also can appear to be retained in nonhumans on tests in which they show apparent memory failure for the specifics of the task. Rats that are trained or overtrained on a black/white discrimination show, more often than not, poor retention of that discrimination in the postoperative tests that follow brain lesions. When rats are trained and tested in a Thompson–Bryant apparatus, preoperative discrimination training must be preceded by pretraining during which rats learn to quickly exit the start box, traverse the choice area, and enter the goal box by knocking over doors that are gradually moved to cover the entire entrance to the goal box. This pretraining can be laborious, but as Meyer and Meyer (1977) noted, "Pretraining was omitted when animals were tested for postoperative retention of the [black/white] habit, for the ancillary skills thus established [during pretraining] were observed to be well retained after most surgical procedures" (p. 134). Bliss (1973) made a similar observation. Monkeys learn in pretraining that they can get and eat peanuts if they sit quietly in a chair and push open little doors. Then, and only then, do they learn that there are peanuts behind green doors but not behind red doors. When these monkeys are later tested for drug-state dependency, they may well fail to discriminate between red and green doors but they demonstrate that failure by pushing open little doors while sitting quietly in a chair. In both of these instances, memory for pretraining habits succeeds at the same time that discrimination performance fails. Because training on the discrimination task neccessarily requires the animal to use the skills and procedures acquired during pretraining, discrimination training may well entail concurrent overtraining on the "simple" habits acquired during pretraining. Hence, the retention of those pretraining habits might well be taken to be the result of overtraining. However, as Bliss also demonstrated, overtraining is not required for the successful transfer across drug states of a "push-open-doors" response. Moreover, deliberate overtraining on a seemingly "simple" task, one-way shock avoidance, does not protect rats from retrieval deficits produced by train-test shifts in apparatus contexts, or shifts from pentobarbital-drugged training to nondrugged testing (Wright, Langley, & Peuser, 1992). If overtraining is not necessary to protect a push-open-doors response and not sufficient to protect a simple one-way avoidance response, it might be more fruitful to suggest that "ancillary" pretraining habits are well retained not because they are overtrained but rather because pretraining represents a categorically different form of learning that is remarkably resistant to retrieval failure, akin, say, to procedural learning. The same suggestion can be made for some components of one-way avoidance learning. During avoidance training rats not only make an increasing number of avoidance responses, they also become increasingly efficient at making short-latency escape responses on nonavoidance trials. A shift from drugged training to nondrugged test results in a dramatic drop in avoidance responding but has no effect on efficient escape responding (Heinzen & Wright, 1992).

The results of a study by Devenport and Cater (1986) also suggest that memory for rules/procedures can be relatively robust in rats. They trained rats on an eight-arm radial maze task with both reference- and working-memory components. Four arms of the maze were never baited and white inserts distinguished them from the remaining black arms. The rats learned to avoid the four arms that were unbaited across sessions (reference memory) and to apply a win-shift rule and enter each of the four baited arms only once in each session (working memory). They then were run on a copy of the maze in a different room. Reference-memory performance was significantly disrupted by the change in environmental context, but working-memory performance was not. That is, these rats continued to use the win-shift rule during the context-shift test, but they dropped markedly in performing the discrimination between black/baited and white/unbaited arms. (As noted later, context change did not disrupt reference- or working-memory performance in rats trained drugged with ethanol.)

"PROTECTING" MEMORY FROM FAILURE

Does Multiple Encoding Produce the Serial Lesion Effect?

Memory also can be remarkably robust even when tested in ways that do indicate that experimental manipulations can result in memory failure. We have been interested in the conditions that prevent memory failure in a variety of circumstances in which memory failure would otherwise be expected. One of the instances in which memory can fail to fail is called the *serial lesion effect*. Table 7.1 outlines typical procedures and findings from studies of the disruptive effects of large posterior cortex lesions on the retention of a preoperatively learned black/white (B/W) discrimination (for review, see Meyer & Meyer, 1977, 1982).

The first group provides the memory-failure baseline. Rats are trained to criterion on a B/W discrimination, subjected to a single-stage bilateral lesion that encompasses the visual cortex of both hemispheres, and are retrained on the task after a recovery interval. As first noted by Lashley (1920), their postoperative retraining scores look like those of task-naive rats. That is, rats with bilateral posterior cortex damage produced by a single-stage lesion (SSL) show no savings from original learning in relearning the task. Rats in the second group also take as many trials to postoperatively relearn the task as they did to preoperatively learn it, and this group demonstrates that preoperative overtraining does not in itself permit postoperative retrieval (Glendenning, 1972; Lashley, 1921), even when preoperative training and overtraining are separated by "sham" surgery (Scheff, Wright, Morgan, & Powers, 1977). If the bilateral damage that precedes final postoperative retraining results from two temporally separated unilateral lesions, the results can be dramatically different. That is, rats with such multiple-

TABLE 7.1
"Typical" Posterior Cortex Lesion Paradigms

Group	Training and Lesion Procedures	Postop Retrain Savings?
1. Single Stage Not Overtrained:	Train---BILAT--R--Retrain	no
2. Single Stage Overtrained:	Train------------Train---BILAT--R--Retrain	no
3. Multiple Stage Not IOI Trained:	Train--UNILAT--R---------UNILAT--R--Retrain	no
4. Multiple Stage B/W IOI Trained:	Train--UNILAT--R--Train--UNILAT--R--Retrain	YES
5. Multiple Stage State Dependent:	UNILAT--R--Train--UNILAT--R--Retrain	no

Note. Typically: "Train" & "Retrain" = to $9/10$ criterion on a black/white task; "BILAT" = posterior cortex removed from both left and right hemispheres; "UNILAT"—"UNILAT" = posterior cortex removed from one hemisphere in one operation and, after an interoperation interval (IOI), from the other hemisphere in a second operation; "R" = a 7–14 day recovery interval.

stage lesions (MSL) may show a serial lesion effect with marked savings in the retraining that follows the second lesion. I think it is unfortunate that this phenomenon is called a serial *lesion* effect because that label does not emphasize the fact that the presence or absence of savings in retraining is critically dependent on what the rat *experiences* in the time interval between the two unilateral lesions. The third group listed in Table 7.1 is the control group that demonstrates that the serial lesion effect is not the consequence of the surgical procedure per se. Despite the fact that bilateral damage occurs in two operations rather than one, MSL rats with no experience with the B/W task in the interval between their two unilateral operations show the same lack of savings that SSL rats show in retraining. However, MSL rats that perform the B/W task during the interoperation interval show marked savings in final relearning (e.g., Bodart, Hata, Meyer, & Meyer, 1980; Glendenning, 1972; Petrinovich & Carew, 1969; Scheff & Wright, 1977; Thompson, 1960).

Is the *engram* lost by SSL rats and MSL rats that show no savings? With some rephrasing this becomes a familiar question in memory research. Is this memory failure a storage failure or is it a retrieval failure? Assuming that loss of engram is failure to have in store and is functionally equivalent to failure to store information, does memory fail because storage fails, or does memory fail because retrieval fails?

There are two straightforward observations that demonstrate that the lack of postoperative savings in SSL rats represents a retrieval rather than storage failure. First, SSL rats do show positive savings on a B/W task if they are post-operatively retrained while receiving amphetamine at dose levels that do not affect the rate at which naive SSL rat learn the task (Braun, Meyer, & Meyer, 1966; Jonason, Lauber, Robbins, Meyer, & Meyer, 1970). That finding indicates that amphetamine does not affect new acquisition in SSL rats but it does permit them to retrieve preoperatively stored information. Second, SSL rats show zero savings when retrained on a preoperatively learned B/W task but show negative transfer when postoperatively trained on a "W/B" reversal of the preoperatively learned B/W task (e.g., LeVere, LeVere, Chappell, & Hankey, 1984). (That negative transfer succeeds where positive transfer fails provides another interesting example of the fact that memory success or failure is not independent of the way in which memory is measured.)

If retrieval also fails in MSL rats that lack appropriate interoperation experience, why does retrieval fail to fail in MSL rats that show the serial lesion effect? The most obvious difference among the first four groups listed in Table 7.1 is that the fourth group, the group that shows the serial lesion effect, is trained on the B/W task while unilaterally brain damaged. The good performance these rats typically show during interoperation interval training demonstrates that information previously stored while intact is successfully retrieved and utilized by the unilaterally damaged MSL rat. This experience could have a number of effects on ultimate retrieval following the second unilateral lesion.

First, information retrieval following the first lesion could affect retrieval as a process. Bjork (1988) argued that "retrieval-practice . . . does not simply strengthen an item's representation in memory, but, rather, enhances some aspect of the retrieval process per se" (p. 397) and asserted that this is a function of the extent to which "processes involved in the initial retrieval overlap the processes required to retrieve that item later" (p. 397). This assertion could be relevant to the serial lesion effect because only the MSL rat with interoperation training following the first unilateral lesion has retrieve-while-brain-damaged practice prior to final, bilaterally damaged, postoperative retraining.

Second, information retrieval doubtless modifies information storage (e.g., Spear, 1981) and interoperation training could result in alterations in what is stored and/or how it is stored. This second consequence of interoperation experience suggests an obvious analogy with drug-state dependent learning. In the standard 2×2 state-dependent learning paradigm, rats are trained drugged or nondrugged and tested drugged or nondrugged. When symmetrical state dependent retrieval obtains, rats for whom the train and test states are the same show excellent retention but the rats for whom the train and test states are different do not (Overton, 1985). If similar results occur with brain-state changes, we can frame a hypothesis that can predict the results for the first four groups in Table 7.1. If we assume that a train-test "mismatch" between intact and brain-damaged "states" will produce the same kind of retrieval failure that we see with a train-test mismatch in drug states, the predictions are straightforward. Prior to final retraining, the SSL rat, or the MSL rat with no interoperation interval training, will have had task experience only as an intact animal. They will fail to show savings when retrained bilaterally damaged because information stored in a normal intact-state will not be readily retrieved in a bilaterally damaged state. If we also assume that the states that result from unilateral damage and bilateral damage are relatively "matched," then MSL rats that have received interoperation interval training will have (re)stored information in a unilaterally-damaged state and this information will be retrievable when the animal is in a bilaterally damaged state. The fifth group in Table 7.1 can be used to test another prediction from this hypothesis. If the serial lesion effect is the result of some relative match between a unilaterally damaged storage state and a bilaterally damaged retrieval state, then there should be no need for any prior intact-state training. Rats that are not trained as normals but receive their first training on the task following a unilateral lesion should store information in a unilaterally damaged state and be capable of retrieving that information when in a bilaterally damaged state. Thus, they should show marked savings when retrained following the second unilateral lesion. This prediction has been tested and contradicted. As outlined by Groups 4 and 5 in Table 7.1, trained rats that receive a unilateral lesion followed by interoperation training 7–14 days later show savings when retrained after the second lesion, but naive rats that are given initial training 7–14 days after a unilateral lesion do not show savings when retrained after their second lesion (Bodart et al., 1980; Wright, Cloud, & Skala, 1988).

There is an obvious difference between all five of the groups in Table 7.1. Only the fourth group is trained both while intact and while unilaterally damaged prior to final retraining. Does this procedure of training in multiple brain states enhance ultimate retrieval because it forces memory to be multiply encoded?

The hypothesis that encoding variablity can enhance ultimate retrieval is a familiar one in the human-learning literature. It is used, for example, to explain the *Melton lag effect*, in which the enhanced retention of an item that appears more than once in a study list depends on the number of other items that separate the repetitions of that item on the study list (Kausler, 1991). Of greater interest here are paradigms in which procedures deliberately designed to result in encoding variability at the time of training reduce the effects of manipulations known to produce retention deficits in humans and nonhumans. For example, studies with human adults (Smith, 1982; Smith, Glenberg, & Bjork, 1978) and infants (Amabile & Rovee-Collier, 1991; Rovee-Collier & Dufault, 1991), and with rats (Logan, Padilla, & Boice, 1968) have shown that when original learning occurs in multiple physical contexts, retention deficits produced by context shifts are attenuated. We reasoned that if encoding variability was responsible for the serial lesion effect, the serial lesion per se might prove to be not only insufficient for the effect, but also unnecessary. If the primary consequence of the serial lesion paradigm is that it forces "multiple-state" encoding prior to final retraining, then any manipulation that forces multiple-state encoding prior to final retraining should result in savings.

In the experiments outlined in Table 7.2 (Wright, Skala, & Cloud, 1987), we used a B/W task and a train/lesion/retrain protocol identical to that described by Meyer and Meyer (1977). In outline the design does not differ from that used for the single-stage lesion groups of Table 7.1 except that our rats were given intraperitoneal injections of saline or sodium pentobarbital during various phases of training, so that some rats would be preoperatively trained in multiple-drug states. All rats were trained for 25 trials/day with a 20-min to 30-min rest interval between Trials 8 and 9 and between Trials 16 and 17. Rats that were trained drugged (D) were given 20 mg/kg of sodium pentobarbital 20 min before the beginning of training and 10 mg/kg of sodium pentobarbital at the beginning of each rest interval. Rats that were trained nondrugged (ND) were given volumetrically equivalent injections of physiological saline. The first experiment included the first two groups listed in Table 7.2; the second experiment was run about a year later and included the remaining four groups. In both experiments rats were preoperatively trained twice to a criterion of 9/10 correct, given a single-stage bilateral lesion of the posterior cortex, and, after a 14-day recovery interval, retrained to criterion. In the group labels the letters separated by a dash indicate whether the rats were drugged (D) or nondrugged (ND) while being trained to criterion the first (Train-1) and/or second (Train-2) time preoperatively. Letter(s) after the slash indicate the D or ND state for postoperative retraining to criterion. The absolute number of trials to a 9/10 criterion differed in the two experiments but the pattern of results was the same. During preoperative training, rats that were D during Train-1 required more trials to criterion than

TABLE 7.2
Multiple-State Preoperative Training Design

| Group (n) | Preop Training Drug State(s) | | Lesion | Postop Retraining Drug State | Postop Retrain Savings? |
	Train-1	Train-2			
D-D/ND (9)	D	D	BILAT	ND	no
ND-D/ND (10)	ND	D	BILAT	ND	YES
ND-ND/ND (10)	ND	ND	BILAT	ND	no
D-D/D (9)	D	D	BILAT	D	no
ND-D/ND (7)	ND	D	BILAT	ND	YES
ND-D/D (8)	ND	D	BILAT	D	YES

Note. "Train-1," "Train-2," and "Retrain" = train to a 9/10 criterion on a black/white task while drugged with pentobarbital (D) or while nondrugged (ND); "BILAT" = posterior cortex removed from both left and right hemisphere in a single operation. From Wright et al. (1987).

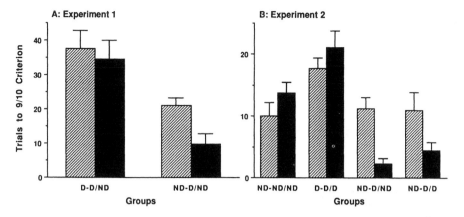

FIG. 7.1. Group mean (error bars = s.e.m.) trials to 9/10 criterion in preoperative training (striped bars) and postoperative retraining (solid bars) in Wright et al., (1987), Experiments 1 (panel A) and 2 (panel B).

rats that were ND trained, and drug-state dependency was seen during preoperative training. Rats that were in the same state for Train-1 and Train-2 (D-D or ND-ND) had fewer trials to criterion in Train-2 then in Train-1, but the Train-1 and Train-2 scores did not differ for rats that were ND during Train-1 and D during Train-2.

Trials to a 9/10 criterion in preoperative Train-1 and postoperative retraining are shown in Fig. 7.1. The results are easy to describe. For groups with single-state ND or D preoperative training, the scores for preoperative Train-1 and postoperative Retrain did not differ (i.e., no savings), regardless of whether the postoperative retraining was in a ND state (Groups ND-ND/ND & D-D/ND) or D state (Group D-D/D).

For rats preoperatively trained both ND and D, the scores for postoperative Retrain were lower than those of preoperative Train-1 (i.e., positive savings), regardless of whether the postoperative retraining was in a ND state (Groups ND-D/ND) or D state (Group ND-D/D).

The train-train/lesion/retrain procedures for Group ND-ND/ND were virtually identical to those of the second group listed in Table 7.1, and the absence of postoperative savings is hardly surprising. The absence of savings for Groups D-D/ND and D-D/D might be surprising if one were to ignore arguments to the contrary (e.g., Meyer & Meyer, 1982), and suggest some form of memory localization hypothesis that asserts that although noncortical memory storage is possible, cortical memory storage is obligatory provided that the cortex is available. Hence, the normal rat will store memory cortically. When the cortex is no longer available because of bilateral lesions, attempts to retrieve information from noncortical storage may be made but they will be fruitless. However, if a drug or a unilateral lesion forces some degree of noncortical storage, retrieval in the absence of the "normal" cortical site may succeed. As it applies here, this notion suggests that it is drugged training, not multiple-state training, that accounts for postoperative savings. However, I find no simple way to make this suggestion compatible with the postoperative failure of rats that were preoperatively trained exclusively drugged (or, for that matter, the absence of savings in the fifth group in Table 7.1). Drugged training produced a memory that failed to fail only when there had been prior nondrugged training. Obviously these results need to be replicated and extended to include, for example, multiple-state groups that are preoperatively trained in the order D-ND. However, the results do strongly suggest that multiple-state preoperative training can "protect" memory from postoperative retrieval failure and that serial lesions may not be necessary to produce results that look astonishingly like those one would expect to see in the serial lesion effect.

If multiple-state training does in fact protect memory from brain damage, it may do so because it ensures encoding variability. Thus it seemed reasonable to ask whether training conditions that foster encoding variability also would protect memory from other manipulations that result in retrieval deficits. We have examined three such manipulations: train-test shifts in drug states, train-test shifts in apparatus contexts, and retroactive interference.

Drug State and Context Interactions

One experiment we conducted suggests that manipulations designed to produce encoding variability do not always prevent memory failure. This study (Wright & Vanover, 1991) was designed to answer two questions. First, would training in multiple physical contexts reduce or eliminate the retrieval deficit we expected to see when drug state was changed between training and test? Second, would multiple-drug state training reduce or eliminate the deficit we expected to see when training and test occurred in different apparatus contexts? We selected a task for which we knew we could produce drug-state or context-shift retrieval deficits, one-way foot-shock avoidance, and trained rats for 10 trials/day either in apparatus/context A or in apparatus/context B. Intraperitoneal injections (15mg/ml/kg pentobarbital for D training, 1ml/kg saline for ND training) were given 20 min before the start of each training day. As can be seen in Table 7.3, either drug state or context was changed in different phases of training.

To address the first question, the training context remained constant on Days 5–9 for all rats, but drug state changed for some groups between Days 6 and 7. As can be seen in Fig. 7.2, there were clear drug-state dependent retrieval deficits. Rats that had no change in drug state from Day 6 to Day 7 (Groups 1 & 4) showed no Day 6–7 difference in performance, but the rats that had a drug-state change between those 2 days did. As is commonly found (Overton, 1985), state dependency was assymetric; the Day 6–7 drop in performance was larger for rats shifted from D to ND (Groups 5 & 6) than for rats shifted from ND to D

TABLE 7.3
Multiple Drug State and Multiple Context Design

Group (n)	Day 1–4 Train-1	Day 5 & 6 Train-2	Day 7–9 Train-3	Day 10 Train-4
1. (8)	ND in A	ND in A	ND in A	ND in B
2. (10)	ND in A	ND in A	D in A	D in B
3. (10)	ND in B	ND in A	D in A	D in B
4. (10)	D in A	D in A	D in A	D in B
5. (10)	D in A	D in A	ND in A	ND in B
6. (10)	D in B	D in A	ND in A	ND in B

Note. From Wright and Vanover (1991).

(Groups 2 & 3). More important, the answer to the question "does multiple-context training eliminate drug-state dependency" appears to be no. The Day 6–7 drop in performance when rats were shifted from ND to D was equivalent for the single-context Group 2 rats that had been ND trained in A on Days 1–6 and the multiple-context Group 3 rats that had been ND trained in B on Days 1–4 and in A on Days 5–6. Similarly, the retrieval deficit that followed the D to ND shift was equivalent for rats D trained only in A (Group 5) and rats D trained in B and A prior to ND training in A (Group 6).

Drug state remained constant across Days 1–6 and Days 7–10. Context changed for some groups between Days 4 and 5, and for all groups between Days

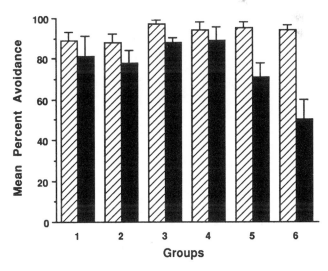

FIG. 7.2. Group mean (error bars = s.e.m.) avoidance responses when drug state changes from Day 6 (striped bars) to Day 7 (solid bars) (Wright & Vanover, 1991).

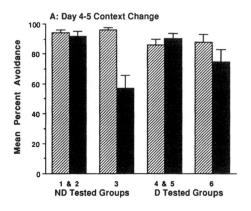

FIG. 7.3. Group mean (error bars = s.e.m.) avoidance responses when context changes in Wright and Vanover (1991). Panel A: Change between Day 6 (striped bars) and Day 7 (solid bars). Panel B: Change between Day 9 (striped bars) and Day 10 (solid bars) (Wright & Vanover, 1991).

9 and 10. Results of the Day 4–5 shift are illustrated in Fig. 7.3, Panel A. The control groups that had no Day 4–5 context change were combined and, as expected, neither those tested ND (Groups 1 & 2) nor those tested D (Groups 4 & 5) showed a Day 4–5 difference in performance.

There were clear context-shift deficits but those deficits were not independent of drug state at the time of the context shift. Context-shifted rats that were ND trained for 4 days in B prior to ND training in A on Day 5 (Group 3) showed a significant Day 4–5 drop in performance, but rats that were D trained for 4 days in B prior to D training in A did not (Group 6). Results of the Day 9–10 context change are illustrated in Fig. 7.3, Panel B. For rats trained in a single-drug state throughout the experiment, the results of the Day 9–10 shift are like those of the Day 4–5 shift. Rats that were ND trained for 9 days in A prior to ND training in B on Day 10 (Group 1) showed a significant Day 9–10 drop in performance, but rats that were D trained for 9 days in A prior to D training in B on Day 10 (Group 4) did not.

With single-state training, ND-trained rats showed context-shift deficits but D-trained rats did not, and the number of preshift training days (4 vs. 9) did not influence the presence of the context-shift deficit in single ND-state trained rats (cf. Wright, Langley, & Peuser, 1992) or its absence in single D-state-trained rats. The absence of a context-shift deficit in pentobarbital-trained rats should not have been surprising given similar results from David Riccio's laboratory (e.g., Gordon, Meehan, & Riccio, 1989), and Devenport and Cater's (1986) demonstration that context change disrupts radial-maze reference memory in non-drugged rats but not in ethanol-trained rats, or given results demonstrating that drug-state cues can overshadow exteroceptive cues when both are used as discriminative stimuli (e.g., Jarbe & Johansson, 1984, 1989). One might suggest that either D-trained rats simply do not attend to context cues, or, as is the case for nominal discriminative stimuli, that the D state is an extremely salient cue that overshadows context cues. The results from Day 10 for the rats in Group 2 do indicate that drugged rats can be sensitive to context change and favor the latter suggestion.

The rats in Groups 2, 3, and 4 were D trained in A on Days 7–9 and D tested for a context-shift deficit in B on Day 10 (Fig. 7.3, Panel B). The Day 9–10 results for Group 3 are difficult to interpret because they had been trained in B prior to the Day-10 shift. However, Groups 2 and 4 were trained only in A prior to Day 10. Group 4 was a single-drug state group that was never ND trained, and Group 2 was a multiple-drug state group that had had ND training on Days 1–6 prior to the D training on Days 7–9 and D context-shift test on Day 10. There was a small but significant Day 9–10 drop in performance for Group 2 but not, as previously noted, for the single-D-state rats of Group 4. Drugged rats were sensitive to context change, but only if they were initially trained nondrugged. Jarbe and Johansson (1989) also reported results that suggest that the use of a blocking paradigm can enhance sensitivity to exteroceptive cues in drugged rats. They trained rats on a sign-differentiated position response in a shock-escape T-maze. In Phase 1 of training, the correct right/left response was designated by drug state (e.g., D go left, ND go right) for some animals and by an exteroceptive cue for others (e.g., Light go left, Dark go right). In Phase 2, the drug state and exteroceptive cues were compounded (e.g., D and Light go left, ND and Dark go right). After Phase 2, they tested the rats with inappropriately mixed compounds (e.g., D and Dark, ND and Light). As might be expected if two exteroceptive nominal stimuli had been used in this blocking paradigm, test behavior was controlled by the blocking stimulus, not the blocked stimulus. Most notably, the presence of the "wrong" exteroceptive cue had little effect on the D-test performance of rats that were trained on the D/ND discrimination in Phase 1, but markedly disrupted the performance of rats given Light/Dark training in Phase 1. Just as in our experiment with context shifts, the exteroceptive stimulus affected performance in drugged rats, but only when ND training with the exteroceptive cue preceded D training.

Although our results suggest that prior ND training can make D-tested rats more sensitive to context change, they also suggest that multiple-state training can eliminate context-shift deficits in ND-tested rats. Groups 1, 5, and 6 were ND trained on Days 7–9, the last 3 days in A, and on Day 10, the shift day in B; thus, they were ND tested for a context-shift deficit (Fig. 7.3, Panel B). The Day 9–10 results for Group 6 support the notion that multiple-state training will protect ND rats from context-shift deficits, but are problematic given that Group 6 was D trained in B on Days 1–4. However, Groups 1 and 5 had never been trained in B prior to Day 10. As previously noted, Group 1 was ND trained throughout the experiment and showed a context-shift deficit. Group 5, a multiple-state group with D training on Days 1–6, showed no significant Day 9–10 drop in performance. Nondrugged rats were insensitive to changes in context, but only if they were initially trained drugged. This is the only result from this experiment that supports the suggestion that encoding variability will make memory fail to fail, and it clearly must be tempered by the finding that a blockinglike phenomenon seems to occur for rats trained in the order ND-D. The finding that training context variability will not affect vulnerability to drug-state changes is intriguing because other studies have demonstrated that training context variability will reduce deficits that result from train-test shifts in context (Amabile & Rovee-Collier, 1991; Logan et al., 1968; Rovee-Collier & Dufault, 1991; Smith, 1982; Smith et al., 1978). Perhaps the effectiveness of encoding variability in preventing memory failure depends on the source of memory failure.

Multiple-State Training and Retroactive Interference

We also examined the effects of multiple-drug-state training on another source of memory failure, retroactive interference (Wright, Langley, & Vanover, 1992). Six groups of rats were trained 10 trials/day on a one-way avoidance task. Two of the four single-state groups, D-D-NE ($n = 6$) and D-D-IT ($n = 10$), had 5 days of training drugged (D, 15 mg/kg pentobarbital); the other two, ND-ND-NE ($n = 7$) and ND-ND-IT ($n = 9$) were trained for 5 days nondrugged (ND). Multiple-state groups were trained for 3 days D and 2 days ND (D-ND-IT, $n = 10$), or for 3 days ND and 2 days D (ND-D-IT, $n = 10$). After a 13-day retention interval during which no injections were given, all rats were tested on the avoidance task in the same D or ND state that was used for the first 3 days of training. Control rats had no scheduled events (NE) during the retention interval. Experimental groups had an *interpolated task* (IT) during the retention interval that included being ND trained, extinguished, and retrained to bar-press for water.

Figure 7.4 shows avoidance responses on training Day 3 and the test day. Control rats showed excellent retention regardless of whether they were trained and tested D (Group D-D-NE) or trained and tested ND (Group ND-ND-NE). As

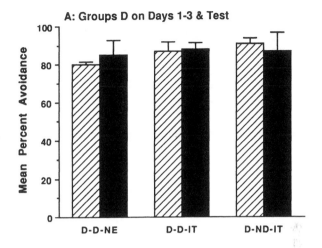

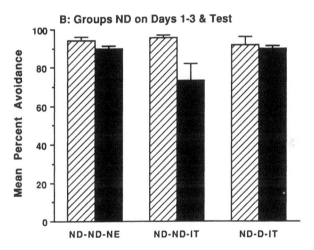

FIG. 7.4. Group mean (error bars = s.e.m.) avoidance responses on training Day 3 (striped bars) and the test day (solid bars) (Wright, Langley, & Vanover, 1992).

was expected from demonstrations of context dependency on retroactive interference effects in humans (Bilodeau & Schlosberg, 1951; Dallett & Wilcox, 1968; Greenspoon & Ranyard, 1957) and rats (Chiszar & Spear, 1969; Moye, Brasser, Palmer, & Zeisset, 1992; Zentall, 1970), the effects of the interpolated task were drug-state dependent. Experimental and control rats that were tested D did not differ in test performance. Interpolated-ND training produced no train-test drop in avoidance responding for experimental groups that were trained and tested D (Groups D-D-IT and D-ND-IT). Experimental rats that were avoidance trained ND for 5 days (Group ND-ND-IT) did show retroactive interference from the interpolated task in ND avoidance testing with a significant train-test drop in performance. However, experimental rats that were avoidance trained ND for 3 days and D for 2 days (Group ND-D-IT) showed no train-test drop in performance when tested ND. Despite the fact that both groups were trained and tested ND, rats that were trained exclusively ND showed retroactive interference from interpolated-ND training but rats that were trained both ND and D did not. It is difficult to argue that the ND-test success of the latter group (ND-D-IT) is the result of D training per se. With this task and drug, rats that are exclusively D trained show substantial performance deficits in ND tests that follow long (or short) train-test intervals, even when no interference tasks are interpolated in that retention interval (e.g., Wright, Langley, & Peuser, 1992).

How Might Multiple-State Training Enhance Retention?

If multiple-state training can protect memory from brain damage, context shifts, and retroactive interference, how might it do so? Two probable consequences of multiple-state training come readily to mind. It may ensure encoding variability by altering what is incorporated in a memory and/or by altering the way in which memory is neurally represented. It also may affect retrieval during training and thereby potentiate the ability of prior retrieval to enhance ultimate recall.

There are several reasons to expect that multiple-drug state training will ensure encoding variability that results, for example, in multiple retrieval paths to that memory. First, both interoceptive stimuli and the perception of exteroceptive stimuli can be expected to differ in D and ND states, and this would alter the population of "cue elements" that could be encoded in a to-be-remembered event. Moreover, even if no such "cue" function is assumed, D and ND states can reasonably be expected to differently bias the momentary availability of neural elements and/or the connectivity patterns among neural elements that constitute the representation of an event.

Multiple-state training also might result in elaborative encoding and/or encoding variability because it makes *intra*training retention difficult. Prior retrieval enhances ultimate recall and, as Spear (1981) noted, the effects of intermediate tests on final retention are a function of the "difficulty" of intermediate tests. Perhaps "easy" intermediate tests do little to enhance ultimate retention because memory items are easily accessible during those test and undergo no additional processing (e.g., Wagner, 1981) or no elaborative processing (e.g., Cuddy & Jacoby, 1982) during those tests. For whatever reason(s), as Kausler and Wiley (1990) put it: "the optimal benefit of prior retrieval as a mnemonic seems to come from conditions that demand some degree of cognitive effort [during prior retrieval] . . . " (p. 185). To put it more colloquially: no pain, no gain. In the lesion paradigm outlined in Table 7.1, for example, simple overtraining, especially massed overtraining, may fail to enhance ultimate retention because retrieval is too automatic or effortless during the overtraining trials. At the least, retrieval of intact/nondrugged memory would be expected to be easier during intact/nondrugged overtraining than during unilaterally damaged or drugged "second" training. Thus, when a unilateral lesion or a drug-state change separates phases of preoperative training, ultimate retention may be enhanced precisely because those manipulations make retrieval difficult during the later phases of preoperative training.

In short, multiple-state training might enhance ultimate retention in a variety of ways: (a) by increasing the variability of the cues that are incorporated in a to-be-remembered event, and/or (b) by increasing the variability of the neural connections that represent that event, and/or (c) by making intratask retrieval "effortful" and thereby facilitating whatever processes are responsible for the "strengthening" of memory by retrieval.

The connectionistic zeitgeist is reasserting a characteristic of the nervous system that Lashley was aware of long ago. It is capable of graceful degradation. Regardless of whether the effects of retrieval and encoding variability are a consequence of that capability or contribute to that capability, I think the implication is clear. Use it or lose it, and use it in a variety of conditions.

Robert Bjork (1988) cited a limerick by Ulric Neisser that cogently summarizes some results from studies on the effects of retrieval on retention: "You can get a good deal from rehearsal / If it just has the proper dispersal. / You would just be an ass / To do it en masse: / Your remembering would turn out much worsal" (p. 399). I cannot overcome the temptation to conclude this essay with a limerick of my own:

> A smiling old rat taught his mates,
> to retrieve at incredible rates,
> Memory's easy he said,
> to pull from your head,
> if you store it in multiple states.

REFERENCES

Amabile, T. A., & Rovee-Collier, C. (1991). Contextual variation and memory retrieval at six months. *Child Development*, *62*, 1155–1166.

Bilodeau, I. M., & Schlosberg, H. (1951). Similarity in stimulating conditions as a variable in retroactive inhibition. *Journal of Experimental Psychology*, *41*, 199–204.

Bjork, R. A. (1988). Retrieval practice and the maintenance of knowledge. In M. M. Gruneberg, P. E. Morris, & R. N. Sykes (Eds.), *Practical aspects of memory: Current research and issues* (Vol. 1, pp. 396–401). Chichester, England: Wiley.

Bliss, D. K. (1973). Dissociated learning and state-dependent retention induced by pentobarbital in rhesus monkeys. *Journal of Comparative and Physiological Psychology*, *56*, 183–189.

Bodart, D. J., Hata, M. G., Meyer, D. R., & Meyer, P. M. (1980). The Thompson effect is a function of the presence or absence of preoperative memories. *Physiological Psychology*, *8*, 15–19.

Braun, J. J., Meyer, P. M., & Meyer, D. R. (1966). Sparing of a brightness habit in rats following visual decortication. *Journal of Comparative and Physiological Psychology*, *61*, 79–82.

Chiszar, D. A., & Spear, N. E. (1969). Stimulus change reversal learning, and retention in the rat. *Journal of Comparative and Physiological Psychology*, *69*, 190–195.

Cuddy, L. J., & Jacoby, L. L. (1982). When forgetting helps memory: An analysis of repetition effects. *Journal of Verbal Learning and Verbal Behavior*, *21*, 451–467.

Dallett, K., & Wilcox, S. G. (1968). Contextual stimuli and proactive interference. *Journal of Experimental Psychology*, *78*, 475–480.

Devenport, L. D., & Cater, N. (1986). Ethanol blockade of context change effects. *Behavioral and Neural Biology*, *45*, 135–142.

Glendenning, R. L. (1972). Effects of training between two unilateral lesions of visual cortex upon ultimate retention of black-white habits by rats. *Journal of Comparative and Physiological Psychology*, *80*, 216–229.

Gordon, T. L., Meehan, S. M., & Riccio, D. C. (1989, May). *The effects of pentobarbital on responding to contextual shifts at testing*. Paper presented at the meeting of the Midwestern Psychological Association, Chicago.

Greenspoon, J., & Ranyard, R. (1957). Stimulus conditions and retroactive inhibition. *Journal of Experimental Psychology*, *53*, 55–59.

Heinzen, C. J., & Wright, D. C. (1992, November). *Train/test drug-state change disrupts avoidance but not escape responding*. Paper presented at the meeting of the Psychonomic Society, St. Louis.

Hine, B., & Paolino, R. M. (1970). Retrograde amnesia: Production of skeletal but not cardiac response gradient by electroconvulsive shock. *Science*, *169*, 1224–1226.

Jarbe, T. U. C., & Johansson, B. (1984). Interactions between drug discriminative stimuli and exteroceptive, sensory signals. *Behavioral Neuroscience*, *98*, 686–694.

Jarbe, T. U. C., & Johansson, B. (1989). Stimulus blocking during compound discrimination training with pentobarbital and visual stimuli. *Animal Learning & Behavior*, *17*, 199–204.

Johnson, M. K., Kim, J. K., & Risse, G. (1985). Do alcoholic Korsakoff's syndrome patients acquire affective reactions? *Journal of Experimental Psychology: Learning, Memory, and Cognition*, *11*, 22–36.

Jonason, K. R., Lauber, S. M., Robbins, M. J., Meyer, P. M., & Meyer, D. R. (1970). Effects of amphetamine upon relearning pattern and black-white discriminations following neocortical lesions in rats. *Journal of Comparative and Physiological Psychology*, *73*, 47–55.

Kausler, D. H. (1991). *Experimental psychology, cognition, and human aging*. New York: Springer-Verlag.

Kausler, D. H., & Wiley, J. G. (1990). Effects of prior retrieval on adult age differences in long-term recall of activities. *Experimental Aging Research*, *16*, 185–189.

Lashley, K. S. (1920). Studies of cerebral function in learning. *Psychobiology*, *2*, 55–135.

Lashley, K. S. (1921). Studies of cerebral function in learning. II. The effects of long continued practice upon cerebral localization. *Journal of Comparative Psychology*, *1*, 453–468.

LeVere, T. E., LeVere, N. D., Chappell, E. T., & Hankey, P. (1984). Recovery of function after brain damage: On withdrawals from the memory bank. *Physiological Psychology*, *12*, 275–279.

Logan, F. A., Padilla, A. M., & Boice, R. (1968). Contextual variability and transfer of discrimination. *Journal of Experimental Psychology*, *76*, 673–674.

McCleary, R. A. (1960). Type of response as a factor in interocular transfer in the fish. *Journal of Comparative and Physiological Psychology*, *53*, 311–321.

Meyer, D. R., & Meyer, P. M. (1977). Dynamics and bases of recoveries of functions after injuries to the cerebral cortex. *Physiological Psychology, 5,* 133–165.

Meyer, P. M., & Meyer, D. R. (1982). Memory, remembering, and amnesia. In R. L. Isaacson & N. E. Spear (Eds.), *The expression of knowledge* (pp. 179–212). New York: Plenum.

Moye, T. B., Brasser, S. M., Palmer, L., & Zeisset, C. (1992). Contextual control of conflicting associations in the developing rat. *Developmental Psychobiology, 25,* 151–164.

Overton, D. A. (1985). Contextual stimulus effects of drugs and internal states. In P. D. Balsam & A. Tomie (Eds.), *Context and learning* (pp. 357–384). Hillsdale, NJ: Lawrence Erlbaum Associates.

Petrinovich, L., & Carew, T. J. (1969). Interaction of neocortical lesion size and interoperative experience in the retention of a learned brightness discrimination. *Journal of Comparative and Physiological Psychology, 68,* 451–454.

Richardson-Klavehn, A., & Bjork, R. A. (1988). Measures of memory. In M. R. Rosenzweig & L. W. Porter (Eds.), *Annual review of psychology* (Vol. 39, pp. 475–543). Palo Alto, CA: Annual Reviews.

Rickett, D. L. (1970). *Transfer of classically conditioned cardiac response.* Unpublished master's thesis, University of Missouri, Columbia.

Rovee-Collier, C., & Dufault, D. (1991). Multiple contexts and memory retrieval at three months. *Developmental Psychobiology, 24,* 39–49.

Scheff, S. W., & Wright, D. C. (1977). Behavioral and electrophysiological evidence for cortical reorganization of function in rats with serial lesions of the visual cortex. *Physiological Psychology, 5,* 103–107.

Scheff, S. W., Wright, D. C., Morgan, W. K., & Powers, R. B. (1977). The differential effects of additional cortical lesions in rats with single- or multiple-stage lesions of the visual cortex. *Physiological Psychology, 5,* 97–102.

Smith, S. M. (1982). Enhancement of recall using multiple environmental contexts during learning. *Memory & Cognition, 10,* 405–412.

Smith, S. M., Glenberg, A., & Bjork, R. A. (1978). Environmental context and human memory. *Memory & Cognition, 6,* 342–353.

Spear, N. E. (1981). Extending the domain of memory retrieval. In N. E. Spear & R. R. Miller (Eds.), *Information processing in animals: Memory mechanisms* (pp. 341–378). Hillsdale, NJ: Lawrence Erlbaum Associates.

Springer, A. D. (1975). Vulnerability of skeletal and autonomic manifestations of memory in the rat to electroconvulsive shock. *Journal of Comparative and Physiological Psychology, 88,* 890–903.

Thompson, R. (1960). Retention of a brightness discrimination following neocortical damage in the rat. *Journal of Comparative and Physiological Psychology, 53,* 212–215.

Wagner, A. R. (1981). SOP: A model of automatic memory processing in animal behavior. In N. E. Spear & R. R. Miller (Eds.), *Information processing in animals: Memory mechanisms* (pp. 5–47). Hillsdale, NJ: Lawrence Erlbaum Associates.

Wagner, A. R., & Brandon, S. E. (1989). Evolution of a structured connectionist model of Pavlovian conditioning (AESOP). In S. B. Klein & R. R. Mowrer (Eds.), *Contemporary learning theories: Pavlovian conditioning and the status of traditional learning theory* (pp. 149–189). Hillsdale, NJ: Lawrence Erlbaum Associates.

Weiskrantz, L., & Warrington, E. K. (1979). Conditioning in amnesic patients. *Neuropsychologia, 17,* 187–194.

Wright, D. C., Cloud, M. D., & Skala, K. D. (1988, November). *State dependent retrieval and the serial lesion effect.* Paper presented at the meeting of the Psychonomic Society, Chicago.

Wright, D. C., French, G. M., & Riley, D. A. (1968). Similarity responding in monkeys in a matching to sample task. *Journal of Comparative and Physiological Psychology, 56,* 1044–1049.

Wright, D. C., Langley, C. M., & Peuser, K. A. (1992). *Functional dissociability of context- and state-dependent retrieval deficits in rats.* Manuscript submitted for publication.

Wright, D. C., Langley, C. M., & Vanover, K. E. (1992). *Multiple encoding and state-dependent retroactive interference in rats*. Manuscript submitted for publication.

Wright, D. C., Skala, K. D., & Cloud, M. D. (1987, May). *Does "multiple encoding" produce the serial lesion effect?* Paper presented at the meeting of the Midwestern Psychological Association, Chicago.

Wright, D. C., & Vanover, K. E. (1991, November). *Drug state and context dependent memory interactions*. Paper presented at the meeting of the Psychonomic Society, San Francisco.

Zentall, T. R. (1970). Effects of context change on forgetting in rats. *Journal of Experimental Psychology, 86*, 440–448.

8 Foraging in Laboratory Trees: Spatial Memory in Squirrel Monkeys

William A. Roberts,
Stephen Mitchell,
and Maria T. Phelps
University of Western Ontario

Spatial Memory in Rats and Birds

Research carried out over the past 15–20 years has shed considerable light on spatial memory in animals. Research with rats on the radial maze (Olton & Samuelson, 1976) indicates that these animals can retain memories of a large number of previously visited spatial locations (Olton, 1977, 1978, 1979; Roberts, 1984) and can retain this information well over retention intervals of several hours (Beatty & Shavalia, 1980b). Although spatial memory is generally resistant to retroactive interference (Beatty & Shavalia, 1980a; Maki, Brokofsky, & Berg, 1979), powerful effects of proactive interference have been found (Roberts & Dale, 1981) that may arise from time discrimination difficulties between recent and earlier visits to spatial locations (Roberts, 1984). Some disagreement exists about the format of rat spatial memory. Several lines of research suggest that rats form maplike representations of space, in which the relationships between visual objects are used to guide navigation (Gallistel, 1990; Mazmanian & Roberts, 1983; Morris, 1981; Suzuki, Augerinos, & Black, 1980). Other investigators argue that rats use a more restricted listlike memory on the radial maze (Brown, 1992; Olton, 1978; see also chapter 9 of this volume). Finally, research by Al Riley and his students suggests that rats may code a sequence of arm visits in the radial maze in a flexible manner, using both retrospective and prospective codes (Cook, Brown, & Riley, 1985).

A variety of findings have emphasized the functional value of spatial memory in animals. Recent research carried out in our laboratory has used the radial maze to study foraging in rats. Several experiments show that rats use spatial memory to maximize the rate of food intake. Multiple food cups were placed along the arms of a four-arm maze, and only selected cups on each arm contained food. When rats had to push a metal cover off these cups to get at their contents, they tended to visit only the baited food cups (Roberts & Ilersich, 1989). When different quantities of food were placed at the end of maze arms or in different food cups placed along maze arms, rats learned to visit food locations with the largest quantities of food before those with smaller amounts (Hulse & O'Leary, 1982; Ilersich, Mazmanian, & Roberts, 1988; Roberts, 1992). All of these findings highlight the functional or survival value of spatial memory processes in rats and suggest that these processes may have evolved because they provided fitness advantages through optimal foraging.

A variety of experiments carried out with birds over the last 10 years is painting a picture similar to that found with rats. Laboratory experiments carried out with pigeons using the radial maze (Roberts & Van Veldhuizen, 1985) or feeders placed on the laboratory floor (Roberts, 1988; Spetch, 1990; Spetch & Edwards, 1986; Spetch & Honig, 1988) show that pigeons can remember a number of food locations over an extended period of time and suffer interference effects similar to those found with rats. The functional aspect of spatial memory in birds has been strongly illustrated by recent studies of food-hoarding species. Clark's nutcrackers bury thousands of pinyon pine seeds in the earth in the late summer and early fall and recover these caches months later in the spring (Shettleworth, 1983). Marsh tits and black-capped chickadees hide seeds in a variety of sites, such as tree crevices, moss, and hollow stems, in their natural environment and retrieve these seeds several hours or days later (Sherry, 1987). Controlled experiments carried out in natural situations and in seminatural laboratory environments have shown that all of these species of birds use spatial memory as the basis for food retrieval (Kamil & Balda, 1985, 1990; Sherry, Krebs, & Cowie, 1981; Shettleworth & Krebs, 1982, 1986). The obvious advantage of hoarding is that it allows a bird to cache food at times when food is abundant and the forager is either satiated or finds it inconvenient to consume food. Because only the bird that hid the food knows its exact location, its chances of recovering a particular food item are substantially higher than those of a random forager. There is evidence that chickadees may store and retrieve food preferentially. Given access to preferred sunflower seeds and less preferred safflower seeds, chickadees cached sunflower seeds in preference to safflower seeds. Given equivalent numbers of stored sunflower and safflower seeds, chickadees preferred to retrieve sunflower seeds (Sherry, 1987). As was the case with the rat experiments, a foraging bird uses spatial memory to optimize its food intake.

Spatial Memory in Nonhuman Primates

Relative to the amount of research done in recent years with rodents and birds, there are surprisingly few studies of spatial memory in monkeys and apes. Numerous investigations of the delayed-response problem with monkeys can be found in the literature, but these experiments typically involved only two-choice spatial discrimination. Studies that involve locomotion over a more complex open field are relatively uncommon. Nevertheless, a few reports do suggest excellent spatial memory in nonhuman primates

An early study by Tinklepaugh (1932) used multiple delayed-response problems to study memory in two common chimpanzees and in two monkeys, one cynomologus macaque and one rhesus macaque. In one version of this experiment, animals were taken through several rooms in the Yale University laboratories. In each room, an animal was shown that one of two containers was baited with food. In another version of the experiment, the subject was placed in the center of a room, with up to 16 pairs of containers placed in a circle around the subject. While the subject watched, food was placed in one member of each pair of containers. After the baiting phase, animals in both tasks were taken to successive pairs of containers and allowed to choose one container. Chimpanzees did well in both tasks, choosing the baited container at 90% accuracy in the room task and at 78% correct in the circle task. The monkeys performed above chance but were not as accurate as the chimpanzees.

Reports of apes in natural habitats suggest that they form spatial maps of their home territory. For example, Savage-Rumbaugh, McDonald, Sevcik, Hopkins, and Rubert (1986) reported that a pygmy chimpanzee had formed a cognitive map containing the locations of over a dozen different food items that extended over a 55-acre farm. Menzel (1973, 1978) carried out experiments in which a common chimpanzee was carried about an open field by one experimenter while another experimenter hid 18 food items in different places. The chimpanzee was then returned to its enclosure and subsequently released to forage for food. Four chimpanzees tested in this manner quickly retrieved about two thirds of the food items and did so in a very purposeful fashion; their paths through the field approached a least-distance-traveled pattern. Subsequent experiments showed foraging preferences between quality and quantity of food. When some hiding places contained fruit and others vegetables, chimps organized their search path to collect as much fruit as possible before vegetables. When one part of the field had a higher density of hidden food than another, the higher density area was visited first. As was the case with rats and birds, chimps remembered not just the locations of food but the properties of food in those locations and then visited food locations according to their food preferences.

Although little work has been done to study monkeys' spatial memory, a recent study by MacDonald and Wilkie (1990) with yellow-nosed monkeys serves as a beginning. MacDonald and Wilkie tested two monkeys within their home enclosure in the Vancouver Stanley Park Zoo. Eight different food sites within the enclosure consisted of food cups buried in straw. On any given trial, four of these food sites were baited, and a monkey was allowed to search all food sites and collect the food. The monkey was then removed from the enclosure and returned later for a retention test. Both win-shift and win-stay tests were made at different stages of the experiment, with food placed in the appropriate four food cups during the retention interval in each case. Both monkeys found all of the food items in four to five visits to food sites and retrieved food accurately at delays up to 1 hr. An analysis of the monkeys' foraging patterns showed that, as was the case with Menzel's chimps, they tended to collect food according to a least-distance-traveled pattern.

SPATIAL MEMORY AND FORAGING
IN THE SQUIRREL MONKEY

Although some research on spatial memory has been carried out with apes and old-world monkeys, little information exists on spatial memory in new-world monkeys. In this chapter, we present some initial studies of spatial memory and foraging in one species of new-world monkey, *Saimiri Sciureus* or the squirrel monkey. *Saimiri* are largely found in the rain forests of Central and South America. They travel in troops that vary in size from 10 to 300 monkeys (Baldwin & Baldwin, 1971). Within troops, squirrel monkeys form age-sex classes consisting of adult females with infants, adult males, and mixed-sex juveniles, and, to a large extent, animals only associate with one another within these classes (Baldwin, 1971; Baldwin & Baldwin, 1972). *Saimiri* are classified as omnivorous frugivores that consume a diet of fruit, leaves, insects, and nuts. Troops of *Saimiri* typically forage at a variety of levels within the rain forest canopy, feeding on fruiting trees. In one study of *Saimiri* foraging in a semi-natural environment (Andrews, 1986), pairs of squirrel monkeys were allowed to search through artificial trees, each containing four opaque food cups baited with miniature marshmallows. The monkeys foraged in an efficient manner, visiting each tree as they encountered it and exhausting its contents. Repeat visits to trees were minimal. The experiments reported here bear some resemblance to the Andrews' study, but we were largely concerned with the spatial memory abilities of squirrel monkeys in a foraging task.

For these experiments, we built a simulated tree environment in our laboratory in which squirrel monkeys could move about freely from tree to tree and search for food in holes made in the trees or in containers attached to the trees. The experiments are concerned mainly with monkeys' ability to form reference memories for food locations and use that memory to forage efficiently. In different experiments, the density of food was varied between trees, and specific food sites were consistently baited on each tree. We anticipated that monkeys would develop preferences for richer trees and for baited locations on trees. As the experiments unfold, however, it will become clear that this was not always the case. Our findings indicate, somewhat surprisingly, that *Saimiri* does not always show evidence of spatial memory. Some possible ecological reasons for this apparent lack of spatial ability are discussed later.

Method

Subjects.   Two 4- to 5-year-old male *Saimiri sciureus*, named Jake and El-wood, served as subjects. Both monkeys had been housed in our laboratory for about 3 years and had previously participated in some delayed matching-to-sample experiments using slide-projected pictures as stimuli. Throughout these experiments, both monkeys were kept at their normal body weight on a diet of monkey chow, fruit, and vitamin supplements. Testing was carried out each day prior to the feeding period, and this procedure guaranteed strong interest in the food bait used in the testing situation.

Apparatus.   The apparatus, shown in Fig. 8.1, consisted of four simulated trees contained within a wire mesh enclosure. The enclosure consisted of four sides and a top, each made of galvanized wire mesh mounted on a pine frame. When connected, they formed a cubic cage measuring 188 cm along each side. A hinged door on one panel allowed the experimenter access to the cage. Within the enclosure, a "tree" was placed in each corner. The trees were vertically placed pieces of 8.75 cm × 8.75 cm cedar timber, each 168 cm high. Runways

FIG. 8.1. A depiction of the monkey foraging apparatus containing four vertical trees connected by runways. The external frame of the enclosure was covered with wire mesh not shown in the figure.

made of 7.5-cm wide × 1.9-cm thick wood connected each tree with the other trees in a cross pattern; the runway was 75 cm above the floor and ran a distance of 140 cm between diagonal trees. The intersection of the runways rested on a center post. A lower runway was attached to the center post at a height of 45 cm and extended 70 cm where it terminated beside the monkey entrance door. The entrance door was a 20-cm high × 15-cm wide opening, covered with a sliding sheet metal guillotine door.

Each tree had 12 1.9-cm diameter holes drilled in it, 4 on each side of the tree. On the side of each tree facing the runway, a hole was drilled at a height of 82 cm above the floor. The initial holes on the other three sides of each tree were drilled 10 cm higher than the prededing hole, moving around the tree in a clockwise fashion. On each side of a tree, the second, third, and fourth holes were drilled 30 cm above the hole immediately below. A piece of 1.9-cm diameter dowling, 12.5-cm long, was mounted 10 cm below each hole to provide a perch for exploration of that hole. Round white acrylic covers, measuring 3.3 cm in diameter and 0.4-cm thick, were placed over each hole. A screw placed through a small hole in the top of each cover allowed it to be attached to the tree just above the hole. The tension on the screw was adjusted so that a monkey with some effort could push the cover aside and inspect the contents of the hole. The hole remained open once the cover was pushed aside.

The apparatus was located within a large testing room, 406 cm × 497 cm, located adjacent to the monkey housing room. The experimenter sat at a computer near the monkey enclosure and recorded the trees and holes visited by entering behavioral codes into the computer. In addition, each monkey's foraging session was recorded on videotape by a camera placed about 3 m from the enclosure.

Procedure.   Each monkey was pretrained on a practice board mounted on the side of an exercise cage kept in the monkey housing room. The practice board contained a number of holes covered with the white covers used on the trees. Raisins were placed in each hole of the board, and a monkey was initially allowed to retrieve raisins from the holes with the covers swung open. The covers were then gradually moved to cover more and more of the holes, until the monkeys learned to push the covers aside and retrieve the raisins. Testing then began in the tree apparatus. A monkey was carried from its home cage in a small carrying cage. The door of the carrying cage was placed flush with the small metal door on the tree enclosure, and both doors were raised simultaneously. The monkey entered the apparatus on the lower runway and from there jumped to the upper runway and proceded to explore the trees. The experimenter recorded each subject's tree and hole visitations until all of the available food items on the trees had been found and consumed. Monkeys then voluntarily returned to the carrying cage through the entrance door and were returned to the home cage.

Experiment 1

Our goal in the first experiment was to find out if squirrel monkeys could learn to forage efficiently among the artificial trees. Before each foraging session, all 12 holes on each of the four trees were baited by placing a raisin in the hole and turning the white acrylic cover into place to cover the hole. Each of the two monkeys then was separately released into the enclosure with the baited trees and allowed to forage for raisins for a period of 10 min. This procedure was continued for 24 daily sessions.

The percentage of holes visited, or raisins consumed (out of 48), is plotted for 24 sessions in Fig. 8.2. Although both monkeys became efficient foragers in this situation, Elwood adjusted to the task much faster than Jake. By Session 2, Elwood was accumulating about 90% of the raisins, and he continued to forage very efficiently, often consuming 100% of the available food. Jake's curve shows that he foraged much more erratically during the early sessions and often visited fewer than 70% of the holes. Eventually, by about Session 16, Jake began to forage as efficiently as Elwood and obtained 90%–100% of the food items on each session.

The difference in foraging efficiency between Jake and Elwood may be partially attributed to different foraging strategies used by the two monkeys. Elwood almost immediately began to use a strategy of searching trees exhaustively from bottom to top. Thus, Elwood began by moving aside the cover on the lowest hole encountered on each tree, eating its contents, and then moving on to the next

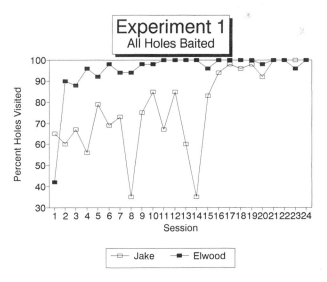

FIG. 8.2. Percentage of holes visited and raisins obtained out of 48 for each monkey plotted over 24 daily foraging sessions (Experiment 1).

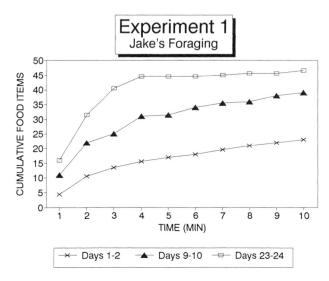

FIG. 8.3. Cumulative raisins obtained by Jake over successive 1-min time periods on selected days (Experiment 1).

highest hole on the tree. He moved around the tree visiting successive holes until he reached the top and then ran along the runway to another tree or often simply jumped to another tree and began foraging in the same manner. By contrast, Jake tended to visit trees for short periods, visiting only a few holes before going to another tree. Jake then needed to revisit trees several times to open holes that had not been opened on previous visits. This less systematic strategy led him to often miss several holes within the 10-min period. By the final sessions, however, Jake's foraging strategy had become more systematic, like that of Elwood; by staying on trees longer to visit more holes, his overall efficiency improved considerably, as seen in Fig. 8.2. As a further demonstration of Jake's improvement, within-session curves are shown in Fig. 8.3 that plot cumulative raisins obtained against minutes spent foraging. It can be seen that both the slope and height of these curves improves considerably from Days 1–2 to Days 23–24. Although Jake accumulated fewer than 25 raisins in 10 min on Days 1–2, by Days 23–24, he accumulated nearly all of the raisins within 4 min.

Experiment 2

Experiment 1 showed clearly that both monkeys had learned to forage efficiently in artificial trees. However, this initial experiment told us little about spatial memory in these animals. The monkeys infrequently revisited a hole previously visited, just as rats infrequently revisit arms on the radial maze. This observation may not indicate memory for previously visited holes, however, because the monkeys may have used the displaced hole covers as a sign that a hole had already been visited. In Experiment 2, we attempted to study reference memory in our subjects by placing food in only a proportion of the holes available. Raisins were placed in 3 holes on one tree, in 6 holes on another, and in 9 and 12 holes on the remaining two trees. Different trees contained the different numbers of baited holes for Jake and Elwood, and the baited holes were randomly distributed over the entire length of the trees in the three-, six-, and nine-baited holes trees. The same holes were baited with food over 20 daily sessions for each monkey.

This procedure allowed us to examine reference memory at two levels, between trees and within trees. Because the trees varied in density of food, we would expect an optimal forager to learn to go to the trees in decreasing order of density, or in the order 12, 9, 6, and 3 baited holes. Within trees, we looked for evidence that the monkeys remembered the specific holes that contained food. Memory for specific food locations should lead the monkeys to visit baited holes before nonbaited holes or to visit only the baited holes.

Figure 8.4 presents data for all holes opened on the initial visit to all four trees. The percent success refers to the percentage of holes visited that contained a raisin. Because 30/48 holes were baited on all four trees, by chance monkeys should have visited 62.5% baited holes. The curves show that both Jake and Elwood fluctuated around this chance level over the 20 days of testing. Over all 20 sessions, Jake visited 61.0% baited holes, and Elwood visited 62.9% baited

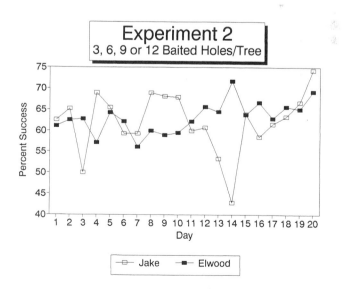

FIG. 8.4.   Percentage of baited holes visited on the first visit to all four trees plotted over 20 daily foraging sessions (Experiment 2).

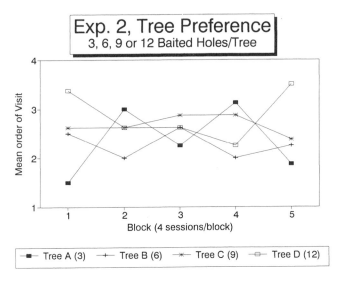

FIG. 8.5.   Mean rank of visit to each of the four trees plotted over five blocks of four sessions each (Experiment 2).

holes. Clearly, there was no evidence here that monkeys had learned the specific locations of food items.

The rank order of visits to trees was used as a measure of tree preferences. In Fig. 8.5, the mean rank of visit to trees varying in food density has been plotted over blocks of four sessions; the curves present the mean data for Jake and Elwood. Because there were four trees, the mean rank is 2.5. All four curves fluctuate around this average. No tree preferences emerge from these data, and we must conclude that monkeys failed to discriminate tree density, as well as food and nonfood locations within trees.

Experiment 3

The failure to find evidence of spatial memory in Experiment 2 might have been a consequence of the fairly high overall density of food. With 62.5% of the holes baited, the payoff for an exhaustive search may have been high enough that monkeys either did not bother to learn and remember food locations or learned food locations but used a very liberal choice criterion. In Experiment 3, the number of baited holes was substantially reduced to one per tree. In this task, the use of reference memory becomes far more important for a forager attempting to maximize rate of food intake. Each tree contained only one raisin placed at varying heights on different trees. Food was located in different holes for Jake and Elwood, but the food locations remained constant for each monkey through-out 24 days of testing.

Mean holes visited per tree until the baited hole was encountered (and includ-ing the baited hole) are shown in Fig. 8.6 as a function of blocks of four sessions. Because each tree contained 12 holes, the baited hole should have been encoun-tered by chance after a mean of 6.5 holes visited. Figure 8.6 shows that both Jake and Elwood started at around this chance level in Block 1. Both curves show a steady decline over blocks, with the baited hole being found in 2.5–3.5 holes visited by Block 6. This improvement in accuracy over blocks of days was significant for both Jake [$F(5,15) = 3.57, p < .05$] and Elwood [$F(5,15) = 5.09, p < .01$]. To measure how close monkeys were searching to the baited hole before finding it, we measured the average distance (in holes) between the baited hole and the hole on the tree visited immediately before the baited hole. Jake's data showed an average distance of 2.31 holes, with the distance expected by chance being 3.95 holes. For Elwood, the average observed distance was 1.82 holes, with the distance expected by chance being 4.70 holes. These data suggest that both monkeys learned and remembered the approximate locations of the baited holes over sessions.

Data for individual trees are shown in Fig. 8.7, which plots mean holes required to find the food item for trees with food located at varying heights. The data for both Jake in the top panel and Elwood in the bottom panel indicate a

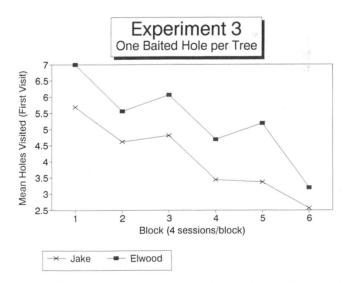

FIG. 8.6. Mean visits to tree holes required to find the baited hole plotted for each monkey across six blocks of four sessions each (Ex-periment 3).

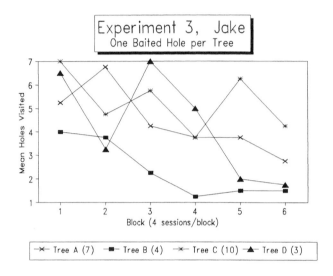

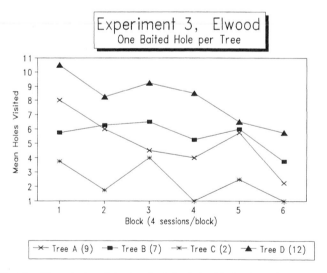

FIG. 8.7. Mean holes visited to find the baited hole plotted separately for each tree across six blocks of four sessions each. The upper panel shows Jake's data, and the lower panel shows Elwood's data. The number after each tree indicates the height of the hole on the tree (Experiment 3).

general improvement in finding food on all trees. The height of the curves appears to correlate with the height of the food item. This relationship depicts the tendency for the monkeys to search for food from the bottom of the trees to the top. Statistical analyses showed significant variation among trees for both Jake $[F(3,9) = 10.60, p < .01]$ and Elwood $[F(3,9) = 43.72, p < .01]$, but no significant interaction between trees and blocks of sessions for either monkey.

Experiment 4

Although both monkeys showed significant improvement at locating food in Experiment 3, their performance was not striking. Both animals still required an average of two to three holes visited before they discovered the correct hole. These results suggest they had learned that food was located in a certain region on each tree but that they had not learned the precise location.

This observation suggests that food locations on the trees may not have been optimally designed to reveal spatial memory. It may have been difficult for monkeys to discriminate precise food locations when there were 12 food holes on each tree placed fairly close together. Past research has shown that retention of the spatial location of food is directly related to the spatial separation of potential food locations (Harrison & Nissen, 1941; Roberts, 1972). In order to improve spatial separation of food locations on our trees, and thus make these locations more distinct, we mounted four food containers on each tree. Each container was made of black acrylic and measured 8.8 cm wide $\times$ 8.8 cm high $\times$ 5 cm deep. One container was placed on each side of each tree at varying heights from the level of the runway to the top. A guillotine door with a handle was placed on the front of each box, and monkeys were trained in a separate cage to raise this door in order to obtain raisins placed in the container. In addition, a wooden peg was placed through the top of the guillotine door and prevented it from opening. Monkeys learned to remove these pegs in order to open the doors. Pegs were used to increase the amount of work monkeys were required to perform to obtain access to empty or baited feeders and thus to sharpen their discrimination between them.

In Experiment 4, we investigated the monkeys' ability to discriminate between trees that contained food or no food. For each monkey, all four boxes on two of the trees contained a raisin, and the four boxes on the other two trees were empty. In addition, the containers on empty trees were glued shut, preventing monkeys from being able to open the doors. This was done to make these containers particularly aversive to the monkeys.

Mean rank of visit to baited and nonbaited trees (counting revisits as a rank) is plotted over five blocks of 4 days each in Fig. 8.8. The curves show that both Jake and Elwood progressively visited baited trees earlier and nonbaited trees later over successive sessions. A types of tree (baited vs. nonbaited) $\times$ blocks of days analysis of variance (ANOVA) on Jake's data yielded a significant effect of

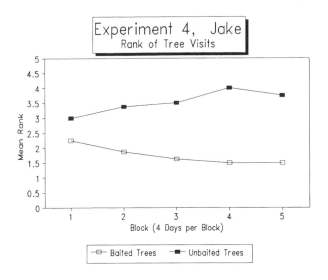

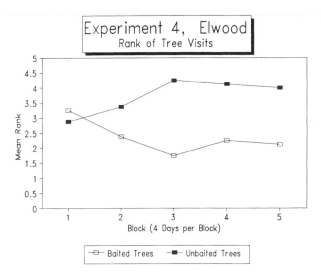

FIG. 8.8.   Mean rank of initial visit to baited and unbaited trees plotted over five blocks of four sessions each. The upper panel shows Jake's data, and the lower panel shows Elwood's data (Experiment 4).

144

type of tree [$F(1,3) = 88.44, p < .01$] but not of block of days [$F < 1.0$] or the tree $\times$ block interaction [$F(4,12) = 3.00$]. In the case of Elwood, the ANOVA produced nonsignificant effects of type of tree [$F(1,3) = 3.37, p > .05$], block of days [$F < 1.0$], and the tree $\times$ block interaction [$F(4,12) = 1.07, p > .05$].

In general, this experiment indicated that squirrel monkeys could discriminate between trees containing food and trees containing no food, although Jake's performance was somewhat stronger than Elwood's.

Experiment 5

In this experiment, we turned again to the question of whether squirrel monkeys could learn specific food locations on each tree. Two containers were randomly chosen on each tree, and these containers were always baited with a raisin throughout 20 days of testing. The other two containers on each tree were empty and the doors could be opened to allow the monkey to see the empty contents of the container. Thus, foraging bouts consisted of visits to baited containers on each tree that yielded a raisin when the door was opened and visits to empty containers on each tree that yielded no food. Learning in this situation would be shown by a reduction in the number of feeders needed to be visited to obtain two raisins on each tree. Perfect performance would be two feeders per tree.

Mean containers visited per tree to obtain two raisins is plotted over blocks of 4 days in Fig. 8.9 for each monkey. Both monkeys show slight improvement

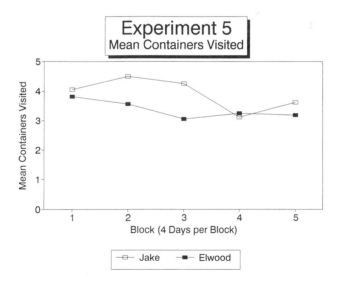

FIG. 8.9. Mean containers visited in order to visit the two containers baited with raisins on each tree. Separate curves are plotted for each monkey across five blocks of four sessions each (Experiment 5).

over days. The drop in Jake's curve was significant $[F(4,12) = 3.73, p < .05]$, but the drop in Elwood's curve was not $[F(4,12) = 1.96, p > .05]$. Assuming that containers are visited without repetition, both rewards should be obtained by chance after a mean of 3.33 containers have been visited. Both monkeys show little deviation from this chance level on the final days of testing and were far above the 2.0 level needed for perfect performance. We must conclude from this experiment that our squirrel monkeys showed little evidence of within-tree spatial memory.

Experiment 6

Our surprising failure to find evidence of discrimination between baited and empty containers that were spatially separated on trees in Experiment 5 led us to consider the possibility that the unbaited containers needed to be made more aversive. Monkeys may have executed an exhaustive search of all containers on a tree because there was little cost involved with a visit to an empty container beyond finding the container empty. In a final attempt to make the empty containers aversive, we again randomly chose two containers on each tree as baited feeders to contain a raisin throughout 20 days of testing. We further established through observation that both Jake and Elwood showed flight reactions when exposed to rubber snake replicas. Each of the two empty containers on each tree contained a rubber snake behind its door. When these doors were opened, the snake often would spring out of the box. The monkeys clearly found this consequence aversive, as they sprang back at the sight of the snake. Jake found the snakes somewhat more aversive than Elwood, and on some occasions would attempt to lift the door to a container just enough to allow him to peek at its contents before deciding to fully open it or go on to another container. Any such contact with a container was counted as a visit.

The mean containers per tree Jake and Elwood required to obtain both raisins is plotted over blocks of days in Fig. 8.10. Jake showed some learning about the locations of the baited feeders early in testing and consistently found both baited containers on each tree in less than the 3.33 feeders predicted by random foraging $[t(19) = >2.45, p < .05]$. However, Jake stayed substantially above the 2.00 feeders visited needed to show perfect retention. Elwood, by contrast, did not achieve a score below the chance level on any of the four trees. Both monkeys showed a slight drop in mean containers visited over blocks of days, but this decline was not significant for either Jake $[F(4,12) = 1.50, p > .05]$ or Elwood $[F(4,12) = 1.34, p > .05]$.

Although neither monkey showed evidence of learning the locations of baited containers in Experiment 5, Jake, but not Elwood, showed clear evidence of learning in Experiment 6. Two factors may be responsible for this differential success. First, Jake had generally performed better than Elwood in several of the preceding experiments, suggesting that he may have a superior spatial memory

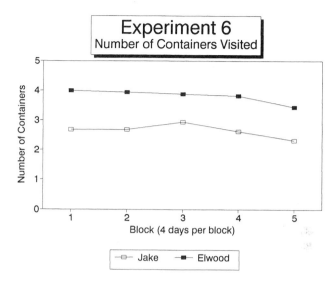

FIG 8.10. Mean containers visited in order to visit the two containers baited with raisins on each tree. Separate curves are plotted for each monkey across five blocks of four sessions each (Experiment 6).

ability. Second, Jake showed more fear of the rubber snakes placed in the unbaited containers than Elwood did. Thus, these results suggest that use of a stimulus that mimics a naturally feared organism may enhance spatial memory in a squirrel monkey.

CONCLUSION

The findings reported in these experiments surprised us considerably. As the review presented in the introduction indicated, evidence of excellent spatial memory has been found in rodents, birds, apes, and old-world monkeys. Further, these animals used spatial memory to forage optimally by maximizing the rate of intake of preferred food. One might expect then that excellent spatial memory would be fairly universal in animals, particularly in primates. In addition, we specifically designed our apparatus to make it ecologically valid. Instead of using a small chamber with limited response possibilities, as is the case in many primate lab experiments, we used a large enclosure in which the monkeys could move about freely from one tree to another. Also, there were a large number of alternative food locations, 48 holes in our initial experiments.

For all of these reasons, we had anticipated that our squirrel monkeys would reveal highly accurate spatial memory. Yet, in a number of these experiments, they showed either no evidence of spatial memory or only mediocre spatial accuracy. A critic's initial reaction to this report might be to question the monkeys' motivation for participating in the experiment. Our monkeys were not reduced in body weight because each one weighs less than 1 kg, and reducing their weight could endanger their health. However, each monkey was tested before its daily feeding, and this procedure ensured the subject was hungry. Beyond these considerations, Jake and Elwood's behavior in the tree apparatus indicated considerable interest in the task. As Experiment 1 showed, both monkeys learned to rapidly move through the trees opening holes and consuming raisins. This level of motivation was maintained throughout the experiments; both monkeys continued to quickly begin searching through food locations when released into the apparatus and to rapidly consume raisins when they were found.

Our initial attempt to demonstrate spatial memory in Experiment 2 met with no success at all. When specific holes in trees were consistently baited, with trees varying in density of holes baited, neither Jake nor Elwood showed any sign of learning food locations on trees or even of learning tree densities. When only one hole on each tree was baited in Experiment 3, both monkeys showed significant improvement in the number of holes they needed to visit on each tree to find the one containing the raisin. Nevertheless, they never reached a level of accuracy at which they went consistently to the precise hole containing the raisin. These findings suggest that monkeys learned the region of the tree at which the raisin was hidden but did not learn the exact hole.

In Experiments 4–6, we attempted to make food locations more discriminable by placing four large food containers on each tree that were separated by substantially more distance than the holes used in the first three experiments. It was found in Experiment 4 that monkeys came to discriminate between trees that varied in food density when two trees had all four containers baited and the other two trees had all four containers empty, with the doors glued in place. However, when we returned to an intratree discrimination between baited and empty containers in Experiment 5, Jake and Elwood again both failed to learn to selectively visit baited containers. Only in Experiment 6 did we find some evidence that Jake learned to selectively visit baited containers. When a feared snakelike object was placed in empty containers, Jake preferred to visit the baited containers beyond a chance level. This manipulation failed to produce evidence of spatial learning in Elwood, who generally showed less fear of these snake replicas than Jake.

It would be rash to jump to the conclusion that *Saimiri* have poor spatial memory from these experiments. Only two monkeys were tested, and their abilities may not be representative of the species in general. Furthermore, although we had anticipated that our tree apparatus would be ideal for demonstrating spatial memory in a laboratory monkey, aspects of our experimental situation may have discouraged the use of spatial memory in ways we do not yet understand. One possibility is that our monkeys learned the locations of baited sites but used such a liberal choice criterion that good discrimination was not revealed (see chapter 9 of this volume). However, several observations suggest that monkeys continued to forage nondiscriminatively even when a price was paid for incorrect choices. For one, measures taken to require the subject to exert some effort to open feeders did not sharpen discrimination, and even placing an aversive snake replica in the unbaited boxes improved spatial memory in one monkey but not the other. In an experiment not reported, we placed raisins laced with quinine at negative food sites. Monkeys continued to visit sites with plain and bitter raisins nondiscriminatively, even though they clearly showed a negative affective reaction to the quinine-laced raisins. One possibility is that the monkeys' foraging was strongly directed by an innate program that dictated variable or exhaustive search of feeding sites and effectively obscured spatial memory (Bernstein, 1984).

On the other hand, these results do force us to entertain the possibility that spatial memory is poorly developed in squirrel monkeys. Nothing in our review of studies of squirrel monkeys in their natural habitat revealed evidence of a need or use of precise spatial memory (Oates, 1987). *Saimiri* forage in large groups that typically move through fruited trees systematically eating all of the food available. It could be that *Saimiri* have evolved sufficient spatial memory to make distinctions between various geographical locations, such as areas with rich versus barren fruit trees, but do not make more precise spatial discriminations between specific locations on a tree with or without food. This position would be in keeping with our observation that our monkeys could discriminate between baited and empty trees but had difficulty finding particular baited feeders on individual trees.

These findings then pose an important research question. Do squirrel monkeys and perhaps other new-world monkeys have poorly developed spatial memory, or is an innate foraging strategy or some other factor masking the behavioral manifestation of spatial memory that is equivalent to that found in other mammals and birds? Hopefully, future research will answer this interesting question.

ACKNOWLEDGMENTS

Support for this research was provided by an operating grant from the National Sciences and Engineering Research Council of Canada awarded to William A. Roberts.

REFERENCES

Andrews, M. W. (1986). Contrasting approaches to spatially distributed resources by *Saimiri* and *Callicebus*. In J. G. Else & P. C. Lee (Eds.), *Primate ontogeny, cognition and social behaviour*. London: Cambridge University Press.

Baldwin, J. D. (1971). The social organization of a semifree-ranging troop of squirrel monkeys (*Saimiri sciureus*). *Folia Primatologia, 14*, 23–50.

Baldwin, J. D., & Baldwin, J. I. (1971). Squirrel monkeys (*Saimiri*) in natural habitats in Panama, Colombia, Brazil, and Peru. *Primates, 12*, 45–61.

Baldwin, J. D., & Baldwin, J. (1972). The ecology and behavior of squirrel monkeys (*Saimiri oerstedi*) in a natural forest in western Panama. *Folia Primatologia, 18*, 161–184.

Beatty, W. W., & Shavalia, D. A. (1980a). Rat spatial memory: Resistance to retroactive interference at long retention intervals. *Animal Learning & Behavior, 8*, 550–552.

Beatty, W. W., & Shavalia, D. A. (1980b). Spatial memory in rats: Time course of working memory and effect of anesthetics. *Behavioral and Neural Biology, 28*, 454–462.

Bernstein, I. S. (1984). The adaptive value of maladaptive behavior, or you've got to be stupid in order to be smart. *Ethology and Sociobiology, 5*, 297–303.

Brown, M. F. (1992). Does a cognitive map guide choices in the radial-arm maze? *Journal of Experimental Psychology: Animal Behavior Processes, 18*, 56–66.

Cook, R. G., Brown, M. F., & Riley, D. A. (1985). Flexible memory processing by rats: Use of prospective and retrospective information in the radial maze. *Journal of Experimental Psychology: Animal Behavior Processes, 11*, 453–469.

Gallistel, C. R. (1990). *The organization of learning*. Cambridge, MA: MIT Press.

Harrison, R., & Nissen, H. W. (1941). Spatial separation in the delayed response performance of chimpanzees. *Journal of Comparative Psychology, 31*, 427–435.

Hulse, S. H., & O'Leary, D. K. (1982). Serial pattern learning: Teaching an alphabet to rats. *Journal of Experimental Psychology: Animal Behavior Processes, 8*, 260–273.

Ilersich, T. J., Mazmanian, D. S., & Roberts, W. A. (1988). Foraging for covered and uncovered food on a radial maze. *Animal Learning & Behavior, 16*, 388–394.

Kamil, A. C., & Balda, R. P. (1985). Cache recovery and spatial memory in Clark's nutcrackers (*Nucifraga columbiana*). *Journal of Experimental Psychology: Animal Behavior Processes, 11*, 95–111.

Kamil, A. C., & Balda, R. P. (1990). Differential memory for different cache sites by Clark's nutcrackers (*Nucifraga columbiana*). *Journal of Experimental Psychology: Animal Behavior Processes, 16*, 162–168.

MacDonald, S. E., & Wilkie, D. M. (1990). Yellow-nosed monkeys' (*Cercopithecus ascanius whitesidei*) spatial memory in a simulated foraging environment. *Journal of Comparative Psychology, 104*, 382–387.

Maki, W. S., Brokofsky, S., & Berg, B. (1979). Spatial memory in rats: Resistance to retroactive interference. *Animal Learning & Behavior, 7*, 25–30.

Mazmanian, D. S., & Roberts, W. A. (1983). Spatial memory in rats under restricted viewing conditions. *Learning and Motivation, 12*, 261–281.

Menzel, E. W. (1973). Chimpanzee spatial memory organization. *Science, 182*, 943–945.

Menzel, E. W. (1978). Cognitive mapping in chimpanzees. In S. H. Hulse, H. Fowler, & W. K. Honig (Eds.), *Cognitive processes in animal behavior* (pp. 375–422). Hillsdale, NJ: Lawrence Erlbaum Associates.

Morris, R. G. M. (1981). Spatial localization does not require the presence of local cues. *Learning and Motivation, 12*, 239–260.

Oates, J. F. (1987). Food distribution and foraging behavior. In B. B. Smuts, D. L. Cheney, R. M. Seyfarth, R. W. Wrangham, & T. T. Struhsaker (Eds.), *Primate societies* (pp. 197–209). Chicago: University of Chicago Press.

Olton, D. S. (1977). Spatial memory. *Scientific American, 236*, 82–98.

Olton, D. S. (1978). Characteristics of spatial memory. In S. H. Hulse, H. Fowler, & W. K. Honig (Eds.), *Cognitive processes in animal behavior* (pp. 341–373). Hillsdale, NJ: Lawrence Erlbaum Associates.

Olton, D. S. (1979). Mazes, maps, and memory. *American Psychologist, 34*, 588–596.

Olton, D. S., & Samuelson, R. J. (1976). Remembrance of places passed: Spatial memory in rats. *Journal of Experimental Psychology: Animal Behavior Processes, 2*, 97–116.

Roberts, W. A. (1972). Spatial separation and visual differentiation of cues as factors influencing short-term memory in the rat. *Journal of Comparative and Physiological Psychology, 78*, 284–291.

Roberts, W. A. (1984). Some issues in animal spatial memory. In H. L. Roitblat, T. G. Bever, & H. S. Terrace (Eds.), *Animal cognition* (pp. 425–443). Hillsdale, NJ: Lawrence Erlbaum Associates.

Roberts, W. A. (1988). Foraging and spatial memory in pigeons (*Columba livia*). *Journal of Comparative Psychology, 102*, 108–117.

Roberts, W. A. (1992). Foraging by rats on a radial maze: Learning, memory, and decision rules. In I. Gormezano & E. A. Wasserman (Eds.), *Learning and memory: The behavioral and biological substrates* (pp. 7–23). Hillsdale, NJ: Lawrence Erlbaum Associates.

Roberts, W. A., & Dale, R. H. I. (1981). Remembrance of places lasts: Proactive inhibition and patterns of choice in rat spatial memory. *Learning and Motivation, 12*, 261–281.

Roberts, W. A., & Ilersich, T. J. (1989). Foraging on the radial maze: The role of travel time, food accessibility, and the predictability of food location. *Journal of Experimental Psychology: Animal Behavior Processes, 15*, 274–285.

Roberts, W. A., & Van Veldhuizen, N. (1985). Spatial memory in pigeons on the radial maze. *Journal of Experimental Psychology: Animal Behavior Processes, 11*, 241–260.

Savage-Rumbaugh, S., McDonald, K., Sevcik, R. A., Hopkins, W. D., & Rubert, E. (1986). Spontaneous symbol acquisition and communicative use by pygmy chimpanzees (*Pan paniscus*). *Journal of Experimental Psychology: General, 115*, 211–235.

Sherry, D. F. (1987). Foraging for stored food. In M. L. Commons A., Kacelnik, & S. J. Shettleworth (Eds.), *Quantitative analyses of behavior* (Vol. 6, pp. 209–227). Hillsdale, NJ: Lawrence Erlbaum Associates.

Sherry, D. F., Krebs, J. R., & Cowie, R. J. (1981). Memory for the location of stored food in marsh tits. *Animal Behaviour, 29*, 1260–1266.

Shettleworth, S. J. (1983). Memory in food-hoarding birds. *Scientific American, 248*, 102–110.

Shettleworth, S. J., & Krebs, J. R. (1982). How marsh tits find their hoards: The roles of site preference and spatial memory. *Journal of Experimental Psychology: Animal Behavior Processes, 8*, 354–375.

Shettleworth, S. J., & Krebs, J. R. (1986). Stored and encountered seeds: A comparison of two spatial memory tasks in marsh tits and chickadees. *Journal of Experimental Psychology: Animal Behavior Processes, 12*, 248–257.

Spetch, M. L. (1990). Further studies of pigeons' spatial working memory in the open-field task. *Animal Learning & Behavior, 18*, 332–340.

Spetch, M. L., & Edwards, C. A. (1986). Spatial memory in pigeons (*Columba livia*) in an open-field feeding environment. *Journal of Comparative Psychology, 100*, 266–278.

Spetch, M. L., & Honig, W. K. (1988). Characteristics of pigeons' spatial working memory in an open-field task. *Animal Learning & Behavior, 16*, 123–131.

Suzuki, S., Augerinos, G., & Black, A. H. (1980). Stimulus control of spatial behavior on the eight-arm maze in rats. *Learning and Motivation, 11*, 1–18.

Tinklepaugh, O. L. (1932). Multiple delayed reaction with chimpanzee and monkeys. *Journal of Comparative Psychology, 13*, 207–243.

9 Sequential and Simultaneous Choice Processes in the Radial-Arm Maze

Michael F. Brown
Villanova University

The radial-arm maze (RAM) task, first described by Olton and Samuelson (1976), has come to be an important paradigm for the study of spatial performance in rats. The task involves sampling with replacement from a set of distinct spatial locations, with the contingencies of reinforcement encouraging a single visit to each location. Subjects must discriminate previously visited arms from those that have not yet been visited in order to perform accurately. It is clear that this discrimination involves memory for extramaze visual cues (e.g., Zoladek & Roberts, 1978; Suzuki, Augerinos, & Black, 1980). The properties of this memory have received a great deal of recent experimental and theoretical attention. The present chapter, however, is not directly concerned with the memory aspects of RAM performance. Rather, what follows examines the structure of the decision process that uses memory in order to determine which maze arm is visited during each choice.

At any particular point in the trial, a rat in the center of the RAM must choose one maze arm from among the 8 (or 12 or 17 or n) arms. Thus, operationally, the animal is confronted with a simultaneous-choice task. The central thesis of this chapter is that the choice process may not, in fact, be simultaneous in nature, but may be more appropriately described as sequential. That is, as the nervous system of the rat determines which arm is to be visited, multiple arms are not simultaneously considered as a possible choice. Rather, arms are considered independently and sequentially.

SIMULTANEOUS MODELS OF CHOICE
AND COGNITIVE MAPS

Although a number of simultaneous models of choice in the RAM could be considered, the one that is focused on here is the theory that performance in the RAM depends on a cognitive map. A cognitive map has been variously defined as (a) a system that "provides the organism with a maplike representation which acts as a framework for organizing its sensory inputs and is perceived as remaining stationary in spite of the movements of the organism" (O'Keefe & Nadel, 1979, p. 488), (b) a "global representation of objects within some manifold or coordinate system from which their mutual spatial relationships can be derived, but that is, to some extent, independent of the objects themselves" (Leonard & McNaughton, 1990, p. 365), and (c) "a record in the central nervous system of macroscopic geometric relations among surfaces in the environment used to plan movements through the environment" (Gallistel, 1990, p. 103). Although there are differences among these various conceptions of cognitive maps, one commonality among them is the idea that the representation guiding behavior is global. When applied to the RAM this means that maze arms are simultaneously represented by the system that determines each choice. A second important defining property of a cognitive map is that it includes the spatial relations among places in the environment.

In the context of the RAM, it is possible that the visited/unvisited status of individual arms might be represented in the medium of a cognitive map and that such a representation might critically determine each arm choice. That is, by consulting a cognitive map that includes the visited/unvisited status of each arm, a rat could determine which arms are possible targets for the next choice. For the present purposes, the critical aspect of this view of choice in the RAM is that information about multiple arms is combined to determine the target of the next arm chosen. Roberts (1984) suggested that a cognitive map may include "markers" that specify the identity of the previously visited arms. According to this view, all maze arms are simultaneously considered as potential choices during at least one stage of the choice process. This view of choice processes in the RAM has been implicit in many, if not most, relevant theoretical discussions.

SEQUENTIAL VIEWS OF SIMULTANEOUS
CHOICE

In contrast to the view that potential choices are considered simultaneously is the view that each observable choice is in fact composed of a number of sequential decisions about individual response alternatives. In the RAM, this means that although the rat's task is to choose from among the arms, the process by which this is accomplished involves a series of "yes/no" or "go/no-go" decisions, the referent of each being an individual maze arm. This conception of choice processes in the RAM was first proposed by Olton (1978).

The idea that choices among simultaneous response alternatives have a microstructure involving decisions about individual response alternatives has a long history in psychological theory, including the notion of "vicarious trial-and-error" in mazes (Muenzinger, 1938; Tolman, 1938). Bower (1959) provided a model of choice in a T-maze that serves as part of the foundation and inspiration for the present analysis. Bower's model assumes that there is an observing response associated with each of the two response alternatives in a T-maze. A subject that is in the observing state associated with one alternative either chooses that alternative or rejects it and enters the observing state associated with the second response alternative. According to the model, subjects are in one of five states during the choice process: a neutral state, in which all subjects are assumed to begin each trial; the observing state associated with the first alternative; choice of the first alternative; the observing state associated with the second alternative; and choice of the second alternative. The relationships among these various states in Bower's "Model B" are illustrated in Fig. 9.1.

A critical aspect of this model for the present purposes is that each response alternative is considered independently. According to the model, choice of a response alternative occurs only when the subject is in the observing state associated with that response alternative. Bower's (1959) model also includes the idea that the choice process is Markovian in nature. This means that the transition probabilities among the states outlined earlier are fixed within each experimental trial. The importance of this is that state-to-state transition probabilities are *path independent*; that is, it does not matter how many times a subject has oriented toward each response alternative, the probability of choosing an alternative, given that orientation to that alternative occurs, is fixed.

Recently, Wright (1991, 1992; Wright & Sands, 1981) developed a model with very similar properties that describes the performance of pigeons in the matching-to-sample procedure. By using a specially constructed apparatus in which pigeons had to make an observable observing response, Wright was able

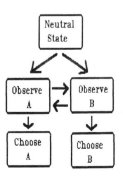

FIG. 9.1. A model of two-alternative choice presented by Bower (1959). A choice begins in a neutral state, from which subjects enter one of two observing states. Choices can be made only from the corresponding observing state. When a choice is rejected, the subject enters the observing state corresponding to the alternative choice (adapted from Bower, 1959).

to specify the frequency of test stimulus observation and the transition probabilities among observing states associated each test stimulus and choice response. Wright provided evidence that these transition probabilities were Markovian in nature. That is, as the pigeon looked back and forth at the available test stimuli, the probability of pecking a test stimulus was independent of the number and nature of previous observing responses; it depended only on the identity of the stimuli. According to this analysis, the information used to determine a choice response is limited to the stimulus currently being observed.

A similar approach to choice behavior was applied to a dolphin by Roitblat, Penner, and Nachtigall (1990). The dolphin was trained to match to sample using three-dimensional shapes as stimuli. Because the dolphin was wearing eyecups and used echolocation in performing this shape discrimination task, it was possible to directly measure the observing responses (ultrasonic emissions) of the dolphin. In contrast to the conclusions of Bower (1959) (regarding rats in a T-maze) and Wright (1991, 1992; Wright & Sands, 1981; regarding pigeons in matching-to-sample), Roitblat et al. argued that the choice behavior of their dolphin was best described by a sequential sampling model that does combine information from separate stimulus observations. This issue of whether or not previous stimulus observations affect choice behavior during an episode of stimulus sampling is returned to later.

CHOICE BEHAVIOR IN THE RAM

Brown and Cook (1986), working in Al Riley's lab, presented an "ethogram" of the behavior of rats in the RAM. In the central arena of the maze, they found the

FIG 9.2.   A rat engaged in visual orientation toward the end of a maze arm. This behavior is classified as a microchoice in the sequential choice model.

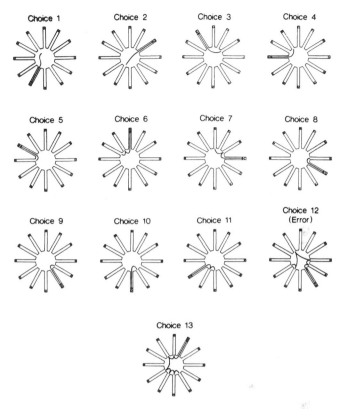

FIG 9.3.   The path of a rat during a representative trial in a 12-arm radial maze. Each panel corresponds to an arm visit (macrochoice). Note that, toward the end of the choice sequence, the rat may orient toward several arms prior to visiting one. Such orientations are termed microchoices.

most predominant behavior to be "arm investigation," defined as "actively examining the entrances to the arms. . . ." Many investigators using the RAM have noticed that rats often traverse the perimeter of the central arena, "exploring" maze arms by visually orienting toward the end of the arm. This behavior increases in frequency and duration late in each trial, when most maze arms have been visited (Brown & Cook, 1986). Figure 9.2 depicts a typical instance of orientation toward a maze arm in a standard RAM in my laboratory. Figure 9.3 shows the path of a rat's snout during a representative trial in a 12-arm maze. Each panel shows the behavior leading up to a single choice. Note that the rat orients toward many arms that are not visited.

Observations like these led Evangeline Wheeler, Al Riley, and me to propose that each observable *macrochoice* in the RAM may in fact be made up of a series of *microchoices*, during which go/no-go decisions are made about individual maze arms (Brown, Wheeler, & Riley, 1989). In rough outline, of course, this view corresponds to the sequential views of simultaneous choice described previously, in that it hypothesizes that when faced with simultaneous choice of arms in the RAM, the rat in fact makes a sequence of decisions about individual maze arms.

SEQUENTIAL DECISION MODEL APPLIED
TO 2AFC PROCEDURE

Brown et al. (1989) presented a model of choice behavior in two-alternative forced-choice (2AFC) tests in the RAM. In this procedure, the rat is first allowed to visit a subset of the maze arms. It is later (usually after being removed from the maze for a retention interval) given a choice between two maze arms, one of which was previously visited (and is therefore the incorrect choice) and one of with was not previously visited (and is therefore the correct choice). The 2AFC choice response in the RAM is formally equivalent to a T-maze choice response.

The model of choice presented by Brown et al. (1989) followed directly from the work of Bower (1959) and Wright and Sands (1981) in that it assumed that rats in the 2AFC procedure independently evaluated each of the two response alternatives, making a series of "go/no-go" decisions until one of the two alternatives was accepted and chosen.

Each of these go/no-go decisions can be described in terms of signal detection theory (SDT). That is, a subject can investigate the correct arm and visit it (hit) or reject it (miss) or investigate the incorrect arm and visit it (false alarm) or reject it (correct rejection). These decision processes might be determined by processes corresponding to discriminability and criterion effects. Figure 9.4 illustrates Brown et al.'s (1989) conceptualization of microchoices in these terms. The distribution on the left represents the psychological effects of observing a previously visited maze arm. The distribution on the right represents the effects of observing an unvisited arm. The metric of this scale might be characterized as "familiarity," with values on the left end of the scale corresponding to arms that are highly familiar due to a previous visit. The dotted line represents the choice criterion, according to which the rat classifies an arm as previously visited (and therefore rejects it) or as unvisited (and therefore visits the arm).

It is important to realize that what is being described in SDT terms here are the processes that occur during each microchoice. As Wright (1991) pointed out, the ultimate choice outcome (macrochoice) is not appropriately described in SDT terms. Rather, the predicted macrochoice outcome is determined by the joint probabilities of the relevant microchoice outcomes. To illustrate, if it is assumed that the target of the first microchoice is randomly determined, and therefore is the correct arm on half of the trials and the incorrect arm on half of the trials, then the probability of a correct macrochoice occurring during the first microchoice is

FIG. 9.4. A theoretical conceptualization of microchoices presented by Brown et al. (1989). The abscissa represents the psychological effects of observing the extramaze cues corresponding to a maze arm. These unspecified effects determine whether or not the arm is visited. Previously visited ("Old" = O) and unvisited ("New" = N)

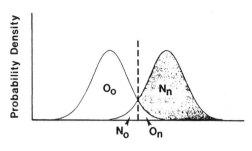

arms produce overlapping distributions of effects. When the effects fall to right of the dashed criterion line, the arm is treated as a new arm (i.e., is visited). When the effects fall to the left of the criterion line, then the arm is treated as an old arm (i.e., is rejected). Thus, four microchoice outcomes are possible: (a) correctly visiting an arm ("hit"; N_n), (b) incorrectly visiting an arm ("false alarm"; O_n), (c) correctly rejecting an arm (O_o), and (d) incorrectly rejecting and arm ("miss"; N_o).

$.5H$, where H = the hit rate at the microchoice level. Similarly, the probability of a correct macrochoice occurring during the second microchoice is $.5CH$, where C = the correct rejection rate at the microchoice level. That is, to be correct on the second microchoice, the rat must first correctly reject the incorrect arm and then correctly accept the correct arm. The probabilities of a correct macrochoice on the third, fourth, and fifth microchoices are $.5(1 - H)CH$, $.5C(1 - H)CH$, and $.5(1 - H)C(1 - H)CH$, respectively. Thus, the predicted probability of a correct macrochoice is the infinite sum,

$$.5H + \sum_{n=0}^{\infty} (.5HC^{n+1}(1 - H)^n). \tag{1}$$

Wright (1991) provides a very useful alternative way of illustrating this prediction.

One implication of the work of Brown et al. (1989) is that choice accuracy in 2AFC tests is sensitive to criterion effects, at least under certain conditions. It is commonly assumed by those applying SDT that 2AFC tests allow a pure measure of discriminability, with choice criterion effects being nullified. When stimulus presentation is under experimental control, this is probably true. Specifically, the experimenter can be sure that the subject is exposed to each stimulus for a constant period of time before the choice response occurs. However, when stimulus presentation depends on the choice behavior of the subject, and when that choice behavior is analyzed in the present terms, it can be seen that the choice criterion can have a profound effect on choice accuracy. A rat with a lax choice criterion will tend to choose the first arm (stimulus) it observes (i.e, the probability of a macrochoice during the initial microchoice is high). A rat with a stricter criterion, on the other hand, will be more likely to observe both arms (i.e., will be likely to make more microchoices) before making a macrochoice. As a result, the rat will be more likely to visit the correct arm. In fact, Brown et al. provided evidence that just such choice criterion differences exist, and that they can, in theory, explain differences in choice accuracy. This finding and one of its implications are discussed in a different context later in this chapter.

SEQUENTIAL DECISION MODEL APPLIED
TO FREE-CHOICE PROCEDURE

More recently, Brown (1992) presented a sequential decision model of performance in the standard "free-choice" RAM procedure. The model assumes that choice decisions are made about individual arms and incorporates empirically determined parameters in predicting several aspects of RAM performance. The parameters used by the model are determined by videotaping the behavior of rats in the central arena of a 12-arm RAM. Episodes of visual orientation toward maze arms are coded as microchoices. Specifically, a microchoice is defined as clear orientation toward the end of a maze arm that is either accompanied by a discernible stop in the motion of the rat's body or by a macrochoice. When a microchoice occurs, it results either in a choice of the maze arm (macrochoice) or in an arm rejection. Furthermore, the maze arm either was or was not previously visited. Thus, four classes of microchoice outcomes can be identified: hits, correct rejections, misses, and false alarms. Hit rates and correct rejection rates are determined separately for each correct macrochoice in the choice sequence (i.e., the 1st through 12th correct macrochoice). These microchoice outcomes form the first class of parameters used by the model. The second parameter used by the model has to do with the movement of the rat from microchoice to microchoice. The spatial separation between one microchoice and the next microchoice is coded in terms of the number of arms separating them. This can range from a microchoice to an arm that is adjacent to the arm that was the target of the last microchoice (separation = 1) to a microchoice to an arm that is directly across the central arena from the arm that was the target of the last microchoice (separation = 6). The distribution of these microchoice spatial separations forms the second set of parameters used by the model.

The model is implemented using a Monte Carlo simulation. Figure 9.5 illustrates the algorithm that represents the model and is used by the simulation program. A maze arm is first randomly selected as the target of the first microchoice. This arm is visited with a probability that matches the empirically determined overall hit rate. After either visiting or rejecting the maze arm (as probabilistically determined by the overall hit rate), a second maze arm is selected as the target of a microchoice. The identity of the target maze arm is determined

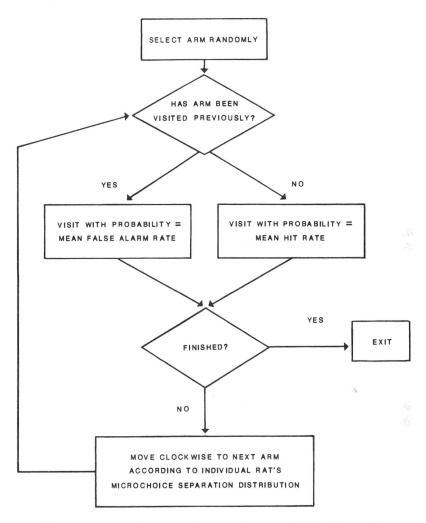

FIG. 9.5. A flow diagram of the algorithm used to implement the sequential choice model.

probabilistically according to the empirically determined microchoice separation distribution of an individual rat. For example, if the target of a microchoice by the rat is adjacent to the target of the last microchoice 75% of the time, then the computer simulation also moves from one maze arm to an adjacent arm 75% of the time as it performs microchoices. If the maze arm so selected was previously visited, then it is visited with a probability that matches the empirically determined false alarm rate. If it was not previously selected, then it is visited with a probability that matches the empirically determined hit rate. This process is iterated until all 12 maze arms have been visited. It is important to emphasize that this model includes no free parameters. The parameters used by the model come directly from the empirical data.

The simulation is run many times, and its average performance is compared to that of the rats. Two major aspects of the performance of the model and rats can be compared. The first is choice accuracy, measured in terms of the number of macrochoices required to complete the maze. The second is the number of microchoices made over the course of the choice sequence. That is, for each macrochoice, how many arms are investigated (microchoices) before one is chosen? This aspect of performance is termed *choice efficiency* for the sake of brevity. Brown (1992) compared the performance of rats in two experiments to this model. One experiment involved the standard free-choice procedure in which rats are placed in a baited maze with access to all 12 arms of the maze. The second experiment involved a forced-choice procedure, in which a subset of the arms was randomly chosen for the rats, and their ability to find the remaining arms was then assessed. In both cases, little or no difference between the rats and the model was found in terms of either choice accuracy or efficiency. The results of this comparison from Experiment 1 of Brown (1992) are shown in Fig. 9.6. The top panel shows the number of choices required to finish the maze by the rats and by the model, as well as an estimate of chance performance. The bottom panel show the efficiency of the rats and the model. The performance of rats in these RAM tasks seems adequately described by the sequential decision model.

The critical feature of the model, in terms of comparisons of sequential and simultaneous models of spatial choice, is the use of the microchoice separation distributions in determining the target of each microchoice. The model moves from one microchoice to the next without regard to the baited/unbaited status of maze arms. In other words, the model does not have available to it any information regarding which arms remain baited at the time the arm-to-arm movement occurs. A cognitive map view of spatial choice, on the other hand, asserts that the representation guiding choice decisions includes global information about the status of multiple maze arms. If so, one straightforward way in which this information should be expressed is in guiding the rat toward arms that are baited. At the extreme, arm rejections should not occur at all because the decision process guiding behavior should not require the presence of the visual cues corresponding to maze arms. Instead, rats would be expected to travel directly from one chosen arm to the next without having to evaluate the arm and make a choice decision upon exposure to the visual cues corresponding to the arm.

Of course, it is possible that the internal spatial representation (the cognitive map) must be verified or calibrated using extramaze cues. For example, one might expect that this must occur at least once at the beginning of each trial in order to provide the cognitive map with the proper "heading" (Margules & Gallistel, 1988). However, if the cognitive map is of any use in guiding choices

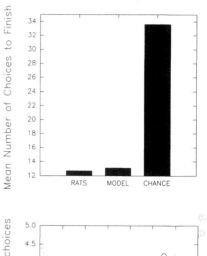

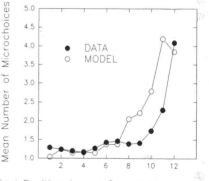

FIG. 9.6. Choice accuracy (top panel) and choice efficiency (bottom panel) of the rats and model in Experiment 1 of Brown (1992). Choice accuracy expected by chance is shown for comparison. Choice efficiency is the number of microchoices leading up to and including each correct macrochoice.

to baited arms, then the target of microchoices should tend to be baited arms more often than would otherwise be expected. The model, which moves from arm to arm according to the same distribution of spatial separations as the rats but without knowledge of which arms are baited, provides the comparison for determining whether this is the case. As indicated earlier, Brown (1992) found little or no difference between the performance of the rats and the model in standard RAM tasks. This indicates that a cognitive map is not necessary to explain accurate and efficient performance in the standard radial-arm maze task.

IS THE CHOICE PROCESS MARKOVIAN?

The sequential model of spatial choice presented by Brown (1992) assumes that the choice process is Markovian within each macrochoice, but not necessarily across macrochoices. That is, between the time that two baited arms are visited, the tendency to visit baited arms (hit rate) and the probability of rejecting previously visited arms (correct rejection rate) do not change. One might expect that as the rat investigates and rejects a series of arms, information acquired during those investigations might affect the outcome of subsequent investigations. For example, discrimination ability might improve as more extra maze cues are observed. Alternatively, the rats' choice criterion (Brown et al., 1989) might change as maze arms are rejected. The fact that the model successfully predicts performance under the assumption that no such changes occur indicates that they do not, and that the choice processes that occur during the microchoices leading up to each macrochoice conform to the Markovian assumption.

More direct evidence for the Markovian assumption is also available. Figure 9.7 shows data collected during one of the experiments reported by Brown (1992). The data come from microchoices that occurred between the penultimate (11th) and final (12th) correct macrochoice of each session. The mean probability of rejecting previously visited arms (correct rejection rate) is shown as a function of the ordinal position of the microchoice among microchoices made between these two macrochoices. There is no evidence that correct rejection rate changes as the rat continues to reject maze arms leading up to the final macrochoice $[F(5,70) < 1]$. Ideally, an analogous analysis of hit rates should be done. However, during late macrochoices, an insufficient number of baited arms are encountered (due to their rarity), whereas during early macrochoices rats make very few microchoices leading up to each macrochoice. These facts make such an analysis impossible. However, the available data are consistent with the view that, during the events leading up to each macrochoice, the choice process is accurately described as Markovian. This supports the characterization of each microchoice as independent.

Although it looks like the choice process is Markovian within each macrochoice, it is less clear whether and how choice parameters change over the course

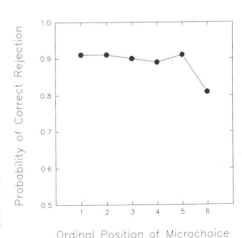

FIG. 9.7. The probability of correctly rejecting arms during microchoices that occur between the 11th and 12th correct macrochoice.

of macrochoices. Brown (1992) measured the mean hit rate and correct rejection rate as a function of ordinal position among the 12 correct macrochoices. These empirical values were used as parameters in the sequential sampling model. Thus, the model makes no assumptions regarding whether microchoice parameters change over the course of macrochoices. In Brown's Experiment 1, no evidence for a change in hit rate over the course of the macrochoice sequence was found. Analysis of correct rejection rates as a function of macrochoice number was not attempted because of the small probability of encountering previously visited arms early in the macrochoice sequence. However, in Brown's Experiment 2, some evidence was obtained that hit rate increases over the course of correct macrochoices. Unfortunately, it is difficult to interpret this effect without being able to perform the corresponding analysis on correct rejection rates.

Some additional data relevant to whether and how choice parameters might change over the course of macrochoices come from the earlier experiment of Brown et al. (1989). They allowed rats to visit a randomly selected set of either 2, 4, 6, 8, or 10 arms in a 12-arm maze, prior to a 15-min retention interval. Upon return to the maze, rats were given a choice between visiting a single available maze arm, or pressing a response manipulandum in the central arena of the maze. If the arm had been in the set of arms visited before the delay, then it was not baited but a response to the center manipulandum was reinforced. However, if the arm had not been visited, then it was baited and a manipulandum response had no effect. Thus, this procedure corresponds to a go/no-go SDT paradigm. Brown et al.'s rats were more likely to press the response manipulandum if the delay occurred following a large number of choices than if it occurred following a small number of choices. This indicated that the choice criterion shifts from a relatively lax state early in the choice sequence to a relatively strict one late in the choice sequence. Data from a more recent unpublished experiment support this conclusion. In this experiment, rats were first allowed to visit 2, 4, 6, 8, or 10 randomly selected maze arms, followed by a 15-min delay. A variety of tests followed the retention interval, but the one of interest for the present purpose was a two-alternative forced-choice (2AFC) test, in which two arms were available, one previously visited and the other not. These two arms were selected so as to be either spatially adjacent or separated by one maze arm. Analysis of videotape was used to measure microchoices to the two available arms. Figure 9.8 shows the number of microchoices that occurred as a function of the number of maze arms that had been visited before the delay. It appears that more microchoices occur when most maze arms had been visited prior to the delay than when few maze arms had been visited. Although this effect is not quite statistically reliable [$F(4, 52) = 2.25$, $.05 < p < .10$], it suggests, in agreement with the earlier data of Brown et al. that the choice criterion becomes more strict as the macrochoice sequence progresses.

Taken together, these data support the assumptions of the model that choice

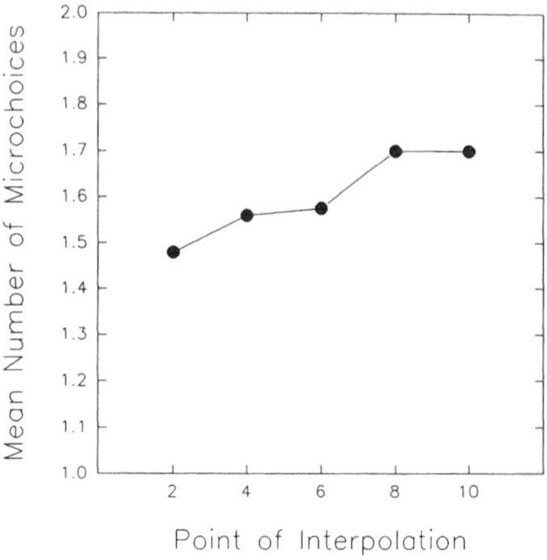

FIG. 9.8. The mean number of microchoices occurring during a two-alternative forced choice test as a function of the number of macrochoices made prior to the test (point of delay interpolation).

processes that occur during each macrochoice are Markovian. The choice parameters guiding behavior (hit rate and correct rejection rate) do not change as maze arms are observed. On the other hand, there is at least some evidence that those same parameters do change over the course of macrochoices. However, the nature of this change is not entirely clear. Data from the go/no-go and 2AFC test procedures indicate a shift in the choice criterion such that arms are less likely to be visited (lower hit rate and higher correct rejection rate) when a larger number of arms have been visited. The data of Brown (1992), however, provide no evidence that this is the case in the standard free-choice procedure. Thus, the resolution of this issue awaits further experimentation.

CONDITIONS UNDER WHICH THE SEQUENTIAL MODEL FAILS

My students and I recently completed a series of experiments similar in design to those of Brown (1992), but in which easy access to the visual cues corresponding to each maze arm is eliminated (Brown, Rish, VonCulin, & Edberg, in press). A door, blocking access to the visual cues, is positioned at the entrance of each maze arm. Furthermore, the central arena of the maze is enclosed by a tall cylinder. Thus, in order to observe the extra maze visual stimuli corresponding to a particular maze arm, the rat must perform an observing response (opening a door). The original purpose of this modification was to provide a more objectively defined microchoice. The act of opening the doors can be unambiguously coded from the videotape records. In the experiments reported by Brown, microchoices were defined by the judgement of coders. Although interrater reliabilities were high, I used the more conservative of the set of ratings obtained for each data set. This means that some microchoices may have been missed using this measurement technique. If so, these omissions depressed the performance pre-dicted by the model, and thus could account for the small but reliable differences between the observed performance of the rats and that predicted by the model. Thus, we hoped that the doors would provide a better measure of microchoices.

In the first experiment of this series, 10 rats were trained in a standard 12-arm RAM. During the critical test sessions, rats were first allowed to visit a randomly selected set of six maze arms. They were then removed from the maze for 15 min. Upon their return to the maze, there was a door constructed of 47 hanging strands of black hematite beads (4 mm in diameter) at the entrance to each maze arm. The strands of beads covered a circular doorway (7.5 cm in diameter) in a metal wall surrounding the central arena. Furthermore, a black fabric curtain surrounded the central arena above the metal wall. A mirror, used to videotape the behavior of the subjects in the central arena, was located above this curtain. A microchoice was defined as the rat's snout sticking through the bead door.

The data of critical importance were analyzed in the same manner as the data of Brown (1992). That is, the overall hit rate and correct rejection rate as a function of ordinal position in the sequence of correct choices was first deter-mined using the videotape records. In addition, the distribution of spatial separa-tions between microchoices was determined for each individual subject. These two sets of parameters are then incorporated into the model illustrated in Fig. 9.5, and the simulation was run 100 times using the microchoice separation distribution of each rat. The top panel of Fig. 9.9 shows the choice accuracy of the rats and model, as well as an estimate of chance performance for comparison. There is a significant difference between the choice accuracy of the rats and the model, although the magnitude of this difference is small when compared to chance performance. There is also a significant difference between the efficiency of the rats and model, which is illustrated in the bottom panel of Fig. 9.9.

These data appear to render the status of the sequential choice model some-what ambiguous. The use of the bead doors was designed to improve the mea-surement of microchoices and possibly eliminate the small but reliable differ-ences in the performance of the model and the rats that were found in the earlier (Brown, 1992) experiments. However, the results of the present experiment replicated the small but reliable differences between the model and the rats in terms of macrochoice accuracy and choice efficiency. One possible interpretation of these results is that, despite our intentions, some visual information might have been available through the bead doors. This possibility was encouraged by the fact that the rats appeared to orient toward bead doors without sticking their snout through the door. To the human eye, there was very little visual informa-tion available through the bead doors, although they were not completely opaque. However, given the large differences between human and rat visual ability, and especially given the importance of low spatial frequency information for rats (Birch & Jacobs, 1979), it is possible that useful information was avail-able to the rats without performance of an operationally defined microchoice.

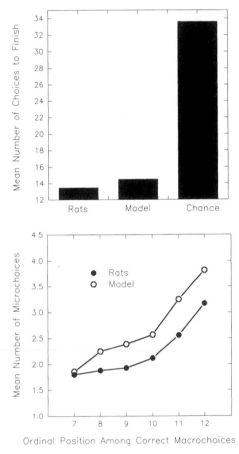

FIG. 9.9. Choice accuracy (top panel) and choice efficiency (bottom panel) of the rats and model when doors constructed of bead strands occlude visual access to the extramaze environment following the sixth macrochoice. Choice accuracy expected by chance is shown for comparison.

Because of this, we designed an improved maze-arm door. These doors were made of plywood (painted black), with a layer of black foam material on the surface facing the central arena. The doors were hinged at the top, so that the rat could push them open (microchoice) and either visit the arm (macrochoice) or reject the arm. Furthermore, to be absolutely sure that no visual information could be obtained through the cylinder that surrounded the central arena, the fabric cylinder was replaced by one made of opaque plastic. Thus, with the exception of the open end of the cylinder (88 cm above the maze surface; mostly taken up by the observation mirror), no visual information entered the central arena of the maze except when a maze door was opened by the rat. Such door openings were defined as microchoices. Figure 9.10 shows the construction of this apparatus.

A second possible explanation for the failure of the sequential choice model to explain performance in the "bead door" experiment is that, although cognitive

FIG. 9.10.   An apparatus in which visual access to the extramaze environment requires pushing open plywood doors, hinged above the entrance to each maze arm.

maps do not to play an important role in the standard radial-arm maze task, they are invoked when easy access to visual cues is not available. In our "bead door" apparatus, rats have access to the visual cues corresponding to maze arms only when they are on the arms of the maze or when they open the door to a maze arm in the central arena. This observing response requires time and effort. Thus, it is possible that use of a cognitive map was not found in the earlier experiments of Brown (1992) even though cognitive maps are, in fact, developed during RAM training. With easy access to visual cues in a relatively undemanding task, a sequential decision process might be used despite the existence of a maplike representation. When easy access to the cues is denied, on the other hand, rats might be encouraged to make use of a cognitive map. This view that a sequential choice process might be only one of several decision processes available to rats has precedence in Wright's (1992) recent proposal that, in the context of pigeon matching to sample, the Markov decision process should be thought of as the limit approached under most conditions, but from which animals deviate in various ways. It may be that rats in the RAM deviate from the sequential choice process by using a cognitive map to guide choice behavior when observations of the visual cues corresponding to individual maze arms require an effortful observing response.

This possibility was examined using rats that had participated in the bead door experiment described previously and the apparatus described earlier and shown in Fig. 9.10. As in the bead door experiment, the rats were first allowed to visit a randomly selected set of six maze arms and removed from the maze for 15 min. During these predelay forced choices, the cylinder was not in place, allowing free access to the extramaze visual cues from the central arena. During the postdelay free-choice test, the degree of access to the extramaze visual stimuli was manipulated over sessions. During all sessions, the cylinder was in place following the delay. During a randomly selected half of the sessions (closed condition), the plywood doors were closed so that access to extramaze visual cues could be gained from the central arena only by opening one of the plywood doors. During the remaining sessions (open condition), the plywood doors were all held in the open position such that visual cues were easily available as in the earlier experiments of Brown (1992). In the open condition, microchoices were coded using Brown's technique.

Microchoice outcomes and microchoice separation distributions were determined separately for the two conditions, and the empirical performance in each condition was compared to the predictions of the sequential choice model. The primary results are shown in Fig. 9.11. As illustrated in the top panel, there was a relatively large (2.7 choices) and statistically reliable difference between the performance of the rats and the model in the closed condition. However, in the open condition, the difference between predicted and obtained performance (1.1 choices) is not reliable. As shown in the bottom panel, neither condition produced a difference between predicted and obtained choice efficiency. These results confirm the conclusion of Brown (1992) that the sequential choice model provides an accurate description of performance in the standard (open) RAM. However, they also confirm the conclusion from the bead door experiment that the model does not accurately describe performance when access to extramaze visual cues requires an effortful observing response.

The failure of the sequential choice model indicates that, in the closed condition, information about the location of baited arms was available to rats in the central arena prior to observation of extramaze visual cues. We considered the possibility that this information was in the form of intramaze visual, olfactory, or tactile cues. However, the results of a probe test in which the maze was rotated during the delay provided strong evidence against this possibility (Brown et al., in press). Thus, we conclude that this information was in the form of a spatially

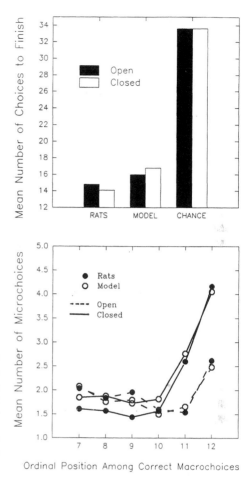

FIG. 9.11. Choice accuracy (top panel) and choice efficiency (bottom panel) of the rats and model when hinged doors constructed of plywood are either open or closed following the sixth macrochoice. Choice accuracy expected by chance is shown for comparison.

organized representation that included the baited versus unbaited status of multiple maze arms. However, this cognitive map was not involved in performance when the maze arm doors were open, allowing easy access to the visual cues corresponding to individual maze-arms. Under those conditions, rats apparently used a sequential decision process to locate baited arms.

SUMMARY AND CONCLUSIONS

Taken as a whole, the work described in this chapter indicates that, at least under the conditions typically used in the radial-arm maze, performance can be adequately described as a sequential choice process, in which each maze arm is independently evaluated and either accepted or rejected. This result may seem at odds with a number of recent lines of evidence providing support for the existence and use of cognitive maps by rats in spatial tasks, including the radial-arm maze (e.g., Cheng, 1986; Dallal & Meck, 1990; Gallistel, 1990). However, our most recent results suggest that the critical issue may not be the existence of cognitive mapping, but rather the conditions under which cognitive maps are utilized. Under relatively undemanding conditions, such as the standard radial-arm maze task, rats apparently do not to make use of cognitive maps, but instead use a (simpler?) sequential choice process. More demanding tasks, however, may elicit the use of more sophisticated choice processes, perhaps including the use of cognitive mapping. The modeling approach outlined in the present chapter promises to help elucidate these issues.

ACKNOWLEDGMENTS

This work was supported by National Institute of Mental Health Grant MH45004. The author gratefully acknowledges the participation of Jackie Edberg, Patti Rish, and Joanna VonCulin in the unpublished work described in this chapter and thanks Ken Cheng, Bill Roberts, Tony Wright, and Tom Zentall for helpful comments on earlier versions of the manuscript or the ideas contained therein.

REFERENCES

Birch, D., & Jacobs, G. H. (1979). Spatial contrast sensitivity in albino and pigmented rats. *Vision Research, 19*, 933–937.

Bower, G. H. (1959). Choice point behavior. In R. R. Bush & W. K. Estes (Eds.), *Studies in mathematical learning theory* (pp. (pp. 109–124). Stanford, CA: Stanford University Press.

Brown, M. F. (1992). Does a cognitive map guide choices in the radial-arm maze? *Journal of Experimental Psychology: Animal Behavior Processes, 18*, 56–66.

Brown, M. F., & Cook, R. G. (1986). Within-trial dynamics of radial maze performance in rats. *Learning and Motivation, 17*, 190–205.

Brown, M. F., Rish, P. A., VonCulin, J. E., & Edberg, J. (in press). Spatial guidance of choice behavior in the radial-arm maze. *Journal of Experimental Psychology: Animal Behavior Processes.*

Brown, M. F., Wheeler, E. A., & Riley, D. A. (1989). Evidence for a shift in the choice criterion of rats in a 12-arm radial maze. *Animal Learning and Behavior, 17*, 12–20.

Cheng, K. (1986). A purely geometric module in the rat's spatial representation. *Cognition, 23*, 149–178.

Dallal, N. L., & Meck, W. H. (1990). Hierarchical structures: Chunking by food type facilitates spatial memory. *Journal of Experimental Psychology: Animal Behavior Processes, 16*, 69–84.

Gallistel, C. R. (1990). *The organization of learning.* Cambridge, MA: MIT. Press.

Leonard, B., & McNaughton, B. L. (1990). Spatial representation in the rat: Conceptual, behavioral, and neurophysiological perspectives. In R. P. Kesner & D. S. Olton (Eds.), *Neurobiology of comparative cognition* (pp. 363–422). Hillsdale, NJ: Lawrence Erlbaum Associates.

Margules, J., & Gallistel, C. R. (1988). Heading in the rat: Determination by environmental shape. *Animal Learning and Behavior, 16*, 404–410.

Muenzinger, K. F. (1938). Vicarious trial and error at a point of choice: 1. A general survey of its relation to learning efficiency. *Journal of Genetic Psychology, 53*, 75–86.

O'Keefe, J., & Nadel, L. (1979). Precis of O'Keefe & Nadel's *The hippocampus as a cognitive map*. *The Behavioral and Brain Sciences, 2*, 487–533.

Olton, D. S. (1978). Characteristics of spatial memory. In S. H. Hulse, H. Fowler, & W. K. Honig (Eds.), *Cognitive processes in animal behavior*. Hillsdale, NJ: Lawrence Erlbaum Associates.

Olton, D. S., & Samuelson, R. J. (1976). Remembrance of places past: Spatial memory in rats. *Journal of Experimental Psychology: Animal Behavior Processes, 2*, 97–116.

Roberts, W. A. (1984). Some issues in animal spatial memory. In H. L. Roitblat, T. G. Bever, & H. S. Terrace (Eds.), *Animal cognition* (pp. 425–443). Hillsdale, NJ: Lawrence Erlbaum Associates.

Roitblat, H. L., Penner, R. H., & Nachtigall, P. E. (1990). Matching-to-sample by an echolocating dolphin (*Tursiops truncatus*). *Journal of Experimental Psychology: Animal Behavior Processes, 16*, 85–95.

Suzuki, S., Augerinos, G., & Black, A. H. (1980). Stimulus control of spatial behavior on the eight-arm maze in rats. *Learning and Motivation, 11*, 1–18.

Tolman, E. C. (1938). The determiners of behavior at a choice point. *Psychological Review, 45*, 1–41.

Wright, A. A. (1991). A detection and decision process model of matching to sample. In M. L. Commons, J. A. Nevin, & M. C. Davison (Eds.), *Signal detection: Mechanisms, models, and applications* (pp. 191–219). Hillsdale, NJ: Lawrence Erlbaum Associates.

Wright, A. A. (1992). Learning mechanisms in matching to sample. *Journal of Experimental Psychology: Animal Behavior Processes, 18*, 67–79.

Wright, A. A., & Sands, S. F. (1981). A model of detection and decision processes during matching to sample by pigeons: Performance with 88 different wavelengths in delayed and simultaneous matching tasks. *Journal of Experimental Psychology: Animal Behavior Processes, 7*, 191–216.

Zoladek, L., & Roberts, W. A. (1978). The sensory basis of spatial memory in the rat. *Animal Learning and Behavior, 6*, 77–81.

10 Representations and Processes in Working Memory

Herbert L. Roitblat
University of Hawaii at Manoa

On the occasion of Al Riley's retirement, it seemed appropriate to look back at some of the work that grew out of my studies with him. My work on delayed matching to sample (DMTS) began in Al's laboratory. He deserves much of the credit for pointing me in the directions this work has taken, but I must accept the blame for inadequacies that may be apparent to the reader. DMTS is a conditional discrimination task in which a sample stimulus is presented at the start of a trial and a set of alternative choice stimuli are presented later in the trial. The subject's task is typically to select the alternative that is the same as or matches the sample. The DMTS performance of many species has been studied, including pigeons (Blough, 1959; Maki & Leith, 1973; Roberts & Grant, 1974; Zentall, Hogan, & Edwards, 1984), various monkey species (D'Amato, 1973; D'Amato, Salmon, & Colombo, 1985), and dolphins (Herman, 1980; Herman & Gordon, 1974; Hunter, 1988).

Delayed matching to sample is an extremely useful task for identifying the representations and processes animals use. The animal must observe the sample stimulus, encode information about its identity, retain that information over a delay, identify the correct matching comparison stimulus, and respond to the correct stimulus. These are the representations and processes with which I have been concerned.

ENCODING DYNAMICS

The most extensive observations of DMTS performance have been conducted with pigeon subjects. One of the most salient features of pigeons' DMTS performance is the typical improvement in choice accuracy observed with increases in the duration of the sample (Maki & Leith, 1973; Roberts, 1972; Roberts & Grant, 1976; Roitblat, 1980). There are a number of theoretical models for why choice accuracy should increase with increasing sample duration. These models can be divided into two general classes: those that suppose that encoding the sample involves a discrete state change and those that suppose that encoding is a continuous process. In the usual two-alternative DMTS task there is no opportunity to distinguish between the various coding models because there is only one informative choice available on any trial and all models make the same prediction. All plausible models predict that choice accuracy should improve with increasing sample duration. The discrete-state models explain this observation as the result of an increasing likelihood that the pigeon's memory will have changed into the appropriate state for a correct response as sample duration increases. The gradual models explain this observation as the result of changes in the strength of the sample representation with exposure.

Although the models all predict the same pattern of performance for single-choice DMTS performance, they make different predictions when three alternatives are available on every trial and the animal is allowed to make a second choice following a first-choice error. Under these circumstances, second-choice as well as first-choice accuracy was found to improve with increasing sample duration (Roitblat, 1980), indicating that the bird had some access to information about the identity of the correct choice even when making an erroneous first choice. These data thereby rule out a simple all-or-none model of pigeon DMTS performance. According to this model, encoding is a simple two-state process in which the bird passes from a state in which it knows nothing about the identity of the sample to one in which it knows perfectly the identity of the sample. Errors, according to this model, occur when the bird is in the know-nothing state and guesses incorrectly. Correct choices result when the bird either knows nothing and guesses correctly, that is before a state transition has occurred, or knows the correct choice and selects it. Hence, this model predicts that second choices following first-choice errors should be accurate no more often than chance because these choices must result from the know-nothing state, and must, therefore, be pure guesses.

A more complicated discrete-state model is based on the assumption that choice accuracy is controlled by two all-or-none processes. According to this model, encoding occurs as an all-or-none process, but on some trials, even when encoding has occurred, the bird does not use the information it possesses, but selects its first choice randomly, perhaps as the result of elicitation processes resulting from the pairing of the comparison stimuli with reinforcement.

Examination of the pattern of first-choice errors does not support this more complicated discrete-state model. First-choice errors were not made randomly; rather the distribution of first-choice errors was found to depend on the sample (Roitblat, 1980). Some errors were more likely to occur than others following one sample, and less likely following another sample. In order for the distribution of first-choice errors to depend on the sample, the bird must have had information about the identity of the sample when making an error.

Other discrete-state models also fail to explain the pigeon data satisfactorily. For example, the nonrandom distribution of first-choice errors means that the bird had some information about the sample when making these errors. This same information could be the basis for making a second choice following a first-choice error. According to this view, the bird's memory entered some particular state following exposure to the sample. The identity of this state was a function of the sample that had been presented and, in turn, controlled both the bird's first, and if relevant its second choice. As a result, second choices are predicted to contain no more information (as measured according to information theory) about the identity of the sample, than that already reflected in the identity of the first-choice error. In contrast to this prediction, the distribution of second choices was found to contain information about the sample that cannot be attributed to the relationship between first-choice errors and samples (Roitblat & Scopatz, 1983). That is, the relationship between the sample and the first-choice errors and that between the identity of the first-choice error and the second choice was not adequate to specify the identity of the second choice. This pattern indicates that the identity of a first-choice error does not exhaustively determine the state of the pigeon's memory. Therefore, if the pigeon does use a discrete-state memory system it must have many more states than stimuli to be remembered. Some of these states represent partial information about the identity of the sample; other states represent more information. As the number of states increases, the model approximates more and more closely a continuous process. See the discussion of dolphin coding dynamics later.

CODES AND CONTENTS

When making its choice at the end of the trial, the subject must compare the alternative comparison stimuli with the information being maintained about the identity of the sample. Theoretical accounts have tended to emphasize the process suggested by the name of the task, that is, to describe the process in terms of "matching" the perceived characteristics of the comparison stimuli against a memory of the characteristics of the sample (Roberts & Grant, 1976). On this view, the content of working memory when a choice is determined is some trace or copy of the sample stimulus. Choice accuracy improves with increasing sample duration because the copy of the sample develops slowly.

There is some support for the hypothesis that pigeons select the matching stimulus on the basis of similarity between the trace of the sample and the perception of a comparison. Pigeons and other animals seem capable of using a generalized concept such as similarity to control their choices (e.g., Herrnstein,

1985; Pisacreta, Redwood, & Witt, 1984; Wilson, Mackintosh, & Boakes, 1985a, 1985b; Zentall, Hogan, & Edwards, 1984). Nevertheless, these same studies also provide substantial evidence that pigeons do not ordinarily rely exclusively on any abstract concept. Similarity between the sample and the correct comparison is not necessary to DMTS performance. A high level of accuracy can be obtained using a so-called "symbolic matching" procedure in which the correct comparison stimulus is only arbitrarily related to the sample (e.g., an orange circle might be the correct choice following a black triangle sample). Experiments have found little difference between symbolic and identity matching performance in pigeons in terms of speed of acquisition or asymptotic accuracy (Carter & Eckerman, 1975; Cohen, Looney, Brady, & Acuella, 1976). Pigeons seem biased toward using stimulus-specific rules, unless the number of training stimuli is very high (Lombardi, Fachinelli, & Delius, 1984; Wright, Cook, Rivera, Sands, & Delius, 1988).

To the extent that an animal uses a generalized matching concept, the transformation from sample to selection of the correct comparison stimulus could be very simple. The pigeon need only retain some "copy" of the sample stimulus until it can be compared with the alternative comparison stimuli. To the extent that pigeons use stimulus-specific rules in the DMTS task, the transformation from sample to selection of the correct comparison stimulus could be very complex. The content of their working memory must be transformed, perhaps via a retrieval process, from a representation of the sample to a representation of the to-be-selected comparison stimulus. The transformation could occur at any time between the onset of the sample and the time the choice is made. Several lines of evidence suggest that this transformation occurs early in the interval.

Identifying the content of the subject's working memory in the DMTS task requires a method to dissociate memories that are related to the sample from memories that are related to the comparison stimulus. The use of a symbolic matching procedure provides this dissociation. In the standard DMTS task, the sample and the correct comparison stimulus are identical, so the effects of any factor that affected one would be indistinguishable from a factor affecting the other. In the symbolic DMTS task, however, different stimuli serve as samples and as comparison stimuli so it is possible to see different effects on one versus the other. A considerable amount of evidence has accumulated showing that pigeons recode the sample into a representation related to the comparison stimulus they will choose early in the interval between the sample presentation and the appearance of the comparison stimuli. The data suggest that the pigeon's working-memory code involves a prospective test code rather than a retrospective sample code (Honig & Thompson, 1982; Roitblat, 1980).

The first line of evidence is that increases in the duration of the retention interval between the sample and the presentation of the comparison stimuli result in increased confusions between similar comparison stimuli, not between similar sample stimuli (Roitblat, 1980). Pigeons were trained in a symbolic DMTS task in which one sample and three comparison stimuli were presented on each trial. The sample set, A,B,D, consisted of two similar stimuli A and B (e.g., red and orange) and one dissimilar stimulus, D (e.g., blue). The comparison stimuli, W, Y, Z, also consisted of two similar alternatives, Y and Z (e.g., vertical and 12.5° from vertical), and one dissimilar stimulus W (e.g., horizontal lines). W was the correct choice following A, Y was the correct choice following B, and Z was the correct choice following D. The pair of similar sample stimuli, A and B, mapped onto a pair of dissimilar comparison stimuli, W and Y; and the pair of dissimilar comparison stimuli, B and D, mapped onto similar comparison stimuli, Y and Z. An analysis of the confusion patterns indicated that as the retention interval increased, pigeons were more likely to confuse similar comparison stimuli than similar sample stimuli. These data and those from a subsequent replication indicate that pigeons maintain the information about the presented sample in a form resembling the comparison stimuli. They seem to translate early from identification of the sample stimulus to a specification of the correct comparison stimulus.

Another experiment by Grant (1982) lends further support to the prospective coding hypothesis. Grant trained pigeons on a version of the symbolic DMTS task in which three different stimuli could serve as samples for the same comparison stimulus. Choice accuracy was higher when three samples were presented at the start of a trial relative to when only one sample was presented, but it did not matter whether the three samples were repeats of the same stimulus or were three different stimuli.

A third line of evidence for prospective coding in DMTS involves the effects of differential outcomes. Pigeons' choice accuracy and retention over delays is enhanced when correct choices of each comparison stimulus are followed by an outcome unique to that stimulus (Brodigan & Peterson, 1976; Peterson, 1984; Urcuioli, 1990a;, Urcuioli, 1990b;, Urcuioli, 1991; Urcuioli & Zentall, 1990; Williams, Butler, & Overmier, 1990). This facilitation suggests that the pigeons code the comparison stimulus and its outcome prospectively. Presumably, the uniqueness of the outcome makes the comparison stimuli more distinct, and hence more memorable. In one experiment pigeons were trained with differential outcomes that were reliably related either to the sample that appeared at the start of the trial or to the comparison stimulus chosen at the end of the trial (Urcuioli, 1990a). Samples were solid blue or yellow, or black and white horizontal, or vertical lines. The comparison stimuli were red and green. The differential outcomes were produced by reinforcing some correct choices 100% of the time and reinforcing other correct choices only 20% of the time. One group received 100% reinforcement for correct choices following red and 20% reinforcement for correct choices following the other comparison stimulus. The other group received 100% reinforcement following either yellow or vertical samples (following which the correct choices were red and green respectively) and received 20% reinforcement following blue or horizontal samples (for which green and red respectively were the correct choices). Hence, for all birds each sample was uniquely associated with a specific outcome. Only birds in the correlated group, however, received a consistent pairing between the correct comparison stimulus and an outcome. Consistent with the hypothesis that the pigeons prospectively coded the outcome as part of their representation of the correct choice on a trial, the birds in the correlated condition acquired the task more quickly and showed better retention over 0–4 s delays than the birds in the uncorrelated condition.

A few experiments have offered evidence suggesting that pigeons may sometimes use a retrospective code during the retention interval. For example, the ease of discriminating or identifying the sample stimuli has a larger effect on choice accuracy and retention than does the ease of discriminating the comparison stimuli (Urcuioli & Zentall, 1986; Zentall, Urcuioli, Jagielo, & Jackson-Smith, 1989). Although this kind of result is perhaps suggestive, in fact, it is not adequate for addressing questions concerning the type of code the animal employs during the retention interval. No matter what code the animal is using, it must discriminate/identify the sample before it can use the correct code. More difficult to discriminate stimuli are likely to be identified less accurately and with less strength or confidence. As a result, more difficult to discriminate stimuli are likely to result in lower choice accuracy and faster forgetting than when performing with easier to discriminate samples. Hence, both the prospective coding hypothesis and the retrospective coding hypothesis predict the same result—lower choice accuracy with difficult to discrimiate samples.

PROACTIVE INTERFERENCE

DMTS trials do not occur in isolation. The data described previously suggest that pigeons employ a gradually strengthening prospective representation for selecting the correct comparison stimulus. To the extent that these representations form and decay gradually, some information may remain from a previous trial when the bird makes its choice on a later trial. Proactive interference (PI) occurs when the events of one trial affect the choice made on the next.

Two types of PI have been observed. DMTS choice accuracy of pigeons (Grant, 1975; Maki, Moe, & Bierley, 1977; Roberts, 1980), monkeys (Jarrard & Moise, 1971), and dolphins (Herman, 1975), among others, is lower when trials are separated by short intertrial intervals (ITI) than when they are separated by long intertrial intervals. In addition, both monkeys (Moise, 1976; Worsham, 1975) and pigeons (Grant, 1975) show a decline in matching accuracy when the stimulus corresponding to the correct choice on one trial is the incorrect choice on the next. Most models attribute both of these effects to the same mechanism (e.g., Roberts & Grant, 1976). The memory from the previous trial is assumed to remain and interfere with the memory from the next trial. Choice accuracy is higher with longer ITI durations because the memory from the previous trial has longer to decay. Choice accuracy is lower when the correct choice from the earlier trial conflicts with that from the later trial because the two memories compete for control of responding.

An alternative account of proactive interference attributes these two effects to different mechanisms. The ITI effect derives from general decrements in the overall efficiency of the animal's information processing system, due, perhaps, to fatigue or temporary satiation. The intertrial conflict effect derives from specific competition or intrusion of the memory from one trial on the choice made on the next.

These alternative accounts were compared using three-alternative matching to sample (Roitblat & Scopatz, 1983). In the standard two-alternative DMTS task, there is only one way to make an error. In contrast, with three alternatives presented on each trial, the animal can err by choosing the comparison stimulus that was correct on the previous trial or by selecting the third stimulus. To the extent that errors are due to intrusions from the previous trial, the pattern of errors should depend on the events of that previous trial (e.g., an erroneous repeat of the choice made on the previous trial should be more likely than an erroneous choice of the third stimulus). To the extent that PI is due to general decrements in the birds' encoding mechanism, the pattern of errors should be independent of the events from the previous trial and repeats of the previous choice should be about as prevalent as selection of the unrelated stimulus. The common mechanism account asserts that these are not separable effects, that choice accuracy improves with longer sample durations precisely because competition from the memories of the previous trial is reduced.

An experiment using three alternative DMTS (Roitblat & Scopatz, 1983) found that choice accuracy was higher with longer than with shorter ITIs as well as evidence for specific intrusions, as expected. The choice made on one trial (the prechoice), and not the sample from the previous trial (the presample), affected the choice made on the next trial (see also Roberts, 1980). In contrast to the predictions of the common-mechanism hypothesis, however, the magnitude of the intrusions did not depend on the duration of the intertrial interval or on the sample or retention interval durations. Increasing the ITI duration increased the strength of the relationship between the sample and the choice (i.e., increased choice accuracy), but did not reduce the relationship between the either the presample or the prechoice and the choice.

The common-mechanism hypothesis for proactive interference also makes another prediction that is not supported by data. Short ITIs are predicted to produce errors because the animal is more likely to base its choice erroneously on its memory of the previous trial. In the case when the same sample is presented on two successive trials, however, the competing memory from the previous trial commands the same response as the sample from the current trial. A choice based on either memory should lead to a correct response. As a result, the common-mechanism hypothesis predicts that choice accuracy should be higher when the same sample is presented on two successive trials and the trials are separated by a short ITI than when the two trials are separated by a long ITI. If relatively strong mismatching memories (following a short ITI) can combine to produce interference, then matching memories with the same strength must combine to produce matching accuracy that is at least as high or higher than that observed when a weak memory from the previous trial combines with a relatively strong memory from the present trial (following a long ITI). In brief, choice accuracy following a repeated sample is predicted to be higher when the trials are separated by a short ITI than when separated by a long duration ITI. To my knowledge, no study has ever reported this effect.

These findings are consistent with a model of pigeon DMTS performance that uses two separate processes as the mechanism for encoding sample information (Edhouse & White, 1988; Roitblat, 1984a, 1984b). One process identifies the sample and then drives a second gradually changing memory process that codes the sample in terms of the correct response to be made. The greater the number of steps taken by this gradual memory process, the greater is the amount of information available to control choice responding (Roitblat, 1984b). Proactive interference consists of two separate affects on these processes. The ITI duration apparently operates on the initial sample-identification process. With short ITI durations the sample-identification process might become fatigued, for example, and hence, become less efficient (see Roitblat & Scopatz, 1983). Intrusions apparently operate on the gradual memory process, perhaps by affecting its starting point. The gradual memory process may not reset completely between trials and may remain partially biased toward the choice the animal made on the previous trial. This bias would correspond to the competition from the choice made on the previous trial.

THE DRIFT MODEL

The dual-mechanism memory hypothesis has been formalized in the drift model, which views representation in terms of a spatial metaphor (Roitblat, 1984a, 1984b). This model consists of two processes. The first is an identification process that determines the identity of the sample. The second is a memory process, driven by the identification mechanism. The memory process operates in a conceptual space in which each of the comparison stimuli is represented as a point. The distance between these points is determined by the dissimilarity among the stimuli. Those stimuli that are more similar are represented as closer together than those stimuli that are less similar. The state of the bird's memory is represented by the location of a "pointer" in this memory space.

During the sample presentation the identification process recognizes the sample after some latency. The more frequently this analyzer is used, the longer is its recognition latency. Sample recognition drives the pointer on a random walk, more or less in the direction of the comparison stimulus that corresponds to that sample. The random walk continues for as long as the sample is present. The pointer begins to drift soon after termination of the sample presentation. Higher choice accuracy is obtained following long rather than short sample presentations because the pointer moves more steps toward the location of the correct comparison stimulus. Similarly, choice accuracy declines during the retention interval because the pointer has time to drift from its position near the correct alternative to some position near one of the other comparison stimuli. All other things being equal, such a drift would tend to move the pointer to a position nearer the comparison stimulus that was more similar to the correct comparison than to the other comparison stimulus because the more similar stimulus is located nearer in the memory space. Between trials, the pointer would tend to reset to a point near the center of the memory space, but biased slightly toward the comparison that was selected. This position ensures that there will be some choice-specific interference on the next trial, but the degree of interference will not depend on the duration of the intertrial interval.

The drift model seems to capture many of the features of pigeon DMTS performance. It also appears applicable to memory processes in other species.

RAT SPATIAL DELAYED MATCHING
TO SAMPLE

Whereas studies of memory in pigeons have focused on the DMTS task, most studies of rat memory have focused on the radial arm maze (e.g., Beatty & Shavalia, 1980; Olton, 1978). The rat's task is to find pieces of food hidden at the ends of arms that radiate from a central platform like the spokes on a wheel. Rats typically retrieve about 7.5 pieces of food in their first eight arm entries. Even after a substantial delay of several hours, rats avoid revisiting arms from which they have already removed the food. A number of studies have ruled out algorithms, odor cues, and other intramaze cues as the basis for this performance. Instead, it appears that rats have excellent memory for spatial information (see Roitblat, 1987; for a review). In a continuing series of studies my students and I applied the DMTS methodology to investigations of the rat's spatial memory using a spatial delayed matching-to-sample (SDMTS) task.

The rats were trained and tested in an elevated three-arm starburst maze. One end of the maze contained a start box from which the rat was released at the start of a trial. The other end contained three goal boxes. At the start of each trial the rat was released from the start box and allowed to run into a preselected goal arm where it received a piece of food, and could see another, larger piece of food still in the goal box. The rat was then returned by hand to the start box and allowed a free choice among the three arms. If it returned to the arm to which it was forced during the sample run it received the remaining food. If it chose a different arm it was returned to the start box, the erroneously chosen goal arm was blocked off, and a second choice was allowed between the remaining two arms. The first run of a trial is called a "sample run." The second, and if necessary, third runs of a trial are called the "choice runs." The rat was rewarded for returning to the arm entered on the sample run and the location of this arm varied randomly from trial to trial.

Rats acquired this task readily averaging around 325 trials to meet an acquisition criterion of about 90% correct (Roitblat & Harley, 1988). Like pigeons', rats' performance on this task declined with increasing retention interval durations. Unlike pigeons, however, the pattern of first-choice errors did not depend on the sample. When errors occurred, they contained no information about the correct location of the sample run. The locations of second choices, on the other hand, were significantly related to the location of the erroneously chosen first choice. If the rats made an error on their first choice, their second choice was likely to be to the adjacent arm. Despite the bias to choose the adjacent arm, second choices were also significantly related to the location of the sample. Hence, second choices demonstrated that the animal had information about the location of the correct choice, even when making errors.

These rats also demonstrated proactive interference. Choice accuracy was higher when trials were separated by long ITIs than when they were separated by short ITIs. Unlike the data obtained from pigeons (Roitblat & Scopatz, 1983), increases in the duration of the intertrial interval resulted in a decrease in the influence of the events of the previous trial. The relationship between the choice made on one trial and the choice made on the previous trial was weaker with longer than with shorter ITI durations. The decrease in intrusions cannot explain all of the improvement in sample-choice relations with increasing ITI durations, however. On some proportion of the trials, the same sample appeared on two successive trials; on other trials, different samples were used. As suggested earlier, a pure intrusion mechanism would predict that choice accuracy should be higher when the same sample is presented twice in a row on trials separated by a short ITI than when the same samples are presented on trials separated by a long ITI. As in pigeons, no evidence was found to support such a prediction. Choice accuracy was higher following long ITI durations whether the samples from two successive trials agreed or not. As in pigeons, proactive interference in rats appears to involve two mechanisms, one for identifying and encoding the sample, and the other for maintaining the information over delays.

The data discussed so far indicate that rat and pigeon DMTS performance is the result of several interacting processes. Samples are apparently identified by one process and encoded by another that varies the strength of the memory representation. Finally, a third process appears to operate to mediate forgetting of the information derived from the sample. These three seem to be separate processes because manipulations can be found that affect one of them without affecting the other. Sample identification seems to be affected by the ITI duration in both rats and pigeons. ITI duration also seems to affect the encoding process for rats, but not for pigeons. Forgetting seems to be independent of the other two processes because the duration of the ITI does not appear to affect the rate of forgetting (Roitblat & Harley, 1988; Roitblat & Scopatz, 1983) and because the drug scopolamine appears to affect encoding or retention of the information without affecting the degree of intrusion by events from the previous trial (Roitblat, Harley, & Helweg, 1989).

The models described so far all assert that the sample is identified and that a comparison is made between the representation and the comparison stimuli. They fail to describe, however, the nature of the comparison mechanism. In fact, there has been little investigation of the strategies and processes animals use to identify and encode the sample and comparison stimuli or about the decision strategies animals employ when selecting a match. Except for the common finding that choice accuracy grows with increasing sample durations (in pigeons at least) and declines with increasing retention interval durations, little is known.

One reason for this limited investigation of the decision processes used by animals is the difficulty of obtaining relevant data. For example, in the typical DMTS experiment, we have no way of measuring the amount of effort expended by the animal in inspecting the sample or of the pattern by which effort is allocated to comparing the comparison stimuli. An ingenious solution to this problem was devised by Wright and Sands (1981). They tested pigeons on a DMTS task in which the stimuli were presented behind small windows where they were invisible unless the pigeon approached and stood directly in front of the window. Hence, by noting the position of the bird's head, one could determine the stimulus that was currently being examined.

Wright and Sands (1981) developed a model for the pigeon's identification of the correct choice. Their model is based on the assumption that the pigeon initially selects at random one of the comparison stimuli for observation. If that stimulus is sufficiently similar to the bird's representation of the sample, then it accepts that alternative as a match and responds accordingly. If the stimulus is not sufficiently similar, the pigeon moves to the other window and repeats the process with the other stimulus. This second comparison stimulus may also fail to meet the similarity criterion, in which case the bird returns to sample the first stimulus again. The bird continues to switch from one comparison stimulus to the other until one of them meets the match criterion. As predicted by their model, and by signal detection theory (Green & Swets, 1974), the probability of a switch from observing one comparison to observing the other depended on the identity of the comparison stimulus first selected for observation and on the similarity between the sample and the comparison stimuli.

Echolocating dolphins provide an excellent means of extending this model. The echolocating dolphin obtains information about objects in its environment by emitting clicks, which emerge from the dolphin's forehead or melon as a highly directional sound beam. Echoes from these sounds reflect from objects near the dolphin (up to more than 100 m) and provide information about the location, structure, and composition of that object (Nachtigall, 1980). The dolphin, hence, provides more precise information about the interrogation of the comparison stimuli by providing a precise measure of the amount of effort exerted by the dolphin in examining the comparison stimuli and by examining the way in which the dolphin combines information obtained from successive looks. We (Roitblat, Penner, & Nachtigall, 1990b) were able to extend the model introduced by Wright and Sands (1981) by examining each click and the information it provides.

The dolphin wore soft removeable eye cups, which occluded his vision. He was signaled to enter a listening hoop at the start of a trial and was allowed to echolocate on a sample that was presented 1 m under water at a distance of about 5 m directly in front of him. He was allowed to echolocate ad lib on the sample. The sample was then removed from the water and three comparison stimuli were lowered. The dolphin was allowed to echolocate on these stimuli as well. He indicated his decision about the location of the matching stimulus by touching a response wand located directly in front of each of the stimuli. Echolocation clicks were detected using hydrophones directly in the path between the listening hoop and the stimuli.

The dolphin was highly experienced at this task and at the time of this experiment averaged 94.5% correct over the course of 48 sessions. He emitted an average of about 37 echolocation clicks in 4.2 scans to the comparison stimuli per trial (Roitblat et al., 1990b). A scan is a train of echolocation clicks directed at one object, terminated either by a cessation of clicking, or by the start of a scan to another object. The number of clicks depended reliably on the stimulus being scanned. More clicks were required to recognize the cone and sphere than to recognize the tube. Furthermore, the dolphin tended to scan multiple stimuli in selecting the matching comparison, and at least some of the targets were scanned more than once in a trial. Therefore, his search strategy was not self-terminating.

The information in a returning echo is a sample from a population of possible echoes. Typically, the echolocating dolphin emits clicks that vary in source level and peak frequency (Moore & Pawloski, 1990). As a result, the echoes returning from objects ensonified by these clicks are also variable (Roitblat, Moore, Nachtigall, & Penner, 1991a). The information derived from any given echo, therefore, varies randomly and there is some probability that the information returning from an echo may be misleading.

When selecting among the comparison stimuli, the dolphin does not need to identify each of the comparison stimuli. It has only to obtain enough information to decide whether a given stimulus is or is not the correct match. The animal can also take advantage of his knowledge that only a limited set of stimuli are used in the experiment, and that one of the three comparison stimuli is always the correct match. Therefore, information obtained from echolocating on one of the stimuli can be used to reduce the uncertainty about the remaining comparison stimuli. To the extent that one item is identified, it reduces the possible set of stimuli that could be present in either of the other two comparison positions. The information obtained from a scan is not independent of the information obtained from preceding scans.

We developed a sequential sampling model that was an extension of that described by Wright and Sands (1981). This model incorporates assumptions that: (a) Each echo is a sample from a random distribution of possible echoes, (b) each echo is obtained at some cost or effort, which the dolphin attempts to minimize, (c) each echo supplies some stochastic information about the identity of the stimulus being scanned, (d) the dolphin combines information from successive echoes, and (e) the dolphin continues to obtain samples of information from additional echoes until the stimulus can be identified with sufficient confidence. Like the drift model, the present model assumes that the subject accumulates information about the identity of the stimulus with increasing exposure to that stimulus. More generally, an observer's confidence in the correctness of an identification grows monotonically with increasing numbers of looks, but at the expense of making those looks. The increasing evidence resulting from multiple examinations of the stimulus tends to be increasingly consistent with one of the alternatives and increasingly inconsistent with the others.

Simulations based on this model provide a reasonably good approximation of the dolphin's performance (Roitblat et al., 1990b). The simulation differed from the dolphin's actual performance, however, in that it was less variable than the live dolphin. The variability in the clicks emitted by the dolphin and in the echoes that he receives provides one possible explanation for the difference in variability between the sequential sampling model and the dolphin. The distribution of echoes is assumed by the model to be stationary and normal. In contrast, we found that spectra and the amplitudes of the outgoing clicks the dolphin used to ensonify the targets varied substantially over a click train (Roitblat, Penner, & Nachtigall, 1990a), but not as a function of the stimulus that was being scanned. In order to model some of this variability, we developed an artificial neural network system with a novel architecture that integrates information over successive echoes (Moore, Roitblat, Penner, & Nachtigall, 1991; Roitblat, Moore, Nachtigall, & Penner, 1991b).

The *integrator gateway network* incorporates features of the sequential sampling model described earlier, including the assumptions that the dolphin averages or sums spectral information from successive echoes and continues to emit clicks and collect returning echoes until it can classify the target producing those echoes with sufficient confidence. It mimics the dolphin's strategy of using multiple echoes to identify each target.

Echoes were presented to the network as a vector (list) of 30 values representing the frequency spectrum of the echo and 1 additional value, the start of train marker, representing whether or not the echo was at the start of an echo train. Recall that the dolphin directs a series or train of clicks to one target at a time, so it seemed plausible to include information marking the start of a click train. Subsequent layers of the network then summed the spectra from successive echoes, resetting the summation at the start of every echo train. The output layer of the network contained three units, each representing one of the sample/comparison stimuli (sphere, tube, cone). The relative activations of these units represented the network's "confidence" that each of the designated objects had been the source of the echoes.

The training set consisted of six sets of 10 successive echoes each, selected from the ends of haphazardly chosen echo trains. An equal number of cone, tube, and sphere echoes were used. The training set was a relatively small subset (4%) of the total set of available echoes (1,335). The test set consisted of 30 echo trains, 9 from the sphere, 11 from the cone, and 10 from the small tube. The network correctly classified 27 of the 30 echo trains; that is, on the basis of the returning echoes, the network was able to correctly determine the stimulus returning that echo. In comparison, another network that was identical in every way to the integrator gateway network except that it did not have the integrator apparatus, correctly classified only 20 of the 30 echo trains successfully. This difference shows that a neural network that implemented the assumptions of sequential sampling theory performed better than an analogous network that classified on the basis of individual echoes. Both networks reached a fixed confidence classification criterion in about the same number of echoes, but the integrator gateway network was more accurate. Although the superior performance of the integrator gateway network does not prove that the dolphin integrated information from successive echoes, it does support the claim.

CONCLUSION

Studies of delayed matching-to-sample performance have been a major factor in the development of our understanding of animal memory because the task is readily learned and lends itself to detailed and controlled analysis. Some early models (e.g., Roberts & Grant, 1976) suggested that matching results from a comparison of the comparison stimuli with a trace resulting from exposure to the sample earlier in the trial. Evidence has slowly accumulated that demonstrates that this simple trace model was inadequate (see Roitblat, 1987). First, correct performance on a DMTS task can occur when the similarity between the sample and the comparison stimuli is irrelevant, as in the so-called symbolic matching procedure. Second, pigeons seem not to make extensive use of a generalized matching rule, when such relations are relevant (e.g., Lombardi et al., 1984; Wright, et al., 1988; Zentall, Hogan, & Edwards, 1984). Third, a number of experiments demonstrate that the identity of the correct comparison stimulus is explicitly part of the representation used by a pigeon during the retention interval between the sample and the comparison stimulus presentations (e.g., Grant,

1982; Roitblat, 1980; Urcuioli, 1990a, 1991). Fourth, proactive interference appears to be the result of at least two separable processes, rather than the one suggested by the trace model (Edhouse & White, 1988; Roitblat, 1984a, 1984b; Roitblat & Scopatz, 1983).

Investigations with other species also reveal the complexity of animal memory. Like pigeons, rats also show evidence of gradual encoding of sample information and dissociable proactive interference mechanisms (Roitblat & Harley, 1988). Similarly, like pigeons (Wright & Sands, 1981), dolphins appear to use a sequential decision process to identify the correct match.

Memory in animals was once thought to be a passive process in which the animal merely recorded sense impressions. In contrast, the research described in the present chapter and elsewhere suggests that animal memory processes are active and quite complex. The animals appear to actively seek information, store, and retrieve it.

REFERENCES

Beatty, W. W., & Shavalia, D. A. (1980). Spatial memory in rats: Time course of working memory and effect of anesthetics. *Behavioral Biology, 28,* 454–462.

Blough, D. S. (1959). Delayed matching in the pigeon. *Journal of the Experimental Analysis of Behavior, 2,* 151–160.

Brodigan, D. L., & Peterson, G. B. (1976). Two-choice conditional discrimination performance as a function of reward expectancy, prechoice delay, and domesticity. *Animal Learning and Behavior, 4,* 121–124.

Carter, D. E., & Eckerman, D. A. (1975). Symbolic matching by pigeons: Rate of learning complex discriminations predicted from simple discriminations. *Science, 187,* 662–664.

Cohen, L. R., Looney, T. A., Brady, J. H., & Acuella, A. F. (1976). Differential sample response schedules in the acquisition of conditional discriminations by pigeons. *Journal of the Experimental Analysis of Behavior, 26,* 301–314.

D'Amato, M. R. (1973). Delayed matching and short-term memory in monkeys. In G. H. Bower (Ed.), *The psychology of learning and motivation: Advances in research and theory* (Vol. 7, pp. 227–269). New York: Academic.

D'Amato, M. R., Salmon, D. P., & Colombo, M. (1985). Extent and limits of the matching concept in monkeys (*Cebus apella*). *Journal of Experimental Psychology: Animal Behavior Processes, 11,* 31–51.

Edhouse, W. V., & White, K. G. (1988). Sources of proactive interference in animal memory. *Journal of Experimental Psychology: Animal Behavior Processes, 14,* 56–70.

Grant, D. S. (1975). Proactive interference in pigeon short-term memory. *Journal of Experimental Psychology: Animal Behavior Processes, 1,* 207–220.

Grant, D. S. (1982). Intratrial proactive interference in pigeon short-term memory: Manipulation of stimulus dimension and dimensional similarity. *Learning and Motivation, 13,* 417–433.

Green, D. M., & Swets, J. A. (1974). *Signal detection theory and psychophysics.* New York: Wiley.

Herman, L. M. (1975). Interference and auditory short-term memory in the bottle nosed dolphin. *Animal Learning and Behavior, 3,* 43–48.

Herman, L. M. (1980). Cognitive characteristics of dolphins. In L. M. Herman (Ed.), *Cetacean behavior: Mechanisms and functions* (pp. 363–429). New York: Wiley.

Herman, L. M., & Gordon, J. A. (1974). Auditory delayed matching in the bottlenose dolphin. *Journal of the Experimental Analysis of Behavior, 21,* 19–26.

Herrnstein, R. J. (1985). Riddles of natural categorization. *Philosophical Transactions of the Royal Society (London), B308,* 129–144.

Honig, W. K., & Thompson, R. K. R. (1982). Retrospective and prospective processing in animal working memory. In G. H. Bower (Ed.), *The psychology of learning and motivation* (Vol. 16, pp. 239–283). New York: Academic.

Hunter, G. A. (1988). *Visual delayed matching of two-dimensional forms by a bottlenosed dolphin.* Unpublished master's thesis, University of Hawaii, Honolulu.

Jarrard, L. E., & Moise, S. L. (1971). Short-term memory in the monkey. In L. E. Jarrard (Ed.), *Cognitive processes of nonhuman primates* (pp. 1–24). New York: Academic.

Lombardi, C. M., Fachinelli, C. C., & Delius, J. D. (1984). Oddity of visual patterns conceptualized by pigeons. *Animal Learning and Behavior, 12,* 2–6.

Maki, W. S., & Leith, C. R. (1973). Shared attention in pigeons. *Journal of the Experimental Analysis of Behavior, 19,* 345–349.

Maki, W. S., Moe, J. C., & Bierley, C. M. (1977). Short-term memory for stimuli, responses, and reinforcers. *Journal of Experimental Psychology: Animal Behavior Processes, 3,* 156–177.

Moise, S. L. (1976). Proactive effects of stimuli, delays, and response position during delayed matching from sample. *Animal Learning and Behavior, 4,* 37–40.

Moore, P. W. B., & Pawloski, D. (1990). Investigations on the control of echolocation pulses in the dolphin (*Tursiops truncatus*). In J. Thomas & R. Kastelein (Eds.), *Sensory abilities of cetaceans* (pp. 305–316). New York: Plenum.

Moore, P. W. B., Roitblat, H. L., Penner, R. H., & Nachtigall, P. E. (1991). Recognizing successive dolphin echoes with an integrator gateway network. *Neural Networks, 4,* 701–709.

Nachtigall, P. E. (1980). Odontocete echolocation performance on object size, shape, and material. In R. G. Busnel & J. F. Fish (Eds.), *Animal sonar systems* (pp. 71–95). New York: Plenum.

Olton, D. S. (1978). Characteristics of spatial memory. In S. H. Hulse, H. Fowler, & W. K. Honig (Eds.), *Cognitive processes in animal behavior* (pp. 341–373). Hillsdale, NJ: Lawrence Erlbaum Associates.

Peterson, G. B. (1984). How expectancies guide behavior. In H. L. Roitblat, T. G. Bever, & H. S. Terrace (Eds.), *Animal cognition* (pp. 135–147). Hillsdale, NJ: Lawrence Erlbaum Associates.

Pisacreta, R., Redwood, E., Witt, K. (1984). Transfer of matching-to-figure samples in the pigeon. *Journal of the Experimental Analysis of Behavior, 42,* 223–237.

Roberts, W. A. (1972). Short term memory in the pigeon: Effects of repetition and spacing. *Journal of Experimental Psychology: Animal Behavior Processes, 6,* 217–237.

Roberts, W. A. (1980). Distribution of trials and intertrial retention in delayed matching to sample with pigeons. *Journal of Experimental Psychology: Animal Behavior Processes, 6,* 217–237.

Roberts, W. A., & Grant, D. S. (1974). Short term memory in the pigeon with presentation time precisely controlled. *Learning and Motivation, 5,* 393–408.

Roberts, W. A., & Grant, D. S. (1976). Studies of short-term memory in the pigeon using the delayed matching to sample procedure. In D. L. Medin, W. A. Roberts, & R. T. Davis (Eds.), *Processes of animal memory* (pp. 79–112). Hillsdale, NJ: Lawrence Erlbaum Associates.

Roitblat, H. L. (1980). Codes and coding processes in pigeon short-term memory. *Animal Learning and Behavior, 8,* 341–351.

Roitblat, H. L. (1984a). Representations in pigeon working memory. In H. L. Roitblat, T. G. Bever, & H. S. Terrace (Eds.), *Animal cognition* (pp. 19–97). Hillsdale, NJ: Lawrence Erlbaum Associates.

Roitblat, H. L. (1984b). Pigeon working memory: Models for delayed matching-to-sample performance. In M. L. Commons, A. R. Wagner, & R. J. Herrnstein (Eds.), *Quantitative analyses of behavior: Discrimination processes* (Vol. 4, pp. 161–181). Cambridge, MA: Ballinger.

Roitblat, H. L. (1987). *Introduction to comparative cognition.* New York: Freeman.

Roitblat, H. L., & Harley, H. E. (1988). Rat spatial delayed matching-to-sample performance: Acquisition and retention. *Journal of Experimental Psychology: Animal Behavior Processes, 14,* 71–82.

Roitblat, H. L., Harley, H. E., & Helweg, D. A. (1989). The effects of scopolamine on proactive interference and spatial delayed matching-to-sample performance by rats. *Psychobiology, 17,* 402–408.

Roitblat, H. L., Moore, P. W. B., Nachtigall, P. E., & Penner, R. H. (1991a). Biomimetic sonar processing: From dolphin echolocation to artificial neural networks. In J. A. Meyer & S. Wilson (Eds.), *Simulation of adaptive behavior* (pp. 66–76). Cambridge, MA: MIT Press.

Roitblat, H. L., Moore, P. W. B., Nachtigall, P. E., & Penner, R. H. (1991b). Natural dolphin echo recognition using an integrator gateway network. In D. S. Touretsky & R. Lippman (Eds.), *Advances in neural information processing systems 3* (pp. 273–281). San Mateo, CA: Morgan Kaufmann.

Roitblat, H. L., Penner, R. H., & Nachtigall, P. E. (1990a). Attention and decision making in echolocation matching-to-sample by a bottlenose dolphin (*Tursiops truncatus*): The microstructure of decision making. In J. Thomas & R. Kastelein (Eds.), *Sensory abilities of Cetaceans* (pp. 665–676). New York: Plenum.

Roitblat, H. L., Penner, R. H., & Nachtigall, P. E. (1990b). Matching-to-sample by an echolocating dolphin. *Journal of Experimental Psychology: Animal Behavior Processes, 16,* 85–95.

Roitblat, H. L., & Scopatz, R. A. (1983). Sequential effects in delayed matching-to-sample. *Journal of Experimental Psychology: Animal Behavior Processes, 9,* 202–221.

Urcuioli, P. J. (1990a). Differential outcomes and many-to-one matching: Effects of correlation with correct choice. *Animal Learning and Behavior, 18,* 410–422.

Urcuioli, P. J. (1990b). Some relationships between outcome expectancies and sample stimuli in pigeons' delayed matching. *Animal Learning and Behavior, 18,* 302–314.

Urcuioli, P. J. (1991). Retardation and facilitation of matching acquisition by differential outcomes. *Animal Learning and Behavior, 19,* 29–36.

Urcuioli, P. J., & Zentall, T. R. (1986). Retrospective coding in pigeons' delayed matching-to-sample. *Journal of Experimental Psychology: Animal Behavior Processes, 12,* 69–77.

Urcuioli, P. J., & Zentall, T. R. (1990). On the role of trial outcomes in delayed discrimination. *Animal Learning and Behavior, 18,* 141–150.

Williams, D. A., Butler, M. M., & Overmier, J. B. (1990). Expectancies of reinforcer location and quality as cues for conditional discrimination in pigeons. *Journal of Experimental Psychology: Animal Behavior Processes, 16,* 3–13.

Wilson, B., Mackintosh, N. J., & Boakes, R. A. (1985a). Matching and oddity learning in the pigeon: Transfer effects and the absence of relational learning. *Quarterly Journal of Experimental Psychology, 37B,* 295–311.

Wilson, B., Mackintosh, N. J., & Boakes, R. A. (1985b). Transfer of relational rules in matching and oddity learning by pigeons and corvids. *Quarterly Journal of Experimental Psychology, 37B,* 313–332.

Worsham, R. W. (1975). Temporal discrimination factors in the delayed matching-to-sample task in monkeys. *Animal Learning and Behavior, 3,* 93–97.

Wright, A. A., Cook, R., Rivera, J., Sands, S., & Delius, J. D. (1988). Concept learning by pigeons: Matching-to-sample with trial-unique video picture stimuli. *Animal Learning and Behavior, 16,* 436–444.

Wright, A. A., & Sands, S. F. (1981). A model of detection and decision processes during matching to sample by pigeons: Performance with 88 different wavelengths in delayed and simultaneous matching tasks. *Journal of Experimental Psychology: Animal Behavior Processes, 7,* 191–216.

Zentall, T. R., Hogan, D. E., & Edwards, C. A. (1984). Cognitive factors in conditional learning by pigeons. In H. L. Roitblat, T. G. Bever, & H. S. Terrace (Eds.), *Animal cognition* (pp. 389–408). Hillsdale, NJ: Lawrence Erlbaum Associates.

Zentall, T. R., Urcuioli, P. J., Jagielo, J., & Jackson-Smith, P. (1989). Interaction of sample dimension and sample-comparison mapping on pigeons' performance of delayed conditional discriminations. *Animal Learning and Behavior, 17,* 172–178.

11
Coding Processes in Pigeons

Douglas S. Grant
University of Alberta

An invitation to author a chapter in a volume honoring an esteemed colleague upon his retirement affords an occasion on which it seems appropriate to take a somewhat reflective look at the field to which Al Riley and the contributors to this volume have devoted such a large portion of their research energies. It has now been two decades since the publication in 1972 of seminal articles by Maki and Leuin (who were at that time graduate students working in Al Riley's laboratory at Berkeley) and by Roberts. Both articles reported experiments using pigeon subjects and the delayed matching-to-sample (DMTS) procedure. In DMTS, accurate performance requires the retention of information derived from the sample stimulus until the test response is executed. Although several prior studies had employed the DMTS procedure and pigeon subjects (e.g., Berryman, Cumming, & Nevin, 1963; Blough, 1959; Cumming & Berryman, 1965; Smith, 1967), the articles by Maki and Leuin (1972) and Roberts (1972) were unique in that the information-processing perspective, which was at that time well entrenched in the literature on learning and memory in humans, was employed to conceptualize and analyze performance in a working memory task (i.e., DMTS) in pigeons. It was those articles that, in large part, gave rise to the now rather substantial field comprising the experimental and theoretical analysis of what is typically referred to as *working memory,* a term coined by Honig (1978).

A substantial data base and considerable theory now constitute the field of working memory in pigeons. Of course, the accumulation of published empirical and theoretical papers is only one indication that an area of scientific endeavor is progressing, and likely not the most important. A second metric that may be used to gauge the vitality of a research enterprise is the level of sophistication of the theoretical concepts employed. In particular, theoretical concepts may be expected to evolve from a relatively few rather global notions to a larger number of more specific notions as a scientific field matures and advances. In concert with this theoretical development, an advancing discipline would be expected to develop more powerful empirical techniques that permit differentiation between the operation of various putative theoretical processes.

One can inquire therefore as to whether the field of working memory in pigeons is advancing by asking (a) whether theoretical concepts are being elaborated, differentiated and specified and (b) whether techniques have been discovered and/or invented that permit the operation of at least some of the specific theoretical processes to be unambiguously differentiated from the operation of at least some others. In my view, the field of working memory in pigeons has made significant advances in both of these ways, both theoretically and empirically.

One purpose of this chapter is to illustrate these developments in one subarea of the field, that concerned with coding processes. An exhaustive review of the literature relevant to coding processes in pigeons is prohibited by limitations on space, and the chapter therefore concentrates on recent research from my laboratory. It is intended that the discussion illustrate two discernable advances in this area. The first is at the theoretical level and involves, in part, the recognition of a larger number of more specific forms of coding and, in addition, the recognition that coding processes are flexible and responsive to task demands. The second is at the empirical level and involves the discovery and development of techniques that can, under some conditions, provide relatively unambiguous evidence of the form of coding.

A CONCEPTION OF THE PIGEON
WORKING-MEMORY SYSTEM

As a prelude to the discussion of coding processes, it is necessary to first consider a general scheme in which to conceptualize processes of working memory in pigeons (for a more extensive elaboration of a similar conceptual scheme, see Grant, 1981). The present scheme endorses the distinction between working memory and reference memory offered by Honig (1978) and acknowledges the interaction between these memory systems in tasks used to study working memory. Working memory may be conceptualized as a repository for dynamic information of temporary relevance, whereas reference memory may be conceptualized as a repository for static information of enduring relevance. In the course of training in a working-memory task, representations of the contingencies are established in reference memory. To-be-remembered items, the sample stimuli in DMTS, activate associated codes stored in reference memory. In this view, working memory consists of a subset of reference-memory codes that is the set of all currently active codes. Test performance in DMTS is held to be controlled by codes that were activated by the sample and remained active throughout the retention interval, possibly through the aid of processes of maintenance rehearsal (for further discussion of rehearsal processes in pigeon working memory, see Grant, 1981, 1984; Maki, 1981).

RETROSPECTIVE AND PROSPECTIVE CODES

Consider a symbolic or arbitrary matching task in which, given a green sample, pecking the horizontal-line comparison is reinforced, whereas given a red sample pecking the vertical-line comparison is reinforced. According to the scheme outlined earlier, in such a task reference memory would contain a representation of the positive relation between green sample and horizontal comparison, and a representation of the positive relation between red sample and vertical comparison. Given such representations, there are two obvious candidates for the type of code that might be activated by the sample stimulus. First, the sample might activate a code that represents features of the sample stimulus: "green sample code" and "red sample code." At the time of testing, the active code in working memory would be used to query reference memory regarding the appropriate comparison response. Alternatively, the sample might activate a code that represents features of the associated comparison stimulus: "peck horizontal code" and "peck vertical code." At the time of testing, the active code in working memory would directly control test responding without any requirement for further inquiry of reference memory.

In the early 1980s, several authors recognized the distinction between these two forms of coding (e.g., Grant, 1981; Honig & Thompson, 1982; Riley, Cook, & Lamb, 1981; Roitblat, 1980, 1982). Although different terms were used to describe these two forms of coding by each of these authors, Honig and Thompson's (1982) descriptors, *retrospective coding* and *prospective coding,* have been employed most typically in the literature. Retrospective coding describes coding processes in which aspects of the sample stimulus are preserved in working memory and prospective coding describes coding processes in which aspects of test responding are preserved in working memory. Early theoretical conceptions of working memory in pigeons tended to emphasize retrospective coding (e.g., Maki, Riley, & Leith, 1976; Roberts, 1972; Roberts & Grant, 1976). By the early to mid-1980s, however, processes of prospective coding were becoming prominent in theoretical discussions of working memory in pigeons (e.g., Grant, 1981; Honig & Dodd, 1986; Honig & Thompson, 1982; Roitblat, 1982; Wasserman, 1986).

ELABORATION AND SPECIFICATION
OF CODING PROCESSES

Although the distinction between retrospective and prospective coding has been useful both conceptually and heuristically, it is now becoming clear that that distinction does not fully capture the diversity and richness of coding processes in pigeons. In particular, recent empirical work from my laboratory (reviewed later), as well as work from the laboratories of Peter Urcuioli and Thomas Zentall (see chapter 12 of this volume, and Zentall, Urcuioli, Jackson-Smith, & Steirn, in press, for further discussion of the work by Urcuioli and Zentall), suggest that the retrospective/prospective coding distinction requires elaboration. Research on memory for samples of food and no food suggests that a distinction orthogonal to the retrospective/prospective distinction should be introduced; that between symmetrical- and asymmetrical-coding processes. Research on memory for event duration suggests a need to distinguish between at least three different types or forms of coding (retrospective, prospective, and intermediate). Finally, research in both areas suggests that coding processes are flexible, controlled processes in that a particular sample stimulus may be coded in different ways depending on other characteristics of the task.

Coding of Samples of Food and No Food

In a typical study of memory for samples of food and no food, trials are initiated by presentation of a preparatory stimulus (e.g., black dot on white field) that terminates either in a brief (i.e., 3 s) presentation of grain (food sample) or an equivalent period of darkness (no-food sample). The comparison stimuli consist of one red and one green keylight; responding to one color is reinforced on trials initiated by a food sample, and responding to the other color is reinforced on trials initiated by a no-food sample.

Colwill (1984), using pigeons, and Wilson and Boakes (1985), using pigeons and jackdaws, examined the effect of retention interval length on accuracy of matching to samples of food and no food. On trials initiated by a sample of food, accuracy decreased markedly as the retention interval was lengthened. On the other hand, accuracy decreased only slightly as the retention interval was lengthened on trials initiated by a sample of no food.

The intriguing aspects of this result, from my perspective, are two. First, we seem to have at hand a diagnostic tool that would permit a determination of whether the same events (food and no-food samples in the present case) can be coded differently in the context of differing task requirements. In particular, suppose it could be shown that markedly different rates of forgetting occur on trials initiated by food and no-food samples in one instance whereas equivalent rates of forgetting on trials initiated by these samples occur in a second instance. Such a result would provide unambiguous evidence that the coding of food and no-food samples is a flexible, controlled process because the same coding process could not be responsible for both results (see Maki, 1981, for further discussion of the distinction between controlled and automatic processing in pigeons). Second, the differential rates of forgetting discovered by Colwill (1984) and by Wilson and Boakes (1985) are inconsistent with my earlier suggestion that samples of food and no food each activate an associated prospective code (Grant, 1981, 1982). Had such a coding process been operating in the experiments conducted by Colwill and by Wilson and Boakes, a sample of food would activate a code to peck one color and a sample of no food would activate a code to peck another color. Because the complexity of the code maintained in working memory would be equivalent on trials initiated by samples of food and no food, equivalent rates of forgetting would be anticipated.

Given these considerations, I recently completed a series of experiments designed to determine whether the way in which samples of food and no food are coded can be controlled experimentally (Grant, 1991). The manipulation chosen was the sample-to-comparison mapping arrangement. In one experiment, the form of coding in a one-to-one (OTO) sample-to-comparison mapping arrangement, similar to that employed by Colwill (1984) and Wilson and Boakes (1985), was compared to the form of coding in a many-to-one (MTO) sample-to-comparison mapping arrangement, similar to that employed by Grant (1981, 1982). Two pairs of samples were employed in each condition; one pair was the occurrence and nonoccurrence of food and the other was the occurrence of a vertical or horizontal line. In the MTO mapping, the food sample and one of the line orientation samples were each associated with the same color comparison (e.g., red), and the no-food sample and the alternate line-orientation sample were each associated with the alternate color comparison (e.g., green). In the OTO mapping, the food sample was associated with one of the color comparisons and the no-food sample was associated with the alternate color comparison, as was the case in the MTO mapping. However, the line-orientation samples were associated with line-orientation comparison stimuli. Thus, two samples were associated with each comparison in the MTO mapping whereas only a single sample was associated with each comparison in the OTO mapping.

The effect of the type of mapping arrangement on the coding of food and no-food samples was assessed using both between-subjects and within-subjects comparisons. Both comparisons produced similar results, and those obtained in the within-subjects comparison are shown in Fig. 11.1. The pigeons were trained and tested first in the OTO mapping condition. When transferred to the MTO mapping arrangement, the contingencies between sample and comparison on food/no-food sample trials were reversed, ensuring that the pigeons would need to relearn the matching task with both pairs of samples.

Inspection of Fig. 11.1 reveals that when trained in the OTO mapping, rate of forgetting was markedly more rapid on trials initiated by a food sample than on trials initiated by a no-food sample. On the other hand, when trained in the MTO mapping, the rate of forgetting was equivalent on trials initiated by each type of sample. Thus, the specific forms in which samples of food and no food are coded in the OTO and MTO procedures differ, revealing that coding processes in pigeons are, at least to some extent, flexible, controlled processes (for additional evidence of flexible coding processes in rats and pigeons, respectively, see Cook, Brown, & Riley, 1985, and Zentall, Steirn, & Jackson-Smith, 1990).

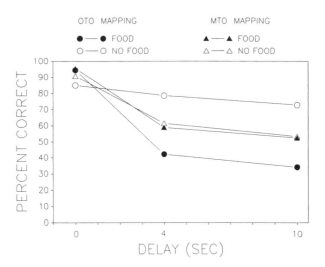

FIG. 11.1. Results of the within-subjects assessment of the effects of sample-to-comparison mapping arrangement on accuracy of matching to food samples and no-food samples as a function of delay. Subjects were trained and tested first in the OTO condition, then were trained and tested in the MTO condition (after Grant, 1991, Figure 3).

Additional empirical work was designed to evaluate a suggestion by Colwill (1984) and Wilson and Boakes (1985) that the differential rates of forgetting on trials initiated by food and no food reflects a failure to code the no-food sample in an OTO mapping arrangement. According to their asymmetrical-coding account, pigeons respond to the comparison associated with a no-food sample by default and, therefore, pigeons respond to the comparison associated with a food sample only when the default is overridden by an active code in working memory. An asymmetrical-coding account of this form would result in little, if any, forgetting on trials initiated by a sample of no food. On the other hand, rapid forgetting on trials initiated by a sample of food would be expected because it is likely that the code activated by a food sample will no longer be active at the time of testing when a substantial delay intervenes between sample and test. On these occasions, the bird should respond to the comparison associated with the no-food sample by default, resulting in a high error rate.

If the asymmetrical-coding account is correct, then any sample-to-comparison mapping arrangement that is not conducive to the operation of a default strategy should result in equivalent rates of forgetting on trials initiated by food and no food. One such condition is the MTO mapping arrangement described earlier, and the asymmetrical-coding account correctly anticipates equivalent rates of forgetting in that procedure. In a subsequent experiment, the asymmetrical-coding account was tested further by training pigeons in one of two different partial MTO procedures in which either the food sample or the no-food sample, but not both, was a component of a many-to-one mapping arrangement. In the MTO-food mapping, the food sample and two additional samples (plus and triangle) were each associated with the same comparison and the no-food sample was associated with the alternate comparison. In the MTO-no food mapping, the no-food sample and the two additional samples were each associated with the same comparison and the food sample was associated with the alternate comparison.

It was anticipated that pigeons trained in the MTO-food mapping, but not those trained in the MTO-no food mapping, would use a strategy in which the no-food sample is not coded and a response to the comparison associated with the no-food sample occurs by default. This anticipation was based on the consideration that a default strategy is readily applied only when the no-food sample is the only event associated with a particular comparison stimulus. When events in addition to the no-food sample are associated with the same comparison, as is the case in the MTO-no food mapping, effective coding strategies based on a default response become cumbersome. For example, one of these strategies involves not coding any of the three samples associated with the same comparison stimulus as the no-food sample (plus, triangle, and no food), and coding the remaining sample (food). Although such a strategy could mediate accurate performance, the fact that pigeons would have to fail to code three of the four samples might well obscure the applicability of a default strategy to solving the MTO-no food task. A second potentially effective default strategy involves coding each of the samples except for the no-food sample. Again, the applicability of such a strategy is likely obscure, in this case because two of the samples (plus and triangle) would activate codes to peck the same comparison as specified by the default.

If, as argued previously, only the MTO-food mapping is conducive to a strategy in which the no-food sample is not coded and a response to the comparison associated with the no-food sample occurs by default, then the asymmetrical-coding account anticipates differential rates of forgetting as a function of sample type only in the MTO-food mapping. The results, shown in Fig. 11.2, support the asymmetrical-coding account. Specifically, pigeons trained in a partial MTO mapping in which the no-food sample did not participate in the MTO mapping (Group MTO-Food) demonstrate differential rates of forgetting on trials initiated by samples of food and no-food. In contrast, pigeons trained in a partial MTO mapping in which the no-food sample did participate in the MTO mapping (Group MTO-No Food) demonstrate equivalent rates of forgetting on each type of trial.

The importance of this research on the coding of food and no-food samples derives from two implications. First, the results necessitate the introduction of a distinction between symmetrical- and asymmetrical-coding processes, a distinction that is orthogonal to that between retrospection and prospection. Combining the two distinctions factorially yields four conceptually distinguishable types of

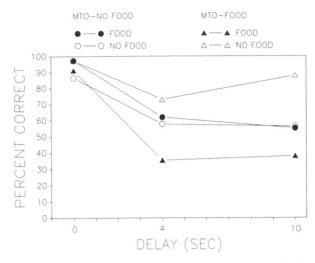

FIG. 11.2. Accuracy of matching to food samples and no-food samples as a function of delay in each partial MTO mapping arrangement.
In Group MTO-No Food, the no-food sample and two additional samples were associated with one comparison stimulus, and the food
sample was associated with the alternate comparison stimulus. In
Group MTO-Food, the food sample and two additional samples were
associated with one comparison stimulus, and the no-food sample
was associated with the alternate comparison stimulus (after Grant,
1991, Figure 4).

coding processes. Second, and more important, the results provide the first
convincing demonstration that the same event (i.e., the occurrence of no food)
may be coded differently depending on aspects of the task. These results suggest
further, therefore, that coding processes are, at least to some extent, flexible,
controlled processes. It could be argued, however, that the degree of flexibility
of the coding system demonstrated is rather minimal in that it was shown only
that an event is not coded under some conditions and is coded under other
conditions. A more powerful demonstration of the flexibility of the coding system would entail a demonstration that a particular event activates qualitatively
different codes in different conditions. Although such evidence has not been
produced in the case of memory for samples of food and no food, such evidence
has arisen from recent research on memory for event duration.

Coding of Event Duration

In a typical study of memory for event duration, trials are initiated by presentation of a stimulus (e.g., overhead houselight) that terminates after either a short
(e.g., 2 s) or long (e.g., 10 s) period. The comparison stimuli consist of one red

and one green keylight; responding to one color is reinforced on trials initiated by a short-duration sample, and responding to the other color is reinforced on trials initiated by a long-duration sample.

Spetch and Wilkie (1982) examined the effect of retention interval length on accuracy of matching to short and long samples. On trials initiated by a long sample, accuracy decreased markedly as the retention interval was lengthened. On the other hand, accuracy decreased only slightly as the retention interval was lengthened on trials initiated by a short sample. Thus, pigeons tended to respond to the comparison stimulus associated with a short sample at longer delays, an effect referred to as the *choose-short effect*. The choose-short effect has now been reported in a number of experiments employing pigeons (Kraemer, Mazmanian, & Roberts, 1985; Spetch, 1987; Spetch & Rusak, 1989; Spetch & Wilkie, 1982, 1983), a result that has been interpreted as reflecting a process of subjective-shortening (e.g., Spetch, 1987; Spetch & Wilkie, 1982, 1983). According to the subjective-shortening account, pigeons code samples of different duration retrospectively in terms of perceived duration and, at the time of testing, remembered duration controls choice between the comparison stimuli. The subjective-shortening account maintains that remembered duration becomes subjectively shorter as time in the absence of the event increases, leading to a tendency at longer delays to remember long-duration events as having been short.

Although research employing both food and no-food samples and long and short samples has revealed asymmetrical retention functions under some conditions, the coding processes mediating this asymmetry are apparently different in the two cases. As suggested earlier, differential rates of forgetting in the case of food and no-food samples is mediated by a coding process in which the no-food sample is not coded. Although a similar asymmetrical-coding process, in which the short sample is not coded, could explain the choose-short effect, the account is contradicted by other results.

Particularly problematic for an asymmetrical-coding account of the choose-short effect is the discovery of the complement of the choose-short effect; the *choose-long effect* (Spetch, 1987; Spetch & Wilkie, 1983). Spetch (1987) found that following training at a fixed delay of either 10 s or 20 s, testing at delays shorter than that employed in training produced a strong tendency to choose the comparison stimulus associated with the long sample stimulus. Had the pigeons employed a coding process in which the short sample is not coded, testing at any delay interval, whether shorter or longer than that employed in training, should reveal that the majority of errors involve selection of the comparison stimulus associated with the short sample. Thus, although an asymmetrical-coding process in which the short sample is not coded can account for the choose-short effect, the account is negated by the discovery of the choose-long effect.

As Spetch (1987) argued, a retrospective and analogical coding process, in which both short and long samples are coded in terms of perceived duration, in combination with a process of subjective shortening, in which remembered duration becomes subjectively shorter as a function of the passage of time since the termination of the event giving rise to the duration, correctly anticipates both choose-short and choose-long errors. Specifically, choose-short errors predominate at delays longer than that employed in training because the duration represented in working memory has foreshortened and therefore should be judged shorter than those represented and associated with particular comparison responses in reference memory. Training at a longer delay should result in foreshortened representations of event duration being associated with particular comparison responses in reference memory. Choose-long errors would predominate at delays shorter than that employed in training because the duration represented in working memory has not foreshortened and therefore should be judged longer than those represented and associated with particular comparison responses in reference memory.

My interest in studying memory for duration derives primarily from considerations similar to those that motivated my study of memory for food and no-food samples. As noted in the preceding section, that consideration is the possibility of obtaining evidence that the same events may be coded differently as a function of task requirements. Such differential coding would be implicated to the extent that conditions could be found under which the choose-short effect does and does not occur (said another way, conditions under which differential rates of forgetting do and do not occur).

Research described in the following sections has revealed two categories of conditions that determine the manner in which samples of different durations are coded. One category is the type of assessment task employed—whether a choice- or successive-matching task is employed. The second category is the type of sample-to-comparison mapping arrangement employed in the choice-matching task—whether a OTO or MTO mapping is employed.

The Effect of Assessment Task.   In the successive task, there are two different test stimuli, but only one of them is presented following the sample on each trial. Pecks to one test stimulus are reinforced following short samples but not following long samples, whereas pecks to the other test stimulus are reinforced following long samples only. Memory is assessed by the discrimination ratio that is the proportion of responses to positive test stimuli. In spite of the fact that the choice- and successive-matching tasks each involve similar memory requirements, Marcia Spetch and I recently obtained evidence that pigeons code samples of event duration differently in the two tasks (Grant & Spetch, 1991; Spetch & Grant, in press).

One experiment involved both a between-subjects and a within-subjects comparison of the effect of assessment task on coding processes (Grant & Spetch, 1991). In both tasks, 2-s and 8-s presentations of houselight served as the samples and red and green served as the test stimuli. The results of the between-subjects comparison are summarized in Fig. 11.3. Pigeons trained and tested in

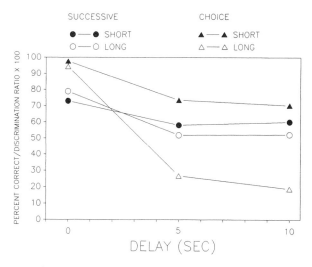

FIG. 11.3.   Accuracy of matching to short and long samples as a function of delay in subjects trained in a successive-matching task and in
subjects trained in a choice-matching task (after Grant & Spetch, 1991,
Figure 3).

the choice task demonstrated differential rates of forgetting as a function of
sample type; accuracy decreased markedly as a function of delay on trials initiated by a long sample, whereas accuracy decreased only slightly as a function of
delay on trials initiated by a short sample. In contrast, equivalent rates of forgetting on short- and long-sample trials was obtained in pigeons trained and tested in
the successive task.

Following original training and testing, the birds assigned to the successive
task were transferred to the choice task, and the birds assigned to the choice task
were transferred to the successive task. The transfer was arranged such that the
reinforcement contingencies were consistent across tasks for each bird. That is,
for birds transferred to the choice task the former positive test stimulus on short-
sample trials became the correct comparison stimulus on short-sample trials, and
the former positive test stimulus on long-sample trials became the correct comparison stimulus on long-sample trials. Similarly, for birds transferred to the
successive task the former correct comparison on short-sample trials became the
positive test stimulus on short-sample trials, and the former correct comparison
on long-sample trials became the positive test stimulus on long-sample trials.

The results of the within-subjects comparison are shown in Fig. 11.4. Birds
who were first trained in the choice task (top portion) demonstrated equivalent
rates of forgetting on trials initiated by a short and long sample when subsequently tested in the successive task. Birds who were first trained in the successive task (bottom portion) did not demonstrate differential rates of forgetting on

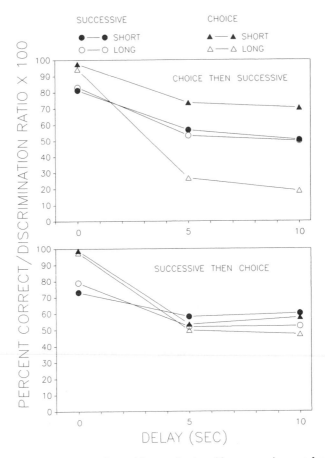

FIG. 11.4. Accuracy of matching to short and long samples as a function of delay in the choice and successive tasks in each group. The top
portion shows accuracy for subjects trained initially in the choice task
and the bottom portion shows accuracy for subjects trained initially in
the successive task (after Grant & Spetch, 1991, Figure 5).

short- and long-sample trials when subsequently tested in the choice task. Instead, accuracy declined at an approximately equivalent rate as a function of
delay on trials initiated by a short and long sample.

Thus, differential rates of forgetting as a function of sample duration are not
obtained when pigeons are tested in the successive task, regardless of whether or
not they had been previously trained in the choice task. On the other hand,
differential rates of forgetting as a function of sample duration are obtained when
naive pigeons are trained and tested in the choice task. However, previous
training in a successive task in which the contingencies are consistent with those

in a subsequent choice task is sufficient to eliminate the differential rates of forgetting that would otherwise obtain.

The different effects of delay in the choice and successive tasks may be interpreted as reflecting different coding strategies that naive pigeons employ in the two tasks. Specifically, naive pigeons might code samples of different durations retrospectively in terms of perceived duration in the choice task. A process of subjective shortening operating during the retention interval then would lead to differential rates of forgetting on the two types of trials. In the successive task, naive pigeons might code the samples prospectively in terms of an instruction to respond and/or not respond to a particular test stimulus. If the appropriate code is forgotten during the delay, responding should be nonsystematic with respect to sample duration, resulting in nondifferential rates of forgetting as a function of sample type.

The view that naive pigeons employ a retrospective coding strategy in the choice task and a prospective coding strategy in the successive task, combined with a consideration of the extent to which each strategy would be expected to be immediately applicable to the alternative procedure, allows one to provide a plausible account of the results from both transfer and delay testing. Specifically, the transfer results should reflect the extent to which accurate performance in the two tasks is readily mediated by coding processes acquired in the alternative task. A prospective coding strategy of the form "respond to stimulus A" and/or "do not respond to stimulus B," acquired in the successive task, is sufficient to generate accurate performance in the choice task. Hence, animals transferred from the successive task to the choice task should, and did, demonstrate substantial positive transfer (78.1% correct on the first session of choice matching). However, because these animals are employing a prospective coding strategy in the choice task, they should not demonstrate a choose-short tendency during delay testing in the choice task. On the other hand, a retrospective coding strategy of the form "if remembered duration is X choose comparison A instead of comparison B" is not sufficient to generate accurate performance in the successive task. In particular, retrieval of a rule of the form "choose comparison A instead of comparison B" would not necessarily lead to either (a) a high rate of key pecking if stimulus A was present and/or (b) a low rate of key pecking if stimulus B was present. Hence, animals transferred from the choice task to the successive task might, and did, demonstrate little positive transfer (the discrimination ratio on the first session of successive matching was .512). Moreover, to the extent that these animals abandon the retrospective strategy and adopt a prospective one in acquiring the successive task, subsequent delay testing should fail to reveal a respond-short tendency.

The point to be emphasized here is that we now have a second demonstration of the flexible, controlled nature of coding processes in pigeons. Moreover, the present case may represent a higher degree of flexibility in that the events (the short and long samples) were actively coded in each case, but were coded in a qualitatively different fashion (retrospectively and prospectively) in the choice and successive tasks. A second demonstration of a similar magnitude of flexibility of the coding system has arisen in recent work in which the sample-to-comparison mapping arrangement has been manipulated in a choice-matching task.

The Effect of Mapping Arrangement in a Choice Task. As noted earlier, naive pigeons trained and tested in the choice-matching task demonstrate differential rates of forgetting on trials initiated by short and long samples. This result is interpretable within the subjective-shortening account, which maintains that naive pigeons code durations retrospectively and analogically in the choice-matching task. The research described in the preceding section suggests that pigeons code durations differently, and probably prospectively, in the successive-matching task. The research described in this section asked whether, under some conditions, naive pigeons might code durations in a nonanalogical form in the choice-matching task.

As in the work involving samples of food and no food described previously, the sample-to-comparison mapping arrangement was manipulated in an attempt to induce different forms of sample coding. In one series of experiments (Grant & Spetch, 1993), the sample on any particular trial was either a short (2 s) presentation of either houselight or keylight (cross) or a long (10 s) presentation of either houselight or keylight. The comparison stimuli on each trial consisted of one red field and one green field. Pigeons received one of two MTO sample-to-comparison mapping arrangements. In the consistent mapping arrangement, the two short samples were associated with one comparison and the two long samples were associated with the alternate comparison. In this arrangement, the event that carried the duration information, houselight or keylight, was irrelevant and only duration was relevant to problem solution. In the inconsistent mapping arrangement, in contrast, both duration and carrier were relevant to problem solution. This was the case because one comparison was correct following a short presentation of one carrier and a long presentation of the alternate carrier (2-s houselight and 10-s keylight), and the alternate comparison was correct following either of the two remaining samples (10-s houselight and 2-s keylight).

It was anticipated that the inconsistent mapping arrangement would be particularly likely to result in nonanalogical, and likely prospective, coding. This anticipation was based, in part, on the fact that only two codes would be required to mediate accurate performance in a prospective-coding process (e.g., peck red and peck green), whereas an analogical-coding process sufficient to mediate accurate performance would require the use of four codes (one for each of the four samples). Moreover, the codes would be rather complex in that information about both duration and carrier would need to be preserved in each code. Honig and Thompson's (1982) suggestion that the coding process that requires less information will operate leads to the expectation of nonanalogical coding in group inconsistent.

In the consistent group, in which only duration is relevant, accurate performance could be mediated by an analogical-coding process in which each of the four samples would be represented analogically as one of two values; one value associated with a 2-s duration and the other value associated with a 10-s duration. However, given that the houselight is brighter than the keylight, Wilkie's (1987) finding that bright events are perceived as longer than equivalent durations of dim events suggests that an analogical-coding process might not be particularly effective. Likely candidates for nonanalogical-coding processes that might operate in the consistent mapping are prospective coding ("peck red" and "peck green") and intermediate coding. In an intermediate-coding process, samples associated with the same comparison stimulus activate a common code, which is in turn associated with a particular comparison response (for similar suggestions see, e.g., Grant, 1982; Maki, Moe, & Bierley, 1977; Roitblat, 1980; Urcuioli, Zentall, Jackson-Smith, & Steirn, 1989). In the most general form of this coding process, samples associated with one comparison activate an "event A" code and samples associated with the alternate comparison activate an "event B" code. At the time of testing, the active code would be used to query reference memory to generate the appropriate comparison response. Although intermediate coding might take a form in which the codes are related arbitrarily to the samples in the consistent mapping condition (as is the case for the codes "event A" and "event B"), intermediate codes might represent a feature common to each of the samples associated with that code, in which case the intermediate codes would be "short" and "long."

A delay test was employed to determine whether either or both of these MTO mappings would result in nonanalogical coding of event duration. Nonanalogical coding, whether prospective ("peck red" and "peck green") or intermediate (either "event A" and "event B" or "short" and "long") would be implicated to the extent that accuracy on both short- and long-sample trials decreased at approximately the same rate as a function of increases in delay. This prediction is predicated on the notion that there is no reason to anticipate that the two codes should be forgotten at different rates.

As shown in Fig. 11.5, neither mapping arrangement produced any evidence of a choose-short effect at longer delays. The equivalent rates of forgetting on short- and long-sample trials suggests that the pigeons in each mapping arrangement employed a nonanalogical-coding process. These results, viewed in conjunction with previous demonstrations of differential rates of forgetting on short- and long-sample trials in the choice-matching task in which a one-to-one mapping was employed, suggest that pigeons are capable of coding samples of event duration in different ways in the choice-matching task and, therefore, provide another demonstration of the flexible, controlled nature of coding processes in pigeons.

In further work concerning the effects of mapping arrangement on the coding of samples of duration in the choice-matching task (Grant & Spetch, 1993), we investigated whether the presence of nontemporal samples can alter the way

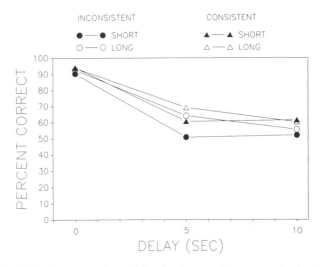

FIG. 11.5. Accuracy of matching to short and long samples in each group as a function of delay. In Group Inconsistent, a short duration of one carrier and a long duration of the alternate carrier were associated with a single comparison. In Group Consistent, a short duration of either carrier was associated with one comparison and a long duration of either carrier was associated with the alternate comparison (after Grant & Spetch, 1993, Figure 6).

in which temporal samples are coded. Two groups of pigeons were trained with two pairs of samples: line samples (a vertical and horizontal line, each 6 s in duration) and duration samples (2- and 10-s presentations of houselight). Color comparisons were always presented following the duration samples. In the MTO mapping, color comparisons were also presented following the vertical- and horizontal-line samples. In the OTO mapping, line-orientation comparisons were presented following the vertical- and horizontal-line samples. A third group served as a control and was trained with 2- and 10-s houselight samples only.

Following acquisition, two delay tests were conducted in succession; delays of 0, 5, and 10 s were employed in the first test, and delays of 0, 10, and 20 s were employed in the second test. As shown in Fig. 11.6, a robust choose-short effect was obtained in both the Control and OTO conditions, as expected. Contrary to our expectations, however, the MTO condition also resulted in a choose-short effect, albeit the effect was not as strong as in either the Control or OTO conditions.

The results from the MTO condition suggest that retrospective and analogical coding of event duration occurred in this condition in spite of the many-to-one mapping arrangement. This result is somewhat surprising in that the work reviewed earlier revealed evidence of nonanalogical coding of event duration in two variations of a many-to-one mapping arrangement in which all of the sam-

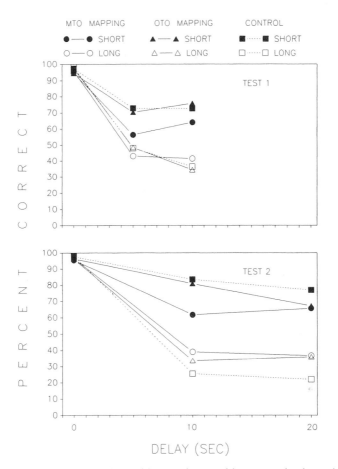

FIG. 11.6. Accuracy of matching to short and long samples in each group as a function of delay during two delay tests. In all three groups, duration samples (2-s and 10-s presentations of houselight) were followed by red and green comparisons. In Group MTO, nontemporal samples (different line orientations) were followed by red and green comparisons. In Group OTO, nontemporal samples were followed by vertical and horizontal comparisons. In Groups MTO and OTO, matching to temporal and nontemporal samples were trained concurrently. In Group Control, nontemporal samples were not employed (after Grant & Spetch, 1993, Figure 2).

ples were temporal. One possible explanation is that the introduction of nontemporal samples has little, if any, effect on the coding of temporal samples. Inspection of the acquisition data in the MTO condition, however, suggested an alternative explanation. That inspection revealed that animals in the MTO condition acquired accurate matching to temporal samples considerably more quickly than they acquired accurate matching to nontemporal (line-orientation) samples. Thus, the failure of nontemporal samples to markedly influence the way in which temporal samples were coded may have occurred because the animals learned to code the temporal samples before they learned to code the line-orientation samples.

To assess this possibility, we conducted a second experiment in which animals in the OTO and MTO conditions first acquired accurate matching with line-orientation samples (the comparisons were line orientations in the OTO condition and colors in the MTO condition). Following acquisition with the lines, half of the trials within each session now involved the temporal samples, which were followed by color comparisons. Following acquisition of accurate matching to the temporal samples, two delay tests, identical with those conducted in the first experiment, were administered.

As shown in Fig. 11.7, a robust choose-short effect was obtained in both the Control and OTO conditions, as was the case in the first experiment. However, the MTO condition revealed no evidence of a choose-short effect and instead accuracy declined at an approximately equivalent rate on trials initiated by short and long samples. At least under some conditions, incorporating nontemporal samples into the memory for duration task can result in nonanalogical coding of the temporal samples.

Although it is clear that a nonanalogical-coding process was employed in the MTO condition, the results do not permit the precise specification of the nonanalogical-coding process that was employed. As noted previously, either a prospective-coding process or an intermediate-coding process would result in equivalent rates of forgetting on trials initiated by short and long samples. To discriminate between prospective- and intermediate-coding processes, Marcia Spetch and I recently applied a technique developed by Urcuioli, Zentall, and their associates (Urcuioli et al., 1989; Zentall, Steirn, Sherburne, & Urcuioli, 1991; Zentall et al., in press; see also chapter 12 of this volume) to the present case. The birds in the OTO and MTO conditions were given a second phase of training in which one pair of samples from the original task (durations for half the birds and line orientations for half the birds) was associated with new comparison stimuli (shapes). In a subsequent transfer phase, the shape comparisons were presented on trials involving the alternate set of samples from the original task. The contingencies during training and transfer testing with the shape compari-sons were identical in both the OTO and MTO conditions and were arranged such that samples that were associated with the same color comparison in the MTO mapping were now associated with different shape comparisons. To illus-trate, suppose during MTO training the 2-s and vertical samples were associated with red, and the 10-s and horizontal samples were associated with green. Then, for example, vertical might be associated with circle and horizontal with dot. During the transfer phase, circle and dot comparisons would follow duration samples; dot would be the correct comparison on 2-s sample trials and circle

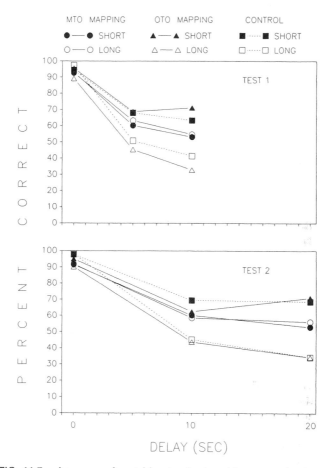

FIG. 11.7. Accuracy of matching to short and long samples in each group as a function of delay during two delay tests. In all three groups, duration samples (2-s and 10-s presentations of houselight) were followed by red and green comparisons. In Group MTO, nontemporal samples (different line orientations) were followed by red and green comparisons. In Group OTO, nontemporal samples were followed by vertical and horizontal comparisons. In Groups MTO and OTO, matching to nontemporal samples was trained prior to training with temporal samples. In Group Control, nontemporal samples were not employed (after Grant & Spetch, 1993, Figure 4).

would be the correct comparison on 10-s sample trials. Thus, if the line sample and the duration sample that were associated with the same comparison during original MTO training were represented by a single intermediate code (e.g., "short" or "sample A"), then accuracy during the transfer test should be lower in the MTO condition than in the OTO condition.

The transfer test did reveal evidence for intermediate coding in the MTO condition in that accuracy during the first 60-trial session of transfer testing was 61.8% correct in the OTO condition and only 43.4% correct in the MTO condition. A more molecular analysis of first session performance in the MTO condition revealed even stronger evidence for intermediate coding; the percentage of correct responses in each of the five 12-trial blocks that constituted the first session of transfer was 34.7, 39.3, 42.0, 52.5, and 48.3, respectively. Thus, when confronted by shape comparisons following samples that had not been associated with either of those comparisons, birds in the MTO condition tended to select the shape comparison that had been correct on trials initiated by the sample that was associated with the same color comparison as the current sample during MTO training.

The transfer result just discussed implicates an intermediate ("short" and "long," or "event A" and "event B") rather than prospective ("peck red," "peck green") coding process (for a similar argument, see Zentall et al., in press). In particular, it is far from obvious through what mechanism a prospective-coding process would give rise to such transfer. To illustrate, given that a vertical line and a 2-s houselight each initially activate the code "peck red," why would, for example, associating the vertical line with a circle comparison result in a tendency to peck circle as opposed to dot following a 2-s houselight? On the other hand, such mediated transfer is readily conceptualized given intermediate coding. To illustrate, given that a vertical line and a 2-s houselight each initially activate the "event A" code, associating the vertical line with a circle comparison might well alter the associative linkage from the "event A" representation in reference memory such that the "event A" representation is now associated with the circle. Thus, presentation of a 2-s houselight should activate the "event A" code (because of Phase 1 training) and, at the time of testing, reference memory rules established in Phase 2 should result in selection of a circle in preference to selection of a dot. The mediated-transfer technique holds great promise as a diagnostic device to discriminate between different forms of common coding in many-to-one mapping arrangements.

CONCLUSION

The selective review of recent research investigating coding processes in pigeons provided here illustrated progress at both the theoretical and empirical levels. At the theoretical level, our conception of coding processes has evolved from a rather simple conception in which a distinction between two forms of coding, retrospection and prospection, was emphasized to a more elaborate, sophisticated conception. In particular, it is clear that at least three broad categories of coding processes should be distinguished: retrospective, prospective and intermediate. Within the category of intermediate-coding processes, moreover, conceptual differentiation can be made between two subcategories; perhaps "arbitrary" and "nonarbitrary" are reasonable descriptors. One example of nonarbitrary, intermediate coding was considered earlier; that is, the possibility that event duration might be coded categorically ("short" and "long"). Another example is provided by Zentall et al. (1991; see also chapter 12 of this volume) who obtained evidence that, under certain conditions, both color and line-orientation samples may be coded as color samples, an instance of what I refer to as "hard-to-easy" coding. Whether these two instances should both be recognized as instances of nonarbitrary, intermediate coding or whether they represent instances of separate subcategories of intermediate-coding processes is unresolved at present.

It is also now clear that it is necessary to distinguish between coding processes that are symmetrical, in that each member of a pair of samples activates a code, and asymmetrical, in that only one member of a pair of samples activates a code. The distinction between symmetrical- and asymmetrical-coding processes is, conceptually, independent of the distinction between retrospective-, prospective-, and intermediate-coding processes. The two distinctions therefore yield six distinct categories of coding processes.

In addition to suggesting new categories of coding processes, recent research has revealed that at least some types of samples are coded differently in different contexts. It is thus becoming clear that processes of coding, as is the case for processes of maintenance rehearsal (Grant, 1981, 1984; Maki, 1981), are flexible, controlled processes. Yet to be determined are the range of applicability of this flexibility and the specific conditions that promote one type of coding over others.

Accompanying, and often inspiring, these theoretical developments have been developments at the empirical level. The discovery of differential rates of forgetting on trials initiated by alternate members of a pair of samples has afforded a unique opportunity to investigate the specific type and general nature of coding processes in pigeons. In addition, the development of techniques involving transfer and reversal has provided an opportunity to assess the form of coding mediating performance in MTO sample-to-comparison mapping arrangements. Further application of these techniques will enhance our understanding of coding processes.

Although our understanding of coding processes has increased significantly in the past few years, our level of understanding is far from complete. Future work could be directed profitably at two general issues. The first issue is whether the current set of distinctions fully capture the richness and diversity of coding processes in pigeons. The second issue involves identifying the conditions that promote one form of coding as opposed to other forms. It has been suggested that, in a OTO mapping arrangement, high sample discriminability promotes retrospection (Zentall, Urcuioli, Jagielo, & Jackson-Smith, 1989) and a successive-matching task promotes prospection (Grant & Spetch, 1991). Are there other conditions that promote one form of coding over others in OTO mapping arrangements? In a MTO mapping, is it generally the case that less discriminable samples activate intermediate codes that represent the more discriminable member of the MTO mapping? Are there other variables that determine the specific form of coding (prospective, intermediate categorical, etc.) in MTO mappings? Addressing these and related issues will result in the substance from which future advances in our understanding of coding processes will be fashioned.

ACKNOWLEDGMENTS

Preparation of this chapter was supported by grant OGP 0443 from the Natural Sciences and Engineering Research Council of Canada. I thank Bill Roberts and Marcia Spetch for commenting on an earlier version of this chapter.

REFERENCES

Berryman, R., Cumming, W. W., & Nevin, J. A. (1963). Acquisition of delayed matching in the pigeon. *Journal of the Experimental Analysis of Behavior, 6*, 101–107.

Blough, D. S. (1959). Delayed matching in the pigeon. *Journal of the Experimental Analysis of Behavior, 2*, 151–160.

Colwill, R. M. (1984). Disruption of short-term memory for reinforcement by ambient illumination. *Quarterly Journal of Experimental Psychology, 36B*, 235–258.

Cook, R. G., Brown, M. F., & Riley, D. A. (1985). Flexible memory processing by rats: Use of prospective and retrospective information in the radial maze. *Journal of Experimental Psychology: Animal Behavior Processes, 11*, 453–469.

Cumming, W. W., & Berryman, R. (1965). The complex discriminated operant: Studies of matching-to-sample and related problems. In D. I. Mostofsky (Ed.), *Stimulus generalization* (pp. 284–330). Stanford, CA: Stanford University Press.

Grant, D. S. (1981). Short-term memory in the pigeon. In N. E. Spear & R. R. Miller (Eds.), *Information processing in animals: Memory mechanisms* (pp. 227–256). Hillsdale, NJ: Lawrence Erlbaum Associates.

Grant, D. S. (1982). Prospective versus retrospective coding of samples of stimuli, responses, and reinforcers in delayed matching with pigeons. *Learning and Motivation, 13*, 265–280.

Grant, D. S. (1984). Rehearsal in pigeon short-term memory. In H. L. Roitblat, T. G. Bever, & H. S. Terrace (Eds.), *Animal cognition* (pp. 99–115). Hillsdale, NJ: Lawrence Erlbaum Associates.

Grant, D. S. (1991). Symmetrical and asymmetrical coding of food and no-food samples in delayed matching in pigeons. *Journal of Experimental Psychology: Animal Behavior Processes, 17*, 186–193.

Grant, D. S., & Spetch, M. L. (1991). Pigeons' memory for event duration: Differences between choice and successive matching tasks. *Learning and Motivation, 22*, 180–190.

Grant, D. S., & Spetch, M. L. (1993). Analogical and nonanalogical coding of samples differing in duration in a choice-matching task in pigeons. *Journal of Experimental Psychology: Animal Behavior Processes, 19*, 15–25.

Honig, W. K. (1978). Studies of working memory in the pigeon. In S. H. Hulse, H. Fowler, & W. K. Honig (Eds.), *Cognitive processes in animal behavior* (pp. 211–248). Hillsdale, NJ: Lawrence Erlbaum Associates.

Honig, W. K., & Dodd, P. W. D. (1986). Anticipation and intention in working memory. In D. F. Kendrick, M. E. Rilling, & M. R. Denny (Eds.), *Theories of animal memory* (pp. 77–100). Hillsdale, NJ: Lawrence Erlbaum Associates.

Honig, W. K., & Thompson, R. K. R. (1982). Retrospective and prospective processing in animal working memory. In G. H. Bower (Ed.), *The psychology of learning and motivation: Advances in research and theory* (Vol. 16, pp. 239–283). New York: Academic Press.

Kraemer, P. J., Mazmanian, D. S., & Roberts, W. A. (1985). The choose-short effect in pigeon memory for stimulus duration: Subjective shortening versus coding models. *Animal Learning & Behavior, 13*, 349–354.

Maki, W. S. (1981). Directed forgetting in animals. In N. E. Spear & R. R. Miller (Eds.), *Information processing in animals: Memory mechanisms* (pp. 199–225). Hillsdale, NJ: Lawrence Erlbaum Associates.

Maki, W. S., & Leuin, T. C. (1972). Information-processing by pigeons. *Science, 176*, 535–536.

Maki, W. S., Moe, J. C., & Bierley, C. M. (1977). Short-term memory for stimuli, responses, and reinforcers. *Journal of Experimental Psychology: Animal Behavior Processes, 3*, 156–177.

Maki, W. S., Riley, D. A., & Leith, C. R. (1976). The role of test stimuli in matching to compound samples by pigeons. *Animal Learning & Behavior, 4*, 13–21.

Riley, D. A., Cook, R. G., & Lamb, M. R. (1981). A classification and analysis of short-term retention codes in pigeons. In G. H. Bower (Ed.), *The psychology of learning and motivation* (Vol. 15, pp. 51–79). New York: Academic Press.

Roberts, W. A. (1972). Short-term memory in the pigeon: Effects of repetition and spacing. *Journal of Experimental Psychology, 94*, 74–83.

Roberts, W. A., & Grant, D. S. (1976). Studies of short-term memory in the pigeon using the delayed matching-to-sample procedure. In D. L. Medin, W. A. Roberts, & R. T. Davis (Eds.), *Processes of animal memory* (pp. 79–112). Hillsdale, NJ: Lawrence Erlbaum Associates.

Roitblat, H. L. (1980). Codes and coding processes in pigeon short-term memory. *Animal Learning & Behavior, 8*, 341–351.

Roitblat, H. L. (1982). The meaning of representation in animal memory. *The Behavioral and Brain Sciences, 5*, 353–406.

Smith, L. (1967). Delayed discrimination and delayed matching in pigeons. *Journal of the Experimental Analysis of Behavior, 10*, 529–533.

Spetch, M. L. (1987). Systematic errors in pigeons' memory for event duration: Interaction between training and test delay. *Animal Learning & Behavior, 15*, 1–5.

Spetch, M. L., & Grant, D. S. (in press). Pigeons' memory for event duration in choice and successive matching-to-sample tasks. *Learning and Motivation*.

Spetch, M. L., & Rusak, B. (1989). Pigeons' memory for event duration: Intertrial interval and delay effects. *Animal Learning & Behavior, 17*, 147–156.

Spetch, M. L., & Wilkie, D. M. (1982). A systematic bias in pigeons' memory for food and light durations. *Behavior Analysis Letters, 2*, 267–274.

Spetch, M. L., & Wilkie, D. M. (1983). Subjective shortening: A model of pigeons' memory for event duration. *Journal of Experimental Psychology: Animal Behavior Processes, 9*, 14–30.

Urcuioli, P. J., Zentall, T. R., Jackson-Smith, P., & Steirn, J. N. (1989). Evidence for common coding in many-to-one matching: Retention, intertrial interference, and transfer. *Journal of Experimental Psychology: Animal Behavior Processes, 15*, 264–273.

Wasserman, E. A. (1986). Prospection and retrospection as processes of animal short-term memory. In D. F. Kendrick, M. E. Rilling, & M. R. Denny (Eds.), *Theories of animal memory* (pp. 53–75). Hillsdale, NJ: Lawrence Erlbaum Associates.

Wilkie, D. M. (1987). Stimulus intensity affects pigeons' timing behavior: Implications for an internal clock model. *Animal Learning & Behavior, 15*, 35–39.

Wilson, B., & Boakes, R. A. (1985). A comparison of the short-term memory performance of pigeons and jackdaws. *Animal Learning & Behavior, 13*, 285–290.

Zentall, T. R., Steirn, J. N., & Jackson-Smith, P. (1990). Memory strategies in pigeons' performance of a radial-arm-maze analog task. *Journal of Experimental Psychology: Animal Behavior Processes, 16*, 358–371.

Zentall, T. R., Steirn, J. N., Sherburne, L. M., & Urcuioli, P. J. (1991). Common coding in pigeons assessed through partial versus total reversals of many-to-one conditional and simple discriminations. *Journal of Experimental Psychology: Animal Behavior Processes, 17*, 194–201.

Zentall, T. R., Urcuioli, P. J., Jackson-Smith, P., & Steirn, J. N. (in press). Memory strategies in pigeons. In L. W. Dachowski & C. F. Flaherty (Eds.), *Current topics in animal learning*. Hillsdale, NJ: Lawrence Erlbaum Associates.

Zentall, T. R., Urcuioli, P. J., Jagielo, J. A., & Jackson-Smith, P. (1989). Interaction of sample dimension and sample-comparison mapping on pigeons' performance of delayed conditional discriminations. *Animal Learning & Behavior, 17*, 172–178.

12

Common Coding and Stimulus Class Formation in Pigeons

Thomas R. Zentall and Lou M. Sherburne
University of Kentucky

Janice N. Steirn
Georgia Southern University

One of the most valuable lessons that I learned while studying with Al Riley was not to let personal expectations or beliefs limit consideration of even the most unlikely explanation of the results of an experiment. Students in Al's lab not only were encouraged to generate possible alternative explanations, they were challenged to do so. The generation of alternative explanations was a goal in itself. The more alternative explanations one could generate the better one would be able to understand the phenomenon being studied. Either one could find a way to rule out each alternative in turn, and hence the more believable one's explanation would be, or one would find better supportive evidence for one of the alternatives, and one would become a convert. Either way we (and science) would be the better for it.

What follows is the description of a line of research that grew out of a curious empirical finding and the willingness to assess what, at the time, seemed to be a rather unlikely explanation for it.

INTRODUCTION

Interest in the development of associations in animal learning has focused primarily on two mechanisms, contiguity (the temporal pairing of stimuli) and contingency (the predictive consequence of stimulus presentation). Based on research that we present, we propose that there exists a third mechanism that we refer to as *common coding* or *stimulus class formation*. Common coding refers to the relation that develops between two events, through their common association with a third. Although the two events are related to the third by a common contingency, the relation between them cannot be anticipated by means of either contiguity or contingency. The processes involved in such common coding are not unknown in learning. For example, they play an integral part in the development of human concept learning because they allow for objects that may not be physically similar to one another (e.g., "apple" and "banana") to be grouped together through their common association with a category label (e.g., *fruit*).

Based on recent findings, this kind of learning apparently also plays a role in the way animals learn about stimulus relations. For example, if stimuli A and B are both associated with an event or outcome X, a relation between A and B may be established (Hall & Honey, 1989). In human conceptual terms, X might be a category label (e.g., *fruit*) and A and B are examples of the concept (e.g., "apple" and "banana"). The question we ask here is, to what extent is a relation established between A and B as a result of being associated with the common event X?

EVIDENCE FOR COMMON CODING

The question of common coding came up quite unexpectedly in the course of conducting an experiment in which we were interested in examining memory mechanisms in pigeons (Zentall, Urcuioli, Jagielo, & Jackson-Smith, 1989; see also chapter 11 of this volume). The task we were using was delayed matching to sample (DMTS), a conditional discrimination in which presentation of an initial (or sample) stimulus indicated which of two test (or comparison) stimuli following a delay was correct. One of the purposes of this experiment was to determine the effect of sample and comparison discriminability (i.e., easy-to-discriminate hues vs. hard-to-discriminate line orientations; see Carter & Eckerman, 1975) on delayed matching performance. In an earlier study we had determined that when sample discriminability and comparison discriminability were manipulated in a between-groups design, pigeons performed significantly better on trials involving hue samples than on trials involving line samples, but that the corresponding variation along the comparison dimension had little effect on delay performance (Urcuioli & Zentall, 1986; see also Fig. 12.1).

In a subsequent study (Zentall et al., 1989), one variable that we manipulated was the "mapping" of samples onto comparison stimuli (i.e., the number of possible samples and comparisons, and the rules governing which samples were associated with which comparisons). For example, for birds in Group 2–2, each of two hue samples was associated with a hue comparison, or each of two shape samples was associated with a shape comparison. In Group 2–4, birds learned to associate each of two samples (either hues or shapes) with a particular hue comparison on some trials and with a particular shape comparison on others. In Group 4–2, pairs of samples consisting of one hue and one shape were mapped onto a single comparison alternative (hues for some birds, shapes for others).

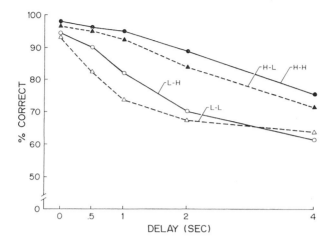

FIG. 12.1.   Retention functions for groups trained with either hue (H) or line (L) samples and either H or L comparisons (after Urcuioli & Zentall, 1986, Fig. 3).

Finally, in Group 4–4, all birds learned to associate two hue and two shape samples with corresponding hue and shape comparisons, respectively.

Following acquisition to criterion levels of performance, all birds then were exposed to delays of variable duration inserted between the sample and comparison stimuli. The results of interest were the rates at which performance dropped with increasing delay for each of the trial types.

Consistent with the results of earlier research, the predominant finding was that when the samples were hues, performance remained at a relatively high level (i.e., the retention functions were relatively flat), whereas when the samples were shapes, performance declined sharply with increasing delay.

The one exception to this finding was in Group 4–2, for which the hue- and shape-sample retention functions were quite similar (see Fig. 12.2). The important difference between this group and the others was that Group 4–2 had a many-to-one mapping of samples onto comparisons (i.e., one hue and one shape were both mapped onto a single comparison; a hue for some birds, a shape for others).

One explanation for this finding was that the two samples associated with a common comparison in Group 4–2 were similarly represented (or "commonly coded") by the pigeons. Such common coding could have been directly based on the common comparisons (i.e., the pigeons could have prospectively coded samples in terms of the anticipated response to a particular comparison stimulus). On the other hand, common coding could have been retrospective, based, for example, on one of the samples (e.g., the sample that was harder to remember could have been coded in terms of the sample that was easier to remember) or on

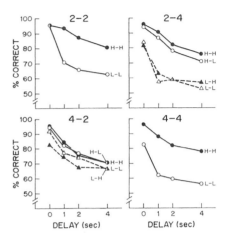

FIG. 12.2. Retention functions for groups trained with either two samples (hues or lines) or four samples (hues and lines) and either two comparisons (hues or lines) or four comparisons (hues and lines). Performance on the different trials types is sometimes within subjects, sometimes between subgroups (after Zentall et al., 1989, Fig. 3).

some unspecified internal representation that was not isomorphic with the common correct comparison stimulus (i.e., as the result of training, the two samples in each associative pair could now belong to a stimulus class "A" or "B"). The data from Group 4–2 presented in Fig. 12.2 suggest that the common codes were probably not prospective in nature. Had the samples been coded prospectively, one might have expected to see some effect of the manipulation of the comparison dimension (hues vs. lines) on the slope of the retention functions. Examination of the data presented in Fig. 12.2 indicates that no such evidence was found.

MANY-TO-ONE MAPPING

Substitution Procedure

Before proceeding further, we wanted to be sure that the greatly reduced sample-dimension effect found during delay testing in the many-to-one group was replicable and could not attributable to sampling error. In our next experiment, we reexamined the effect on retention of hue versus line-orientation samples when a many-to-one (i.e., 4–2) procedure was used (Urcuioli, Zentall, Jackson-Smith, & Steirn, 1989). Again, we found a greatly reduced sample-dimension effect (i.e., performance on hue sample trials was only somewhat better than that on line sample trials).

The muted effect of sample dimension on retention with the many-to-one procedure provides suggestive evidence for some kind of common-coding effect, however, stronger support would be provided by evidence from a transfer of training experiment.

If a hue and shape sample were commonly coded through their association with a common comparison, it should be possible to show evidence for such coding by then training the birds on a new DMTS association involving one of the samples and a new comparison, and then testing the birds for the emergence of an association between the remaining sample (from original training) and the new comparison. For example, after training on a many-to-one DMTS task in which red (R) and vertical-line (V) samples were associated with a R comparison, and green (G) and horizontal-line (H) samples were associated with a G comparison, pigeons then were trained to match just the R and G samples to (new) circle (C) and dot (D) comparisons, respectively (the design of this experiment is presented in Table 12.1). If, for example, during original training, R and V samples were commonly coded because of their common association with the R comparison, and then the R sample was associated with a new comparison, C, then evidence for an emergent V-C association should be evidenced. When the pigeons were tested with the V and H samples and the C and D comparisons, transfer consistent with common coding was found (Urcuioli et al., 1989). Specifically, pigeons for which the reinforced associations were consistent with the hypothesized common codes performed significantly better than those for which the associations were inconsistent with those codes. The transfer data for individual birds are presented in Fig. 12.3.

Total Reversal Versus Partial Reversal Procedure

Many-to-One DMTS. An alternative means of demonstrating the establishment of common coding was suggested by Lea (1984), who was concerned that many reported findings of concept learning in animals actually may have involved individually learned responses (i.e., to peck or not to peck) to a large

TABLE 12.1
Design of Many-To-One Common Coding Substitution Experiment

Group	Phase 1	Phase 2	Phase 3 (Test)
Consistent	R → V+ C → H+ V → V+ H → H+	R → C+ G → D+	V → C+ H → D+
Inconsistent	R → V+ G → H+ V → V+ H → H+	R → C+ G → D+	V → D+ H → C+

Note. R = red homogeneous field. G = green homogenous field. V = vertical lines. H = horizontal lines. C = circle. D = dot. Samples are shown to the left of the arrows, and correct comparisons are shown to the right. Incorrect comparisons, and balancing of sample-comparison associations in Phases 2 and 3, have been omitted for clarity. From Urcuioli et al., 1989.

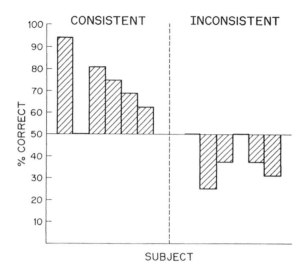

FIG. 12.3.   The first 16 trials of Phase 3 transfer from the many-to-one common coding substitution experiment for individual birds in the Consistent and Inconsistent groups (after Urcuioli et al., 1989, Fig. 4).

number of stimuli (see, e.g., Herrnstein & Loveland, 1964). According to Lea, common coding would be indicated if one could demonstrate the following: If after training an organism to make a particular response to one set of stimuli (e.g., peck) and to make a different response to another set of stimuli (e.g., not peck), one reversed the response associations for a subset of each stimulus set and found (without further training) that the remaining stimuli also had their response associations reversed.

Vaughan (1988) provided such a demonstration with a variation of Lea's (1984) suggested procedure. Pigeons were trained to respond to one (arbitrarily selected) set of stimuli (A+) and to not respond to another stimulus set (B−). Following acquisition, the contingencies associated with each set were reversed (i.e., A− B+), and then reversed again, and again, repeatedly. After a large number of reversals, the pigeons were able to discriminate the positive from the negative stimuli on any reversal following presentation of the first few stimuli in each set. Apparently, the pigeons had developed a common code for the stimuli in each set, and once the first few stimuli in each set had been presented, thus allowing the animal to determine the current valence of each set (i.e., + or −), appropriate responding occurred to the remaining members.

We took a different approach to Lea's (1984) proposal for the assessment of concept development. After training pigeons on a many-to-one DMTS task, we asked if it would be easier for them to reverse both pairs of sample-comparison associations, than for them to reverse only one pair (Zentall, Steirn, Sherburne, & Urcuioli, 1991). We reasoned that if training on a many-to-one DMTS task resulted in the development of common sample codes, then it should be easier for pigeons to reverse two common-code/comparison-response associations, than it would be for them to regroup the stimuli into new coding classes (see Nakagawa, 1986).

All birds were initially trained to match, for example, a circle comparison to red and vertical samples, and a dot comparison to green and horizontal samples. Birds assigned to Group Hue-Line were then given a total reversal, such that a circle comparison response was correct following green and horizontal, whereas a dot comparison response was correct following red and vertical. Birds assigned to Group Hue were given a partial reversal (i.e., a reversal involving only the hue dimension), such that a response to the circle was now correct following green and a response to the dot was now correct following red. The remaining associations were the same as in training. Birds assigned to Group Line also were given a partial reversal (i.e., a reversal involving only the line dimension), such that a response to circle was now correct following a horizontal sample, and a response to dot was now correct following a vertical sample. The design of this experiment (Zentall et al. 1991) is presented in Table 12.2.

We found that birds in Group Line took significantly longer to reverse their line-sample associations than did birds in Group Hue-Line (see Fig. 12.4). A corresponding difference was not found for Group Hue, however. We return to this asymmetry in results later.

Simple Discriminations.   If the common-coding effect found in many-to-one DMTS is a general phenomenon, then one should be able to find similar partial-versus total-reversal results with a simpler task involving two successive simple discriminations.

In this experiment (Zentall et al., 1991), pigeons were trained, for example, to peck red and vertical stimuli for food, and were trained not to peck (extinction) green and horizontal stimuli (i.e., red+, green−, vertical+, horizontal−). If the pigeons learned to commonly code red and vertical stimuli because of the com-

TABLE 12.2
Design of Many-To-One Common Coding Partial- Versus
Total-Reversal Experiment

Phase 1	Phase 2		
All Groups	Group Hue	Group Line	Group Hue-Line
R → C+	R → D+	R → C+	R → D+
G → D+	G → C+	G → D+	G → C+
V → C+	V → C+	V → D+	V → D+
H → D+	H → D+	H → C+	H → C+

Note. From Zentall et al., 1991.

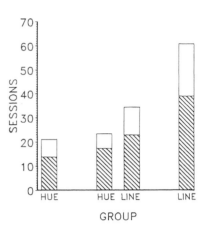

FIG. 12.4. Acquisition of partial reversal (Groups Hue and Line) and total reversal (Group Hue-Line) in the DMTS, many-to-one, common-coding experiment (after Zentall et al., 1991, Fig. 1). Lined bars represent sessions to a performance criterion of 80% correct; open bars, a criterion of 90% correct.

mon association of these stimuli with food, and to commonly code green and horizontal stimuli because of their common association with the absence of food, then the pigeons should be able to reverse those common-code/outcome associations more easily than to regroup those stimuli into new coding classes.

Following acquisition, birds in Group Hue-Line received a total reversal (i.e., red−, green+, vertical−, horizontal+), whereas birds in Group Hue and Group Line received partial reversals (i.e., red−, green+, vertical+, horizontal−, and red+, green−, vertical−, horizontal+, respectively). The design of this experiment is similar to that of the preceding experiment except that the circle and dot comparisons from that experiment were replaced by food on S+ trials and by no food on S− trials.

Again, results indicated that birds in Group Line took significantly longer to reverse their line discrimination than did birds in Group Hue-Line (see Fig. 12.5), whereas birds in Group Hue did not take significantly longer to reverse their hue discrimination than did birds in Group Hue-Line. The results of this experiment look remarkably similar to the results found with the DMTS task (Zentall et al., 1991). Thus, common coding appears in simple as well as conditional discriminations.

Let us return for a moment to the lack of symmetry in the two partial- versus total-reversal experiments. Although there was evidence of common coding involving the line stimuli in both experiments, there was little evidence of common coding involving the hue stimuli (i.e., there was little difference in rate of hue-dimension reversal between Groups Hue and Hue-Line). This consistent asymmetry may provide a clue concerning the nature of the common code. The pattern of results obtained is compatible with the assumption that all of the sample or initial stimuli (i.e., both hues and lines) are coded in terms of the red and green hues. According to this view, reversals for both Groups Hue-Line and Hue would require reversal of the common-code/response associations. The line reversal for Group Line, however, would not require reversal of the common-code

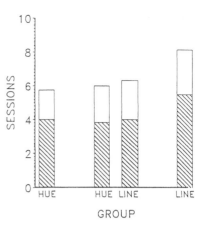

FIG. 12.5. Acquisition of partial reversal (Groups Hue and Line) and total reversal (Group Hue-Line) in the simple-discrimination, many-to-one, common-coding experiment (after Zentall et al., 1991, Fig. 2). Lined bars represent sessions to a discrimination-ratio criterion of 0.80; open bars, a discrimination-ratio criterion of 0.90.

associations, but rather would require reconfiguration of which line stimuli go with which hue code. A more complete description of this argument appears in Zentall et al. (1991).

RELATION BETWEEN COMMON CODING AND STIMULUS EQUIVALENCE

According to Sidman (1986), stimuli are equivalent when they satisfy three mathematically defined conditions: *reflexivity, symmetry,* and *transitivity.* Reflexivity, also referred to as identity, is indicated when an organism demonstrates that it recognizes the relation between two identical stimuli (i.e., Stimulus A = Stimulus A). Symmetry, also referred to as bidirectionality or the presence of backward associations, means that if Stimulus A is associated with Stimulus B, then stimulus B is also associated with Stimulus A. Transitivity, also referred to as mediation, means that if Stimulus A is associated with Stimulus B, and Stimulus B is associated with Stimulus C, then Stimulus A is also associated with Stimulus C.

The results of the common coding experiments described earlier certainly imply that stimuli that are commonly coded belong to the same stimulus class or can be interchanged for one another. But are they truly equivalent? Although no direct test of stimulus equivalence was carried out in the research described earlier, there is evidence from other experiments that pigeons are capable of demonstrating each of the stimulus relations that defines equivalence.

Reflexivity

The identity relation between two stimuli can only be assessed indirectly. It is not sufficient to show that an organism can learn to respond to Comparison A when

FIG. 12.6. Transfer performances for birds trained
with either red-green (identity)
matching or oddity and transferred to blue-yellow matching
or oddity. For Nonshifted birds
the training and transfer tasks
were the same (both matching
or both oddity); for shifted birds
the two tasks were different (either training with matching and
transfer to oddity or the reverse; after Zentall & Hogan,
1978, Fig. 4).

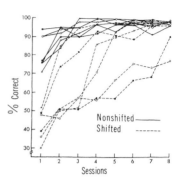

the sample is A and to respond to Comparison B when the sample is B, because such learning indicates nothing about the subject's sensitivity to the identity relation between them. Instead, the most accepted way to show that the identity relation has been learned (or recognized) is by demonstrating positive transfer when the subject is presented with new stimuli involving the same relation. Zentall and Hogan (1978), for example, found that when pigeons had been trained on either a matching or an oddity task involving red and green hues, and then were transferred (in a 2 × 2 design) to a matching or to an oddity task involving blue and yellow hues, they transferred at significantly higher levels of performance when the training and transfer tasks were the same (matching-matching and oddity-oddity) than when the training and transfer tasks were different (oddity-matching and matching-oddity). The transfer results from this experiment are presented in Fig. 12.6.

Similar findings have been reported when the training dimension (e.g., shape or brightness) was orthogonal to the dimension used in transfer (e.g., hue; Zentall & Hogan, 1974, 1975, 1976, 1978). Although some disruption in perfor-mance often is observed during transfer, at least some of this disruption can be attributed to the use of novel transfer stimuli (Zentall, Edwards, Moore, & Hogan, 1981; Zentall & Hogan, 1978).

Symmetry

The existence of a symmetry relation or a bidirectional association assumes that the establishment of an association between Stimulus A and Stimulus B also results in the formation of an association in the opposite direction (i.e., between Stimulus B and Stimulus A). The study of backward associations has a long history in animal learning, though most of this research has involved simple Pavlovian conditioning procedures (see e.g., Spetch, Wilkie, & Pinel, 1981; but see also Hearst, 1989).

One can examine backward associations in the context of the DMTS task by initially training pigeons on symbolic matching (i.e., a task in which there is no physical match between samples and correct comparisons), and then switching the roles of the original samples and comparisons. For example, pigeons could be trained in a DMTS task to peck the red comparison when the sample is a circle and to peck the green comparison when the sample is a dot. Backward associations (or symmetry) would be demonstrated if birds would then peck a circle comparison when the sample was red and peck a dot comparison when the sample was green.

Unfortunately, interchanging the samples and comparisons in a DMTS task not only alters their temporal relation to one another, but it also alters (among other things) the location of those cues (e.g., the samples that originally appeared on the center key now appear on the side keys). These other changes could be responsible for the typically poor performances that pigeons have shown when tested for backward associations (see Gray, 1966; Hogan & Zentall, 1977; Richards, 1988; Rodewald, 1974).

Recently, we have overcome that problem by training pigeons on a symbolic matching task with differential outcomes (Zentall, Sherburne, & Steirn, 1992). When the sample was, for example, a circle, a correct response to the red comparison was followed by food (peas); when the sample was a dot, a correct response to the green comparison was followed by the feeder light alone (i.e., no food). To get the pigeons to peck the green comparison, incorrect comparison responses caused the trial to repeat. Then, in a test session, the circle and dot samples were replaced with food and no-food samples.

Although in test the pigeons showed a consistent bias to peck the comparison associated with the food outcome, they were significantly more likely to peck that comparison when the sample was food, and to peck the other comparison when the sample was no food, than the reverse (see Fig. 12.7). For birds in Group Positive, responses defined as correct were to comparisons consistent with the backward associations presumably learned to food and no food during original training, whereas for Group Negative, correct responses were inconsistent with the presumed backward associations. Thus, for example, the forward pairing of red with food and green with no food apparently resulted in the develop-

FIG. 12.7. Performance on the first 16 trials of the first transfer session (left) and on all trials from the first transfer session (right) by the positive and negative transfer groups.

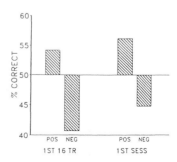

ment of a backward association or a symmetry relation between food and red, and between no food and green.

Transitivity

The transitivity requirement associated with stimulus equivalence can be restated as follows: If Stimulus A leads to Stimulus B, and Stimulus B is associated with a response to Stimulus C, then Stimulus A should lead to a response to Stimulus C.

Here too, there is some evidence that such mediated learning does occur in animals, although, again, most of the research has involved Pavlovian conditioning procedures. At least two lines of research that fit this paradigm can be identified: sensory preconditioning and second-order conditioning.

The sensory preconditioning procedure involves the repeated pairing of two "neutral" stimuli, S1 and S2, in Phase 1, followed by the pairing of one of the stimuli (e.g., S2) with an unconditioned stimulus (US) in Phase 2. In Phase 3, elicitation of the conditioned response (CR) developed during Phase 2 upon presentation of S1 (relative to the appropriate control procedures) is evidence for the mediated association of S1 with the US (see Seidel, 1959, for a review).

In the second-order conditioning procedure, Phase 1 involves multiple pairings of S1 with the US. In Phase 2, S2 is repeatedly paired with S1. In Phase 3, when the S2 is presented alone, elicitation of the CR (relative to the appropriate control procedures) is an indication of second-order conditioning (see Rescorla, 1980, for a review).

One also could examine such mediated learning in DMTS by first training pigeons on a symbolic task (Phase 1), and then in Phase 2, training them with a task in which the comparisons from Phase 1 are used as samples and are associated with novel comparisons. Then in Phase 3, evidence of transitivity or (mediated learning) would be indicated by positive transfer when the samples from Phase 1 were presented with the comparisons from Phase 2. Again, however, as in the case of assessing backward associations, using the comparisons from Phase 1 as the samples in Phase 2 alters not only their temporal position in the trial, but also their spatial location (i.e., instead of being presented on the side keys, they are now presented on the center key).

Recently, we have overcome this methodological problem by holding constant the spatial location of the stimuli common to the two training phases (Steirn, Jackson-Smith, & Zentall, 1991). We first trained pigeons in a simple successive (response independent) discrimination, to associate responses to a red stimulus, for example, with food and responses to a green stimulus with no food. We then trained the pigeons on a DMTS task, to respond to a vertical comparison if food was the sample, and to a green comparison if no food was the sample. Finally, we presented them with a red sample on some trials and a green sample on others and provided a choice between vertical and horizontal comparisons.

In Phase 3, a significant difference in transfer performance was found be-

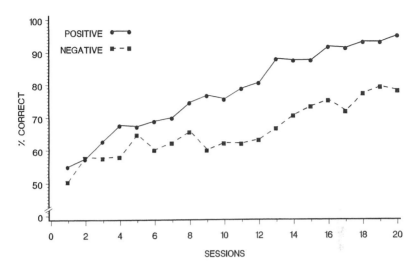

FIG. 12.8. Phase 3 transfer effects when initial stimuli from Phase 1 are presented with the comparison stimuli from Phase 2 for the positive and negative transfer groups (Steirn et al., 1991).

tween the positive transfer group (for which correct responses were consistent with the hypothesized mediated associations) and the negative transfer group (for which correct responses were inconsistent with those associations). The results are presented in Fig. 12.8. Thus, although no direct association between hues and lines was explicitly trained, the pigeons apparently were able to use the food and no-food events to mediate those associations.

As can be seen in Fig. 12.8, transfer effects were not immediate, but developed only with continued training on the transfer task. Furthermore, all pigeons showed a strong preference for the comparison that had been associated with the no-food sample during training. This comparison preference is not surprising when one considers that all transfer trials involved no-food samples (i.e., samples of red or green hues). In a follow-up experiment (Steirn et al., 1991), we used a more salient no-food event (presentation of the hopper light with an empty food tray) that could be more easily distinguished from the red and green samples, and found both a smaller preference for the comparison associated with the no-food sample and, more important, a significant difference in performance between the positive and negative groups on the first transfer session.

Thus we have provided evidence in pigeons for a derived (emergent) transitivity relation in the context of a delayed matching task. In describing this emergent relation one can propose that the expectation of food or no food generated by the red and green hues can, at least in part, substitute for the actual occurrence of food and serve as the basis for choice of the line orientation comparison.

Conclusions

Taken as a whole, the research cited in the preceding sections indicates that pigeons are capable of demonstrating each component of stimulus equivalence: reflexivity, symmetry, and transitivity. Although these phenomena have been demonstrated individually in separate experiments, in principle there is no reason to believe that all three could not be demonstrated in a single animal.

RELATED RESEARCH

If common coding is a general phenomenon that occurs when two unrelated stimuli are treated similarly, then one should be able to find evidence for common coding under procedural variations that have been used to demonstrate the acquisition of equivalence sets in humans. Furthermore, if one can determine the boundary conditions under which the common coding phenomenon occurs, it should be possible to better specify the nature of the underlying mechanisms. The research designs that follow (only some of which we have begun to explore) are offered as a first step in determining the defining conditions for common coding.

One-to-Many Common Coding

Earlier, we demonstrated that if two samples, A and B, are associated with the same comparison, C, then A and B will be commonly represented, such that if A is now associated with a new comparison D, B also will be associated with D. In the human literature, mapping samples onto more than one comparison (i.e., a one-to-many mapping) also can produce stimulus equivalence of the type described by Sidman (1986; cf. Sidman & Tailby, 1982). If these common-coding results with humans reflect underlying processes that are similar to those observed in pigeons, it should also be possible to demonstrate common coding in pigeons using a one-to-many procedure analogous to the many-to-one procedures described earlier.

To test this hypothesis, we used the total- versus partial-reversal procedure (following initial acquisition). During training, the samples were circle and dot shapes and the comparisons were either red and green hues or vertical and horizontal lines. A response to red or vertical was correct when the sample was a circle, whereas a response to green or horizontal was correct when the sample was a dot. Following acquisition, all sample/correct-comparison associations were reversed for Group Hue-Line (i.e., red and vertical were now correct following the dot sample, and green and horizontal were now correct following the circle sample). By contrast, Groups Hue and Line received partial reversals involving the hue- and line-comparison dimensions, respectively.

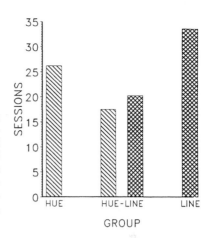

FIG. 12.9. Acquisition to a criterion of 80% correct performance of partial reversal (Groups Hue and Line) and total reversal (Group Hue-Line) in the DMTS, one-to-many, common-coding experiment (after Zentall, Sherburne, Steirn, Randall, & Roper, in press, Fig. 1).

During this test phase, acquisition of the line reversal took significantly longer for Group Line than for Group Hue-Line. Although the difference in the hue reversal between Groups Hue and Hue-Line did not reach statistical significance (due to unusually high within-group variability in Phase 2 acquisition for Group Hue), the magnitude of the effect was similar to that for the line reversal (Zentall, Sherburne, Steirn, Randall, Roper, & Urcuioli, in press). These results are presented in Fig. 12.9.

In a follow-up experiment (Zentall et al., in press), we asked if similar results could be found if the comparisons in a one-to-many task were different spatial locations, rather than hues and lines. In this experiment, the samples were red and green hues and the comparisons were white keys to the left and right of the sample key on some trials, and above and below the sample key on others. For a given pigeon, the correct comparison following a red sample might be the left spatial location on some trials and the top spatial location on others, whereas the correct locations following a green sample would be the right and bottom spatial locations.

As in the earlier study, there were three transfer groups. For birds in the total-reversal group (Group Vert-Hor), all the sample-comparisons associations during training were now reversed. For the horizontal-reversal group (Group Hor), the hue-sample/left-right-comparison associations were reversed but the hue-sample/top-bottom associations remained as they were during training. For the vertical-reversal group (Group Vert), associations involving the top and bottom locations were reversed but the associations involving the left and right locations remained as they were during training.

Acquisition in the reversal phase indicated that the two partial-reversal groups acquired their reversals significantly slower than did the total-reversal group (see Fig. 12.10). The results of these two experiments indicate that common coding can be found with one-to-many training as well as with many-to-one training.

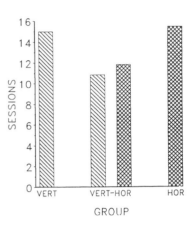

FIG. 12.10. Acquisition of partial reversal (Groups Vert and Hor) and total reversal (Group Vert-Hor) in the spatial-DMTS, one-to-many, common-coding experiment (after Zentall, Sherburne, Steirn, Randall, & Roper, in press, Fig. 2).

However, unlike the many-to-one task, in which there was an indication that the common code was some representation of the easier-to-discriminate samples, the common code that develops when the one-to-many procedure is used does not appear to be a representation of the more discriminable comparison.

Common Coding Through Bidirectional Training

According to Dixon and Spradlin (1976), bidirectional training, in which Stimulus A is associated with Stimulus B and Stimulus B is associated with Stimulus A, should be sufficient to produce an equivalence relation between the two stimuli. Dixon and Spradlin demonstrated such equivalence in humans by first establishing bidirectional associations between two stimuli, A and B, then training a new association between Stimuli A and C, and then showing that this training resulted in the development of an association between Stimuli B and C.

On the one hand, it is possible that such training would result in the development of two independent and unrelated associative "rules" (if A then B, and if B then A). On the other hand, the bidirectional nature of the A-B association may be sufficient to result in a common representation of these two stimuli.

To test these hypotheses, one could first train pigeons to associate red and green samples with vertical and horizontal comparisons, respectively, and to associate vertical and horizontal samples with red and green comparisons, respectively. All pigeons then could be trained to associate red and green samples with circle and dot comparisons, respectively. Finally, the pigeons could be tested with vertical and horizontal samples and circle and dot comparisons. If such initial training results in the formation of two equivalence relations, then pigeons for which reinforced associations in the test phase were consistent with the presumed equivalence relation established in original training should perform significantly better than pigeons for which the associations in the test phase were inconsistent with the equivalence relation.

Interference/Facilitation Paradigm

If common coding of samples occurs in a many-to-one DMTS task, then one would expect commonly coded samples to be compatible with each other, but differently coded samples to be incompatible. One could test this hypothesis by training new associations involving one member of each presumably commonly coded pair with new comparisons (as in the substitution procedure described earlier). Then, after training with extended delays between offset of the sample and onset of the comparisons, one could insert either of the remaining sample stimuli into the delay. If the inserted stimulus on a given trial had been commonly coded with the sample (during original training), facilitated performance should occur. On the other hand, if the inserted stimulus had been commonly coded with the alternate sample, performance should be disrupted (Zentall, Sherburne, & Urcuioli, in press).

Direct Associations Between Commonly Coded Stimuli

In the pigeon research described earlier, the common code is inferred from the effect produced when one stimulus of an associative pair replaces the other in an independently trained task (see, e.g., the substitution procedure). Alternatively, one can ask whether commonly coded stimuli will be "matched" to one another more readily than stimuli that do not share a common code. Two designs follow directly from this question, one involving many-to-one DMTS, the other, one-to-many DMTS.

Many-to-One DMTS.   If pigeons are trained to associate Samples A and B with Comparison C, can one demonstrate that an association has developed between A and B by showing positive transfer to a task involving A as a sample and B as one of the comparison stimuli. The problem with asking this question of pigeons is that it requires changing not only the designation of B from sample to comparison, but also changing the location of B from center key to side key. As noted earlier, such a change may result in sufficient disruption of performance to obscure transfer effects (see Hogan & Zentall, 1977).

A solution to this problem is to use food and no food as one element of each pair of commonly coded stimuli, as was done in testing for the presence of backward associations (see Zentall, Sherburne, & Steirn, 1992). In original training, red and food samples, for example, would be associated with a response to a circle comparison stimulus, whereas green and no-food samples would be associated with a response to a dot comparison. In the transfer phase, pigeons would be trained on a simple successive discrimination to respond to red but not to green (positive transfer), or the reverse (negative transfer). Better transfer by the positive transfer group than by the negative transfer group would suggest that common coding of the red hue and food had occurred in original training. An additional measure of common coding of the samples would be indicated if, during original training, pigeons pecked the hue sample associated with the same comparison as the food sample more than the hue sample associated with the same comparison as the no-food sample.

This many-to-one design (e.g., train A-C and B-C associations) with a test for the association between hypothesized commonly coded samples (e.g., test for the presence of the A-B association) has some interesting properties that, according to Sidman (1986), allow it to be used as a direct test for stimulus equivalence. According to Sidman, the presence of an A-B association following training with A-C and B-C, demonstrates two important properties of stimulus equivalence, transitivity and symmetry. Transitivity, is demonstrated to the degree that C serves as a mediator between A and B (A-C, C-B). Symmetry (or bidirectionality) is demonstrated because transitivity can only operate through the C-A association (i.e., the backward associate of the trained A-C association).

One-to-Many DMTS.   A similar question can be asked using the one-to-many design. In other words, if pigeons are trained to associate Sample A with Comparisons B and C, will tests reveal the presence of a B-C association? Again, food and no food would be used as one member of each of the proposed commonly coded stimulus pairs. During original training, red samples would be followed sometimes by a choice between vertical and horizontal lines (responses to vertical would result in reinforcement), and at other times by food presentation directly. Green samples, on the other hand, would be associated with horizontal lines on some trials and the absence of food on others. Then, if the hedonic events were commonly coded with the appropriate hues, the pigeons' rate of acquisition of a simultaneous line-orientation discrimination in Phase 2 should depend on whether the Phase 2 S+ is the hue that was presumably commonly coded with food (i.e., the positive transfer group) or no food (i.e., the negative transfer group) in Phase 1.

Again, the presence of significant differences in Phase 2 acquisition between the positive and negative transfer groups would indicate the pigeons' capacity for both transitivity (i.e., A serving as a mediator between B and C), and symmetry (i.e., the establishment of a backward B-A association, given A-B training).

The results of these proposed experiments should suggest whether it is appropriate to view common coding as an example of stimulus equivalence. Furthermore, they should clarify the conditions under which common coding can be found in pigeons.

CONCLUSIONS

The research presented indicates that relations develop between stimuli that have a common association with some other event. Furthermore, we suggest that these common representations or common codes are analogous to the equivalence relations between stimuli that are purported to be a necessary component of human concept learning, as well as referential meaning and the semantic relations needed for language acquisition (see Sidman & Tailby, 1982). If further research confirms these assertions, it will offer considerable support for the assumption that there is a continuum of cognitive ability across species, and that in many cases capacity differences among species may reflect quantitative rather than qualitative differences in ability.

ACKNOWLEDGMENTS

The research presented here was supported by NIMH Grants MH 35376 and MH 45979, by NSF Grants BNS 8418275, RII 8902792, and BNS 9019080, and by a grant from University of Kentucky Research Foundation. The authors thank Peter J. Urcuioli for his important contribution to the research reported here and both Peter J. Urcuioli and Douglas S. Grant for their helpful comments on an earlier version of this chapter.

REFERENCES

Carter, D. E., & Eckerman, D. A. (1975). Symbolic matching by pigeons: Rate of learning complex discriminations predicted from simple discriminations. *Science, 187*, 662–664.

Dixon, M. H., & Spradlin, J. E. (1976). Establishing stimulus equivalence among retarded adolescents. *Journal of Experimental Child Psychology, 21*, 144–164.

Gray, L. (1966). Backward association in pigeons. *Psychonomic Science, 4*, 333–334.

Hall, G., & Honey, R. C. (1989). Contextual effects in conditioning, latent inhibition, and habituation: Associative and retrieval functions of contextual cues. *Journal of Experimental Psychology: Animal Behavior Processes, 15*, 232–241.

Hearst, E. (1989). Backward associations: Differential learning about stimuli that follow the presence versus the absence of food in pigeons. *Animal Learning and Behavior, 17*, 280–290.

Herrnstein, R. J., & Loveland, D. H. (1964). Complex visual concept in the pigeon. *Science, 146*, 549–551.

Hogan, D. E., & Zentall, T. R. (1977). Backward associations in the pigeon. *American Journal of Psychology, 90*, 3–15.

Lea, S. E. G. (1984). In what sense do pigeons learn concepts? In H. L. Roitblat, T. G. Bever, & H. S. Terrace (Eds.)., *Animal cognition* (pp. 163–176). Hillsdale, NJ: Lawrence Erlbaum Associates.

Nakagawa, E. (1986). Overtraining, extinction, and shift learning in a concurrent discrimination in rats. *Quarterly Journal of Experimental Psychology, 38*, 313–326.

Rescorla, R. A. (1980). *Pavlovian second-order conditioning.* Hillsdale, NJ: Lawrence Erlbaum Associates.

Richards, R. W. (1988). The question of bidirectional associations in pigeons' learning of conditional discrimination tasks. *Bulletin of the Psychonomic Society, 26*, 577–580.

Rodewald, H. K. (1974). Symbolic matching-to-sample by pigeons. *Psychological Reports, 34*, 987–990.

Seidel, R. J. (1959). A review of sensory preconditioning. *Psychological Bulletin, 56*, 58–73.

Sidman, M. (1986). Functional analysis of emergent verbal classes. In T. Thompson & M. D. Zeiler (Eds.), *Analysis and integration of behavioral units* (pp. 213–145). Hillsdale, NJ: Lawrence Erlbaum Associates.

Sidman, M., & Tailby, W. (1982). Conditional discrimination vs. matching to sample: An expansion of the testing paradigm. *Journal of the Experimental Analysis of Behavior, 37,* 5–22.

Spetch, M. L., Wilkie, D. M., & Pinel, J. P. J. (1981). Backward conditioning: A reevaluation of the empirical evidence. *Psychological Bulletin, 89,* 163–175.

Steirn, J. N., Jackson-Smith, P., & Zentall, T. R. (1991). Mediational use of internal representations of food and no-food events by pigeons. *Learning and Motivation, 22,* 353–365.

Urcuioli, P. J., & Zentall, T. R. (1986). Retrospective memory in pigeons' delayed matching-to-sample. *Journal of Experimental Psychology: Animal Behavior Processes, 12,* 69–77.

Urcuioli, P. J., Zentall, T. R., Jackson-Smith, P., & Steirn, J. N. (1989). Evidence for common coding in many-to-one matching: Retention, intertrial interference, and transfer. *Journal of Experimental Psychology: Animal Behavior Processes, 15,* 264–273.

Vaughan, W., Jr. (1988). Formation of equivalence sets in pigeons. *Journal of Experimental Psychology: Animal Behavior Processes, 14,* 36–42.

Zentall, T. R., Edwards, C. A., Moore, B. S., & Hogan, D. E. (1981). Identity: The basis for both matching and oddity learning in pigeons. *Journal of Experimental Psychology: Animal Behavior Processes, 7,* 70–86.

Zentall, T. R., & Hogan, D. E. (1974). Abstract concept learning in the pigeon. *Journal of Experimental Psychology, 102,* 393–398.

Zentall, T. R., & Hogan, D. E. (1975). Concept learning in the pigeon: Transfer of matching and nonmatching to new stimuli. *American Journal of Psychology, 88,* 233–244.

Zentall, T. R., & Hogan, D. E. (1976). Pigeons can learn identity, difference, or both. *Science, 191,* 408–409.

Zentall, T. R., & Hogan, D. E. (1978). Same/different concept learning in the pigeon: The effect of negative instances and prior adaptation to the transfer stimuli. *Journal of the Experimental Analysis of Behavior, 30,* 177–186.

Zentall, T. R., Sherburne, L. M., & Steirn, J. N. (1992). Development of excitatory backward associations during the establishment of forward associations in a delayed conditional discrimination by pigeons. *Animal Learning and Behavior, 20,* 199–206.

Zentall, T. R., Sherburne, L. M., Steirn, J. N., Randall, C. K., Roper, K. L., & Urcuioli, P. J. (in press). Common coding in pigeons: Partial versus total reversals of one-to-many conditional discriminations. *Animal Learning and Behavior.*

Zentall, T. R., Sherburne, L. M., & Urcuioli, P. J. (in press). Common coding in a many-to-one delayed matching task as evidenced by facilitation and interference effects. *Animal Learning and Behavior.*

Zentall, T. R., Steirn, J. N., Sherburne, L. M., & Urcuioli, P. J. (1991). Common coding in pigeons assessed through partial versus total reversals of many-to-one conditional discriminations. *Journal of Experimental Psychology: Animal Behavior Processes, 17,* 194–201.

Zentall, T. R., Urcuioli, P. J., Jagielo, J. A., & Jackson-Smith, P. (1989). Interaction of sample dimension and sample-comparison mapping on pigeons' performance of delayed conditional discriminations. *Animal Learning and Behavior, 17,* 172–178.

IV

PERCEPTUAL PROCESSES

13 Attention: Neurocognitive Analyses

David S. Olton
Kevin Pang
Fred Merkel
Howard Egeth
The Johns Hopkins University

Both neural and cognitive systems have considerable plasticity. The same stimulus at different times may elicit different responses depending on the influence of other variables on the system. The characteristics of this plasticity differ markedly in terms of many parameters: the variables that produce it, how quickly it occurs, how long it lasts, what stimulus and response systems can be influenced, and so forth.

The major purpose of this chapter is to compare analyses of plasticity that have been conducted in the context of memory and attention, with the primary goal of suggesting ways in which lessons learned from the analysis of the brain mechanisms involved in mnemonic processes can be applied to the analysis of the brain mechanisms involved in attentional processes. Riley has made substantial contributions in the analysis of both memory and attention. Our approach builds on those contributions, relying heavily on the ways in which cognitive processes can be analyzed in animals, and extending that approach to the integration of neural and cognitive analyses of psychological processes.

COMPARATIVE COGNITION

Although comparative cognition is used most often as a term to describe comparisons between different species, it is also appropriate for the comparison of two individuals within the same species because these individuals may differ substantially in their neural mechanisms, genetic background, experience, or other variables that may influence cognitive processing. Historically, the discussion of comparative cognition often has begun with a substantial bias toward one particular view of the extent to which the individuals being compared are similar or different. This bias may reflect a particular view of evolution, species, cognitive processes (especially language), genetic background, or any other influential variable. A major contribution of the recent developments in comparative cognition is to provide a theoretical and experimental basis to choose empirically among the many different alternative explanations of the differences among individuals. The general issues involved in comparative cognition have been discussed extensively elsewhere, both by Riley and by others, and are not reviewed here (Riley, Brown, & Yoerg, 1986; Roitblat, 1987). Rather, the emphasis is placed on the application of this approach to the analysis of plasticity in memory and attention, with special emphasis on the importance of comparative cognition to determine the neural mechanisms underlying cognitive processes.

For the understanding of neural mechanisms involved in cognitive processes, comparative research with animals is important for two reasons. First, the brains of animals differ substantially in many characteristics, suggesting that some of the neural mechanisms involved in a given cognitive process must differ from one species to the next. For example, the hippocampus of both mammals and birds is involved in memory, but the gross structure of the hippocampus, and its cellular organization are substantially different in the two species (Olton, 1989; Sherry & Vaccarino, 1989; Sherry, Vaccarino, Buckenham, & Herz, 1989). An appreciation of the ways in which the similarities and differences of this brain area contribute to the similarities and differences in memory between the two species is critical to determine the ways in which neural systems can mediate memory. Many characteristics of the brain vary substantially across species. Consequently, any attempt to describe the general principles involved in the neural mechanisms of cognitive processes must include a comparative approach to encompass this variability.

A second reason for using a comparative approach to study cognitive processes in animals is to obtain general principles that may hold for the neural mechanisms of cognitive processes in several species of animals, including humans. Direct access to the human brain, both to manipulate it and measure it, is extremely limited, both in current practice and in the foreseeable future. As long as the welfare of the individual being studied is more important than the acquisition of basic knowledge (an ethical position that is so strongly developed that it is unlikely to change), detailed information about the neural systems involved in human cognition cannot be obtained in sufficient detail to resolve fundamental issues about the relation between mind and brain. The focus of this chapter is on this second approach, the use of animals to identify the general neural mechanisms involved in cognitive processes such as memory and attention.

MEMORY

The integration of comparative cognition and neural analyses to understand the neural systems involved in memory in humans is well established. An empirical indicator of the extent to which the analysis of memory has been integrated between humans and other mammalian species is the number of cross-references in articles using one species to data obtained from other species. Information about the three mammalian species most often tested (rats, monkeys, and humans) is frequently brought together to address important issues concerning the neural mechanisms of memory. For example, recent memory in rats, monkeys, and humans is impaired following lesions of the hippocampus. These results suggest a common function of the hippocampus across species. This conclusion could be obtained only by the free exchange of information between investigators studying recent memory in the different species. This free interchange of information may both reflect the benefits of a comparative approach, and be a catalyst to stimulate it. Once this interaction is begun, of course, it often continues easily and may be refined to allow even further integration. This article suggests that the steps taken to integrate studies of the neural mechanisms of mnemonic processes may be beneficial to produce a similar integration of the neural mechanisms of attentional processes, and develops this line of thought by first reviewing the relevant information in memory, and then extending the principles to the analysis of attention.

The term "model" has been used often when examining memory in animals. Sometimes, this term is appropriate in the sense that the procedure is designed to replicate that used in a human. In other cases, it is probably inappropriate because the memory process itself is of direct interest, whether or not it has been studied in humans. The term "model" often implies that a particular standard is the only criterion for being correct. This attitude is inappropriate in most analyses of comparative cognition because each analysis may be legitimate in its own right. For the present discussion, the major purpose is to analyze the commonalities in memory across different mammalian species, particularly, humans, rats, and monkeys. In this context, the goal is to provide accurate assessment of a particular memory process in all three species rather than to adopt a particular procedure used for a particular species as a model for the other species.

To provide some specific examples of the general points raised in this discussion, consider the analysis of recent memory in mammals. Recent memory involves the recall of previously presented information, usually maintaining it within a specific context or episode. It is often assessed by some specific variation of a delayed conditional discrimination. In this type of discrimination, some information is presented to the individual at the beginning of the trial. After a delay, two or more response alternatives are made available, and the response that is correct depends on the information that was presented at the beginning of the trial. In humans, recent memory can be assessed in many ways; some examples include reading a specific story and asking for recall of it as in the Wechsler Adult Intelligence Scale, or by presenting three specific nouns and asking for recall of them as in the "Mini-Mental" test (Folstein, Folstein, & McHugh, 1975). In animals, some version of a match-to-sample or nonmatch-to-sample is often used. For example, in an operant box, the stimulus might be either a red or green circle, and the response might be pressing a lever under either a black square or a black triangle. For a symbolic match-to-sample test, a response to the black square is correct following the red circle as the stimulus; a response to the black triangle is correct following a green stimulus.

An important step in the integration of studies from humans and nonhuman mammals has been the development of nonverbal tests of recent memory for humans (Bartus & Dean, 1981; Freedman & Oscar-Berman, 1986; Kesner, Adelstein, & Crutcher, 1987; Olton, 1990b; Wright, 1989). A comparison of the results from these nonverbal tests in humans with those from the verbal tests in humans and with the nonverbal tests in animals can assess the generality of the conclusions drawn from experiments with verbal material in humans. Again, the ability to adapt procedures across different species, without giving primacy to any one given test in any one given species, is an important step in integrating the relevant information (Wright, 1989).

Indeed, a major contribution of comparative cognition has been to emphasize the importance of abstract cognitive and computational analyses that are conducted independently of particular test procedures. A complication in any comparative analysis arises when the concepts or data used for comparison are tied so closely to a particular experimental procedure that the only way to test the generality of results from one species to another is to replicate the procedure exactly. A fundamental goal in assessing cognitive processes is to provide multiple, converging operations so that variables influencing the empirical processes of manipulation and measurement can be separated from those influencing the primary cognitive process being studied (Garner, Hake, & Eriksen, 1956; Olton, 1990a; Platt, 1964; Rescorla, 1988; Waldrop, 1990a, 1990b).

The comparison across mammalian species for analyses of recent memory has been subject to considerable discussion, often in the context of animal models of procedures used for humans. The measurement of normal memory in humans, and the description of the amnesic syndromes following different types of brain damage, has been extensive, and has stimulated considerable research with animals. Conferences, books, and articles typically include references to four different areas: experimental psychology (memory in normal humans), clinical neuropsychology (amnesic syndromes in patients), comparative cognition (memory in animals), and physiological psychology (the brain mechanisms involved in memory). This continued cross-fertilization has led to considerable agreement about the criteria and dimensions that should be used to compare analyses and memory in different species (Olton, Wible, & Markowska, 1991).

Independent Variables.   The demand on recent memory can be manipulated parametrically by altering the *task demand,* the extent to which a particular component of a task is required for successful performance (Olton, 1989). Task demand for recent memory can be manipulated by procedures that alter interference among the items to be remembered. Several empirical manipulations are commonly used. The first is increasing the delay/retention interval between presentation of the stimuli and the opportunity to make responses. The longer this delay interval, the greater the demand on recent memory, the worse the performance following the delay. The second manipulation is the number of items to be remembered. The greater the number of items presented as sample stimuli at the beginning of the trial, the greater the demand on recent memory, the worse the performance following the delay. The context in which a given stimulus is presented can have a marked effect on the ability to recall that stimulus. Both the type of trial preceding the current one, and the temporal intervals in all of the components of the test session, not just the current trial, may influence interference, and choice accuracy. For example, choice accuracy on the current trial with a given delay and a different correct response than on the previous trial may be greater if the intertrial interval is longer, rather than shorter, because the longer interval reduces interference from the previous trial. Interference may be produced by stimuli and responses, and probably other components of the trial (Olton & Shapiro, 1992).

In summary, comparative analyses of the neural and cognitive mechanisms involved in recent memory are helped immensely by agreement on the parametric variables that can be used to manipulate task demand for recent memory. These are sufficiently empirical to reach consensus about what should be done in a given experiment, yet sufficiently general that the actual manipulations are easily adapted to testing with any given species.

Conceptual Framework. Considerable discussion concerns the appropriate taxonomic categorization of memory, and many different theoretical frameworks have been suggested. Although many of the disagreements are substantial and require resolution to provide a complete analysis of mnemonic processes, sufficient agreement is available that an appropriate conceptual framework for the analysis of recent memory can proceed. For example, both proactive and retroactive interference are present in recent memory. Proactive interference is the disruptive effect of previously presented information on the ability to remember currently presented information. Retroactive interference is the disruptive effect of currently presented information on the ability to remember previously presented information. Serial order effects are also present in recent memory. Primacy and recency, respectively, refer to the more accurate memory for items at the beginning and end of a list than for items in the middle. Tests of recent memory in different species can compare the extent to which these different psychological processes are involved (Wright, 1989).

In summary, in addition to the empirical manipulations of recent memory described in the previous section, theoretical descriptions of recent memory also can be used for comparative purposes. The agreement on the kinds of cognitive processes that should be observed in recent memory permits assessment of the effectiveness of manipulations of independent variables on the desired cognitive processes.

Evolutionary Factors. When searching for food, animals can use many different strategies. Discussions of optimal foraging have identified strategies that vary in the extent to which they require recent memory. Both birds and rodents have been studied in the natural habitat to determine the extent to which recent memory guides their behavior. For example, some birds hoard food by placing seeds in a cache for a short period of time to hide the food, and then return to obtain it. This behavior is similar to that in an experimentally controlled delayed conditional discrimination, and these birds use recent memory to find their food. In contrast, other birds do not store food in caches, and use other kinds of strategies to obtain it. Similar differences exist among rodents. These analyses are important because they demonstrate that the natural habitat for an animal can put demands on recent memory, and that the types of procedures used to assess recent memory in laboratory tests can involve naturally occurring foraging strategies rather than being some unnatural never-before-experienced challenge to memory (Kamil & Balda, 1985; Sherry, 1984; Shettleworth, 1985). Involvement of the hippocampus in natural foraging as well as in laboratory tests of recent memory provides additional evidence that laboratory procedures can elicit natural mnemonic processes (Harvey & Krebs, 1990; Krebs, Sherry, Healy, Perry, & Vaccarino, 1989; Sherry & Vacarino, 1989; Sherry et al., 1989).

ATTENTION

Several model systems have been developed to assess the neural basis of specific kinds of attention. However, no large-scale integration of the neural analysis of attention has yet been undertaken. As one indicator of the difference in the maturity of neural analyses of memory and of attention, consider the presentations at the Society of Neuroscience in 1991. In the slide and poster sessions, 21 were titled "learning and memory," and these had subheadings referring to electrophysiology, anatomy, neurochemistry, and other topics. No single session was labeled "attention." In the list of keywords, many presentations had the word *learning* or *memory* listed, whereas only a few had the word *attention* listed. Presentations at the Society for Neuroscience are not the only criteria to judge the amount of work being done in a given area, but even if these data are biased for some unknown reason, they demonstrate such a substantial difference that even significant quantitative adjustments would not change the picture. The point is clear: Much more research is being conducted on the neural analysis of mnemonic plasticity than on the neural analysis of attentional plasticity.

The reasons for this difference may be many, but one important difference in the study of memory and attention is the availability of experimental procedures for animals to assess attentional processes in ways that are homologous to those used in humans. In the study of attention, no analysis equivalent to recent memory, with appropriate independent variables, psychological constructs, and ethological considerations, has yet been developed. The seminal work of Riley and colleagues has made an important step in this direction, showing how an approach using the principles of comparative cognition can be applied to the study of attention in animals (Bond & Riley, 1991; Brown, Cook, Lamb, & Riley, 1984; Riley, 1976; Riley & Leith, 1976; Riley & Roitblat, 1976). The object of the research program that we have just begun is to extend the analysis of attention and develop procedures that can be used for rats and that incorporate the same features used in experiments for humans. If tests of attention in humans and rats can be integrated in the same way that tests of memory have been integrated, then the rich conceptual and empirical body of information developed in the study of human attention can be applied to the study of attention in animals, which then opens the door to an examination of the neural bases of these attentional mechanisms.

EXPECTANCY AND ATTENTION

Expectancy is preparation for a specific event. If the expectancy is accurate and the event occurs, responding can be more rapid and more accurate (Hick, 1952; Hyman, 1953). The conceptual framework to describe the cognitive processes involved in expectancy have been developed primarily in analyses of the performance of humans in choice reaction time tasks. For the same reasons outlined previously in the study of memory, facilitation of the comparative analysis to examine the brain mechanisms involved in expectancy can be accomplished by using experimental procedures that have the same conceptual framework, operational procedures, and independent variables in both animals and people. The experiment described here was a first attempt to begin that unification of procedures (Pang, Merkel, Egeth, & Olton, 1992).

Conceptually, the experiment was designed to examine the effects of expectancy on information processing. As described previously, accurate expectancy for an event should facilitate responding to that event, decreasing reaction time, and increasing choice accuracy. The independent variable was the relative probability of two stimulus/response events within each test session. The dependent variables included reaction time, choice accuracy, discriminability, and bias. The procedures used variations of a two-choice reaction time task. For the rats, the procedures were made as similar as possible to those used in previous experiments with people. For humans, the procedures were as similar as possible to those used for the rats including the minor variations that were necessary to train the rats in the task. The procedures for the two species were identical except for three components: the increased amount of training given to the rats prior to testing, an explicit water reinforcement given to the rats but not the humans, and the specific equipment used for the testing.

For both species, a trial was initiated by depressing two response keys. After a variable delay, one of two stimuli, a tone or a light, was presented. For the light, releasing one response key was correct. For the tone, lifting the other response key was correct. Within each session, the probability of the two stimuli remained constant. Across sessions, these probabilities varied (probability of light/probability of tone): 100/0, 90/10, 50/50, 10/90, 0/100. Reaction time and choice accuracy were measured, and discriminability and bias were calculated.

The predictions were as follows. If the probability of a stimulus within a session alters expectancy, responses to the expected stimulus should be faster and more accurate, producing a bias toward the expected stimulus. If rats, like humans, respond to changes in stimulus probability by altering expectancy, and expectancy has similar effects on information processing, then the pattern of results for rats should be similar to those for humans.

The pattern of results was similar to that predicted. As the probability of a specific stimulus increased, reaction time decreased, choice accuracy increased, response bias increased, and discriminability did not change. This pattern was similar for both rats and humans, with one exception: For rats, expectancy had little effect on responses to the auditory stimulus.

NEURAL MECHANISMS OF ATTENTION

The success of this comparative approach in the study of attention offers the opportunity to pursue neural analyses of attention in the same way that the neural analyses of memory have been developed. With conceptual and operational similarity in the approaches taken for the two species, additional behavioral experiments can be designed to test the extent to which rats and humans use similar cognitive mechanisms for other types of attentional processes, and then the neural bases of these can be examined.

The cholinergic projections from the nucleus basalis magnocellularis (NBM) to the frontal cortex may have an important influence on some types of attention. Lesions in this system disrupted divided attention for the temporal discrimination of two stimuli (Olton, Wenk, Church, & Meck, 1988; Olton, Wenk, & Markowska, 1991). Changes in the cholinergic innervation of cortical areas altered receptive fields to somatosensory stimuli (Dykes, Tremblay, Warren, & Bear, 1991). Microinfusion of drugs that altered cholinergic activity in the NBM altered attentional processes for stimuli in the procedures described here. The cingulate cortex also may be involved in certain types of divided attention as indicated by scanning with positron emission tomography (PET) in humans (Corbetta, Miezin, Dobmeyer, Shulman, & Petersen, 1990). Other analyses have emphasized specific types of attention, such as visual attention related to eye movements and spatial location (Spitzer, Desimone, & Moran, 1988). Together, all of these analyses provide clues about the kinds of neural systems that may underlie attentional processes in humans and other mammals.

PLASTICITY

Both attention and memory are associated with plasticity in the nervous system, changing pathways from receptors to effectors. Currently, the conceptual frameworks to describe the cognitive mechanisms involved in attention and memory are very different, and as compared to the substantial amount of information currently available about the neural mechanisms of memory, little information is available about the neural mechanisms of attention. What are the similarities and differences in these neural systems? The likelihood of them differing in fundamental neurobiological mechanisms, such as the flow of ions across membranes or synaptic currents, is very small, although possible. A more likely difference is in localization of function, the conjunction of neuroanatomical areas and neurochemical systems that respond to differences in attentional demands and mnemonic demands (Olton, Givens, Markowska, Shapiro, & Golski, 1992).

Considering the similarities and differences in the neural mechanisms of attention and memory can stimulate some useful thinking about attentional processes in a general cognitive/systematic/computational framework that is independent of the specific experimental procedures and limited theories that have been proposed so far. In both cases, the input/output functions of the nervous system have been changed by the addition of some other variable, attentional or mnemonic. How do these changes take place? What do the similarities and differences in neural mechanisms tell us about the distinction in the cognitive processes of memory and attention? Can this approach help to unify and integrate two fields that have developed relatively independently of each other, and most important, can it help to take the success that has characterized the neural analyses of memory and be equivalently successful in the neural analyses of attention?

In summary, the comparative analysis of attention in humans and other mammals provides opportunities to investigate the neural systems underlying attention. The seminal studies of Riley examining attention in animals provide an important building block for this enterprise, and if the goals of the research enterprise outlined here are achieved, Riley will have provided major links in the long chain of events leading to this success.

ACKNOWLEDGMENTS

Preparation of this chapter was supported in part by Air Force Office of Scientific Research, AFOSR-89-0481 and AFOSR-91-0110. The authors thank A. Dürr for preparation of this manuscript, and S. Hulse and T. Zentall for helpful comments.

D. O. especially acknowledges the personal and professional contributions of Al Riley, both directly from him, and indirectly through the students he has educated and the other people that he has influenced. Many happy interactions have taken place with Al at Berkeley and other places, resulting in substantial professional growth and personal pleasure. May we all be so successful in leaving behind such a delightful legacy.

Ode to Al

A retiring professor named Riley,
Like the coyote, had an intellect wiley.
With a voice deep and low,
His wisdom he'd show
With a humor expressed most dryly.

REFERENCES

Bartus, R. T., & Dean, R. L. III (1981). Age-related memory loss and drug therapy: Possible directions based on animal models. In S. J. Enna (Ed.), *Brain neurotransmitters and receptors in aging and age-related disorders* (pp. 209–223). New York: Raven.

Bond, A. B., & Riley, D. A. (1991). Searching image in the pigeon: A test of three hypothetical mechanisms. *Ethology, 87,* 203–224.

Brown, M. F., Cook, R. G., Lamb, M. R., & Riley, D. A. (1984). The relation between response and attentional shifts in pigeon compound matching-to-sample performance. *Animal Learning and Behavior, 12,* 41–49.

Corbetta, M., Miezin, F. M., Dobmeyer, S., Shulman, G. L., & Petersen, S. E. (1990). Attentional modulation of neural processing of shape, color, and velocity in humans. *Science, 248,* 1556–1559.

Dykes, R. W., Tremblay, N., Warren, R. A., & Bear, M. F. (1991). Cholinergic of synaptic plasticity in sensory neocortex. In R. T. Richardson (Ed.), *Activation to acquisition functional aspects of the basal forebrain cholinergic system* (pp. 325–345). Boston: Birkhauser.

Folstein, M. F., Folstein, S. E., & McHugh, P. R. (1975). "Mini-Mental State"—A practical method for grading the cognitive state of patients for the clinical. *Journal of Psychiatric Research, 12,* 189–198.

Freedman, M., & Oscar-Berman, M. (1986). Bilateral frontal lobe disease and selective delayed response deficits in humans. *Behavioral Neuroscience, 100,* 337–342.

Garner, W. R., Hake, H. W., & Eriksen, C. W. (1956). Operationism and the concept of perception. *The Psychological Review, 63,* 149–159.

Harvey, P. H., & Krebs, J. R. (1990). Comparing brains. *Science, 249,* 140–146.

Hick, W. E. (1952). On the rate of gain of information. *Quarterly Journal of Experimental Psychology, 4,* 11–26.

Hyman, R. (1953). Stimulus information as a determinant of reaction time. *Journal of Experimental Psychology, 45,* 188–196.

Kamil, A. C., & Balda, R. P. (1985). Cache recovery and spatial memory in Clark's nutcrackers (*Nucifraga columbiana*). *Journal of Experimental Psychology: Animal Behavior Processes, 11,* 95–111.

Kesner, R. P., Adelstein, T., & Crutcher, K. A. (1987). Rats with nucleus basalis magnocellularis lesions mimic mnemonic symptomatology observed in patient with dementia of the Alzheimer's type. *Behavioral Neuroscience, 101,* 451–456.

Krebs, J. R., Sherry, D. F., Healy, S. D., Perry, V. H., & Vaccarino, A. L. (1989). Hippocampal specialization of food-storing birds. *Neurobiology, 86,* 1388–1392.

Olton, D. S. (1989). Dimensional mnemonics. In G. H. Bower (Ed.), *The psychology of learning and motivation* (pp. 1–23). San Diego: Academic.

Olton, D. S. (1990a). Experimental strategies to identify the neurobiological bases of memory: Lesions. In J. L. Martinez & R. Kesner (Eds.), *Learning and memory, a biological view.* New York: Academic.

Olton, D. S. (1990b). Mnemonic functions of the hippocampus: Past, present and future. In L. R. Squire & E. Lindenlaub (Eds.), *The biology of memory* (pp. 427–440). New York: Schattauer Verlag.

Olton, D. S. Givens, B. S., Markowska, A. L., Shapiro, M., & Golski, S. (1992). Mnemonic functions of the cholinergic septohippocampal system. In: L. R. Squire, N. M. Weinberger, J. L. McGaugh (Eds.), *Memory: Organization and locus of change* (pp. 250–269), New York: Oxford University Press.

Olton, D. S., & Shapiro, M. (1992). Mnemonic dissociations: The power of parameters. *Journal of Cognitive Neuroscience, 4,* 200–207.

Olton, D. S., Wenk, G. L., & Markowska, A. L. (1991). Basal forebrain, memory, and attention. In R. Richardson (Ed.), *Activation to acquisition: Functional aspects of the basal forebrain.* Boston: Birkhauser.

Olton, D. S., Wenk, G. L., Church, R. M., & Meck, W. H. (1988). Attention and the frontal cortex as examined by simultaneous temporal processing and lesions of the basal forebrain cholinergic system. *Neuropsychologia, 26,* 307–318.

Olton, D. S., Wible, C. G., & Markowska, A. L. (1991). A comparative analysis of the role of the hippocampal system in memory. In R. G. Lister & H. J. Weingartner (Eds.), *Perspectives on cognitive neuroscience* (pp. 186–196). New York: Oxford University Press.

Pang, K., Merkel, F., Egeth, H. & Olton, D. S. (1992). Expectancy and stimulus frequency: A comparative analysis in rats and humans. *Perception & Psychophysics, 51,* 607–615.

Platt, J. R. (1964). Strong inference. *Science, 146,* 347–353.

Rescorla, R. A. (1988). Behavioral studies of pavlovian conditioning. *Annual Review of Neuroscience, 11,* 329–352.

Riley, D. A. (1976). Information processing and attention in animals. *International Congress in Psychology, 22.*

Riley, D. A., Brown, M., & Yoerg, S. (1986). Understanding animal cognition. In I. J. Knapp & L. C. Robertson (Eds.), *Approaches to cognition* (pp. 111–136). Hillsdale, NJ: Lawrence Erlbaum Associates.

Riley, D. A., & Leith, C. R. (1976). Multidimensional psychophysics and selective attention in animals. *Psychological Bulletin, 83,* 138–160.

Riley, D. A., & Roitblat, H. (1976). Selective attention and related cognitive processes. In S. Hulse, W. K. Honig, & H. Fowler (Eds.), *Cognitive processes in animal behavior* (pp. 249–276). New York: Macmillan.

Roitblat, H. L. (1987). *Introduction to comparative cognition.* New York: Freeman.

Sherry, D. F. (1984). Food storage by black-capped chickadees: Memory for the location and contents of caches. *Animal Behavior, 32,* 451–464.

Sherry, D. F., & Vaccarino, A. L. (1989). Hippocampus and memory for food caches in black-capped chickadees. *Behavioral Neuroscience, 103,* 308–318.

Sherry, D. F., Vaccarino, A. L., Buckenham, K., & Herz, R. S. (1989). The hippocampal complex of food-storing birds. *Brain, Behavior and Evolution, 34,* 308–317.

Shettleworth, S. J. (1985). Food storing by birds: Implications for comparative studies of memory. In N. M. Weinberger, J. L. McGaugh, & G. Lynch (Eds.), *Memory systems of the brain* (pp. 231–250). New York: Guilford.

Spitzer, H., Desimone, R., & Moran, J. (1988). Increased attention enhances both behavioral and neuronal performance. *Science, 240,* 338–340.

Waldrop, M. M. (1990a). Hubble: The case of the single-point failure. *Science, 249,* 735–736.

Waldrop, M. M. (1990b). Hubble hubris: A case of "certified" blindness. *Science, 249,* 1333.

Wright, A. A. (1989). Memory processing by pigeons, monkeys, and people. In G. H. Bower (Ed.), *The psychology of learning and motivation* (pp. 25–70). New York: Academic.

14 Gestalt Contributions to Visual Texture Discriminations by Pigeons

Robert G. Cook
Tufts University

One of the fundamental questions of animal cognitive psychology concerns the nature and processes determining the effective stimulus in any discrimination. In his 40 years at Berkeley, this question, in one form or another, has been the theoretical and experimental foci for Donald Riley, his many students, and collaborators. These experiments have examined a broad range of topics, with inquiries into transposition in both children and rats, conditioning in the octopus, selective and divided attention in pigeons, echolocation in rats, and verbal learning and memory in humans (Postman & Riley, 1959; Riley, Goggin, & Wright, 1963; Riley & Leith, 1976; Riley & McKee, 1963; Riley & Rosenzweig, 1957; Warren, Riley, & Scheier, 1974).

On the surface these topics appear to be quite diverse. But a more careful inspection reveals an underlying unity to these investigations, reflected in a recurring concern with how discrimination learning is influenced by the gestalt properties of the stimulus. This interest in how the configuration, component relations, or structural organization of the stimulus can modify learning and behavior is present throughout his work. These include: how attention in pigeons is influenced by the structure of a compound stimulus, (Cook, Riley, & Brown, 1992; Lamb & Riley, 1981; Riley & Leith, 1976; Riley & Roitblat, 1978), the importance of stimulus relations in visual and auditory transposition in children and rats (Riley, 1965; 1968; Riley et al., 1963), and the contribution of figural simplification or "Pragnanz" to the human retention of form stimuli (Riley, 1962).

Discrimination Learning (Riley, 1968) and *Memory for Form* (Riley, 1962) in particular are insightful essays concerning the potential role of gestalt components in discrimination learning and memory. Riley's careful delineation of the basic arguments and his critical examination of the literature are exemplary illustrations of using evidence to evaluate competing accounts of experimental data—and make for instructive and highly recommended reading for all students of learning and memory.

This persistent interest in how animals perceive and process discriminative stimuli has naturally infected many of his students. I am no exception. This influence is reflected in my research aimed at understanding the mechanisms of visual cognition in birds (Cook, 1983, 1992a, 1992b, 1992c; Cook et al., 1992; Cook, Wright, & Kendrick, 1990; Riley, Cook, & Lamb, 1981). Over the last several years, I have been particularly interested in the visual perception and processing of multidimensional texture stimuli in pigeons (Cook, 1992a, 1992b, 1992c). In keeping with D. A. Riley's lifelong interest in the influence of gestalt stimulus properties on animal behavior, this chapter examines the contribution of emergent stimulus properties to the avian perception of textured visual stimuli. This topic is not only appropriate, but is an acknowledgment of the important and long-standing scientific tradition involving the intergenerational transmission of questions and ideas from mentor to student.

This chapter has three sections. In the first part I discuss the general rationale for comparing visual processing in different animals, particularly mammals and birds. The second section elaborates on this issue, reviewing briefly how textured visual stimuli are perceived by humans and pigeons, and the contribution of emergent stimulus properties to human texture perception. The third section describes three experiments recently conducted in my lab suggesting that similar emergent factors are involved in pigeon texture perception and processing.

THE COMPARATIVE PSYCHOLOGY OF VISUAL PERCEPTION

Birds behave as if they readily and accurately perceive an object-filled visual world. Their unerring avoidance of obstacles during flight, pinpoint landings on small limbs, and superb precision in collecting food items—be it pigeons searching out and pecking up small cryptic grains or flycatchers hawking for insects— are all examples of their apparent keen awareness of the visual objects surrounding them. Apropos to the importance of vision to these animals, birds have enormous eyes for their size, and a considerable portion of their central nervous system is devoted to the processing of visual sensory information. Because of the severe weight limitations of muscle-powered flight, birds have simultaneously been subjected to strong evolutionary pressures to limit their overall size. The competing requirements for acute vision and limited size have caused birds to evolve small visual and central nervous systems that are nonetheless both powerful and efficient (Donovan, 1978; Pearson, 1972). This remarkable package of size and power creates something of a paradox given current thinking about the computational complexity of visual processing (Arbib & Hanson, 1987; Fischler & Firschein, 1987; Landy & Movshon, 1991). For example, no visually-directed robot can match the diminutive chickadee's ability to rapidly move through any environment, much less one as visually noisy as the tangled web of limbs that is this species' forest habitat.

This puzzle naturally leads one to compare avian visual mechanisms with other biological and nonbiological visual pattern recognition systems, such as diurnal mammals and computers. For example, humans and pigeons can both discriminate many types of form stimuli. Of greater theoretical importance, however, is that both species often manifest the same difficulties and make similar errors in discriminating these stimuli (Blough, 1982; 1985; Blough & Franklin, 1985; Cook, 1992b; Cook et al., 1990; Fujita, Blough, & Blough, 1991). This suggests that the common final perception of these stimuli are similar for both species, a conclusion that raises several interesting questions. Are these similar perceptions derived from comparable visual and cognitive processes? How do the large differences in size and anatomical organizations of the pigeon and human visual systems influence these processes? Are pigeons visually more efficient than humans? If not, what aspects of avian visual perception and performance have been curtailed in the interest of brain size or processing speed?

The answers to such questions require detailed comparisons of the two species at the different processing stages occurring in the journey of visual information from retina to behavior. With this approach in mind, I have recently been examining the early stages of avian visual processing through investigations of how pigeons perceive and process visual textures. In humans, such textured stimuli seem to straightforwardly reveal some of the earliest and most fundamental operations of the human perceptual system (Beck, 1982; Grossberg & Mingolla, 1985; Julesz, 1981). As a result, they seemed to be an excellent starting point for comparing the mechanisms of visual perception in these two different, but highly visual, species.

TEXTURE PERCEPTION AND EARLY VISION IN HUMANS AND PIGEONS

Texture stimuli are composed of perceptually distinct spatial regions derived by the visual grouping of smaller elements. Humans quickly perceive the global differences in many different kinds of texture stimuli (Beck, 1966, 1982; Julesz, 1975, 1981). The effortless and rapid nature of human texture segregation has suggested that its perceptual mechanisms are located early in the visual processing stream, are parallel and preattentive in character, and are used to identify extended areas with similar visual attributes (Beck, 1982; Broadbent, 1977; Grossberg & Mingolla, 1985; Julesz, 1981; Marr, 1982; Treisman, 1986; Treisman & Gelade, 1980; Watt, 1988). This rapid grouping and segregation of the different parts of the visual input are considered fundamental for the subsequent identification of objects, their surfaces, and contours. Thus, the study of human texture discrimination has been extremely useful in revealing the organization and structure of the early visual processes related to the determination of fig-ure/ground relations, the separation and combination of dimensional informa-tion, the identification of potential visual primitives and features, and the dis-crimination of object surfaces and edges.

The human perception of textured stimuli also seems to invoke and involve many of the organizational and grouping principles outlined by the early Gestalt psychologists (e.g., Kohler, 1947). For instance, the similarity, proximity, and arrangement of component elements all influence the ease of texture segregation in humans. Such findings have suggested to some that the gestalt "laws" of perceptual organization (good continuity, proximity, similarity, and closure) are products of the early visual system, and its attempts to group the luminance and hue differences in the visual input for the purpose of deriving edge, contour, and surface information (Beck, 1982; Marr, 1982; Uttal, 1981). By grouping similar or closely spaced features that form smooth continuous or closed organizations, the visual system can take advantage of the regular structure of objects in order to determine the relations between adjacent or occluding object surfaces and their respective edges. Similar grouping algorithms for detecting and identifying edge and surface information are critical first steps in many computerized vision programs as well (see Cohen & Feigenbaum, 1982; Fischler & Firschein, 1987).

Are similar mechanisms for processing textured information operating in the avian visual system? My texture discrimination experiments have suggested that pigeons also rapidly perceive the regional organization of textured displays (Cook, 1992a; 1992b; 1992c). In those initial experiments, the pigeons detected and pecked at a *target* region in order to receive food reinforcement. This target region, formed by a group of small, identical, colored forms or elements, was randomly located within a larger array of *distractor* elements. The elements of these two visually distinct regions systematically differed in either color, shape, or both dimensions. These stimuli were computer generated and presented direct-ly to the pigeons on color monitors, with their pecking responses detected by an infrared LED touchscreen.

The pigeons easily acquired these texture discriminations, quickly learning to peck at the target of both color and shape displays (Cook, 1992a). Additionally, the rate of acquisition was not influenced by the number of local elements composing the set of training texture displays. After learning the task, the pi-geons also readily transferred this discrimination to new texture displays com-posed of novel elements. Collectively these findings seemed best explained by the idea that the pigeons readily perceived the global organization of the target and distractor regions, and that this acted as the effective stimulus mediating this visual discrimination.

A further set of experiments found that pigeons and humans appear to similarly process dimensional information in textures (Cook, 1992b). In these experiments, both species were tested with feature and conjunctive textured stimuli directly analogous to those previously tested with humans (Treisman & Gelade, 1980). Both species were better at discriminating feature displays in which the target and distractor regions consistently differed in color or shape (e.g., shape: red and green *squares* embedded within red and green *circles;* color: *red* circles and squares embedded within *green* circles and squares), than conjunctive displays where the regions were derived from conjunctive combinations of the same elements (e.g., *red squares* and *green circles* embedded within *red circles* and *green squares*).

The effortless discrimination of feature displays and the corresponding difficulty of segregating conjunctive displays has suggested that early human preattentive processes parse visual information into separate dimensional channels, where it is then grouped into distinct visual regions (Treisman, 1986; Treisman & Gelade, 1980; see also Broadbent, 1977; Cave & Wolfe, 1990; Hoffman, 1979; Marr, 1982; Neisser, 1967, for similar ideas about the early parsing and combination of dimensional information). These analogous findings with pigeons suggest that their early visual mechanisms parse and then group dimensional attributes in a similar manner. Thus my experiments have so far suggested that humans and pigeons are comparable in their perception and processing of textural information—with both species immediately perceiving contrasting global regions of color or form through the grouping and segregating of information within separate dimensional channels.

These similarities naturally lead one to ask whether these early avian grouping processes are also important in edge and object contour discrimination, as seems to be the case in humans. This chapter provides an opportunity to draw together three independent lines of research that have converged to suggest that the edgelike emergent visual discontinuities at the boundaries of textured regions are indeed critical in the pigeons' discrimination of such displays. Each experiment used a different type of procedure and textured stimulus. The first used a target detection discrimination procedure and examined performance with various arrangements of the distractor elements. The second used a go/no-go procedure to examine the discrimination of aggregated and distributed mixtures of different elements. The third used a two-alternative forced-choice procedure to examine the discrimination of dotted textured stimuli. Despite these differences, each one seems to point toward a common conclusion—that like humans, the detection and discrimination of edges and related visual features may be an important operation of the avian early visual system.

The Effects of Distractor Organization

The purpose of the first experiment was to investigate how the organization of the distractor elements in textured displays influenced choice accuracy. Three different organizations of distractors were used to surround the small block of target elements. These involved either randomly scattering the distractor elements about the display, spreading the distractors out over the display so that they were not spatially contiguous, or arranging them into randomly placed vertical and horizontal strings of distractor elements.

The testing procedures were identical to those described earlier for the first set of texture discrimination experiments (Cook, 1992a; 1992b). The six highly experienced pigeons from those experiments also participated in this study. Each trial began with a peck to a randomly located "ready" signal, which was then replaced by a textured stimulus containing a small randomly located target region. The elements of this target region differed in shape (e.g., red triangles within red circles) or color (e.g., green squares within red squares) from those of the distractor area of the display. The pigeon's task was to locate and peck at the target region five times in order to be rewarded with food. If the distractor region of the display was pecked five times, however, the trial was scored as incorrect, causing the display to disappear and the overhead houselight to extinguish for 15 s.

A total of 1,452 shape and 1,320 color texture displays were available for testing, as formed by the pairwise combination of 132 different colored shape elements (derived from 11 colors and 12 shapes). All displays were formed within a 24 × 16 grid of spatial locations. The target region was a 7 × 6 block of elements randomly located within this grid. This target region was then surrounded by different numbers of distractor elements (34, 68, or 102 elements) arranged in one of three configurations. In the *Random* condition, the distractors were positioned in nonoverlapping random locations within the overall array. In the *Spaced* conditions, the distractors were positioned at random, but with the constraint that each distractor have at least one open grid interval to its right, left, top and bottom. In the *Linear* condition, strings of five vertically or horizontally arranged distractors were positioned at random within the overall grid. If a string prematurely reached the edge of the grid, the remaining distractors were carried over to form a shorter line of distractors elsewhere in the display. Illustrative examples of these three distractor organizations, as formed by 68 distractor elements, are displayed in the upper part of Fig. 14.1.

The graph in Fig. 14.1 shows the results from the 10 daily sessions of testing with these three distractor organizations. Only the linear arrangement of distractors significantly reduced target detection accuracy relative to the other two conditions. This was true for both color and shape targets, and for each of the three different quantities of distractor elements that were tested (Fig. 14.1 is averaged over this latter factor). The greater interference produced by linear arrangements indicates that it contained features more in common with those used to locate and identify the target region than the other organizations. The long straight edges, their frequent right angles, or the rectangular shape overall, all consistently present in both target and linear distractor arrangements, seem the most likely features. Although the present results do not discriminate be-

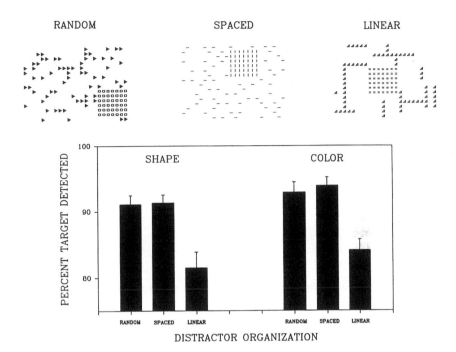

FIG. 14.1. Representative examples of the Random, Spaced and Linear distractor organization conditions are displayed at the top of the figure. The bottom graph depicts mean accuracy for shape and color target detection trials when tested with the three distractor organization conditions.

tween these alternatives, all of them suggest a feature closely tied to the linear global geometry of the target region as being the most critical to its accurate detection and discrimination by the pigeons.

The Effects of Element Spatial Organization

Experiments using the directed *target detection* procedure, like the one just described, require that the target area be formed by a small group of blocked elements, as it is this feature that tells the pigeons where to peck to obtain food. This very useful procedure unfortunately prevents the investigation of other potentially important spatial configurations of textured stimuli. For instance, displays with randomly distributed individual elements represent an important comparison condition for evaluating the effects of element grouping on texture discrimination, but cannot be tested with the target detection procedure.

The next experiment, done in collaboration with Karen Beal and Werner Honig, specifically addressed this limitation. In this experiment, textured displays of two different organizations were tested using a go/no-go mixture positive discrimination procedure (Honig, 1991). In our go/no-go discrimination, the pigeons were rewarded with food for pecking at textured displays containing any mixture of differently colored elements, but were not reinforced for pecking at a uniform display of any one color. The simpler and undirected response requirement of the go/no-go procedure, in comparison to the target detection task, allows displays with different kinds of element organizations to be tested. As a result, the pigeons' responses to two different textural configurations were examined.

In the *Aggregate* condition, a set of colored elements was grouped into a rectangular area and randomly placed within a set of elements composed of a different color (see Fig. 14.2). The *Random* condition was the same in every respect, except the two kinds of colored elements were randomly distributed throughout the entire display. Pairwise combinations of three different types of colored elements (red, green, and cyan squares) were used to form these *mixture* displays. Mixture displays of either organization and any combination of colors were treated as positive discriminative stimuli in this experiment and pecks to

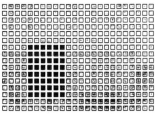

AGGREGATE MIXTURE

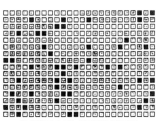

FIG. 14.2. Representative examples of the displays used to test the effects of spatial organization on element mixture discrimination. The Random and Aggregate displays are shown for only one of the six different proportions of mixed elements tested. Colors are shown as different gray levels. The contours around the elements were added only for illustrative purposes.

RANDOM MIXTURE

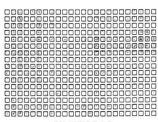

UNIFORM

these were reinforced with food. In the *Uniform* condition, the entire display was composed of square elements of only one color (cyan, red, or green), and pecks to these displays were never reinforced.

The relative proportion of the two different colored elements used to compose the positive mixture displays was also varied. Six proportions in all were tested, ranging from displays having only one different element (.26% of the 384 total elements), to those having equivalent numbers of the different colored elements (a 50%/50% mixture). At the more asymmetrical proportions, the Aggregate displays (see Fig. 14.2) were highly similar to the textures examined in the target detection experiments (Cook, 1992a, 1992b; the previous section).

Each discrimination trial started with a peck to a "ready" signal, which produced a uniform or mixture display on the computer screen. Each display was presented for a minimum of 20 s. On mixture trials, food reinforcement was delivered immediately after the first peck following the completion of this 20-s interval. On uniform trials, the displays were turned off after twenty s. Both trial outcomes were immediately followed by a 10-s intertrial interval. Daily sessions consisted of a set of 48 uniform negative trials (16 of each color) and 72 mixture positive trials (six color combinations × two display organizations × six proportions). Four pigeons that had already participated in an earlier mixture positive experiment were tested, and thus were readily performing the basic discrimination when these data were collected.

Before presenting the results, consider for a moment an observer that examined these displays in the most limited manner imaginable, by comparing only adjacent elements two at a time over the entire display. An observer with such a small viewing and comparison "aperture" then searches each display until two different colored elements are encountered (indicating a positive trial that should be pecked at) or until the entire display has been exhaustively searched (indicating a uniform negative trial).

Computer programs designed to model such a small-aperture observer using this type of serial, self-terminating response rule generate several interesting predictions about the nature of responding to these different displays. First, the number of successive "scans" required to locate a color difference should increase as the proportion of the two colors becomes increasingly more asymmetrical. That is, as the number of different element pairs decrease, it takes longer and longer to find a difference. Second, element differences should be found sooner in distributed Random displays than in the grouped Aggregate displays. That is, the number of scans needed to find a difference when the elements are distributed is less than when the elements are grouped into two regions. The size of this display difference decreases with evenly proportioned mixtures, but is never completely eliminated. These predictions held whether the model was programmed to scan the displays in a systematic manner or at random.

On the other hand, an observer simultaneously looking for color differences over large areas of the display containing many elements generates a different set of predictions. For such a large-aperture observer, neither variations in element proportion nor the organization of the colored elements should have a great effect on performance. This is because any element differences present within any form of mixture display should be immediately detected and readily discriminated within only a very few scans.

Presented in Fig. 14.3 are the results collected from the four pigeons tested. Consider first the top panel, which displays the average peck rates to uniform and mixture displays over the first 20 s of display presentation, and prior to any feedback given by food reinforcement. These birds clearly had little trouble discriminating the uniform and mixed nature of these displays as revealed by the considerable differences in peck rates. More important, there was little to no effect of element proportion on the discrimination of the positive mixture displays (the right 6 data points of Fig. 14.3). Even at the smallest proportion, in which only one element was different, birds responded almost as much as when the display contained an even mixture of the two element types. Analyses of the time to first peck also showed this same pattern. This overall insensitivity to the

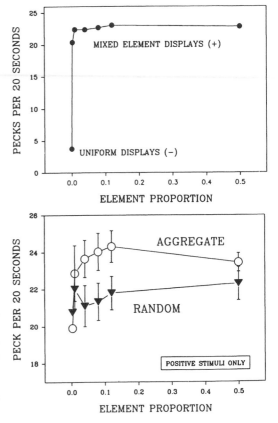

FIG. 14.3. The top panel displays the mean number of pecks to Uniform and Mixture displays conditions (Random and Aggregate displays combined) over the 20 s prior to any trial outcome. The labels for Element Proportion are referenced to the less numerous colored element in the display, and range from the Uniform condition (0) to the evenly mixed displays (.5). The bottom panel displays the mean number of pecks per 20 s for Aggregate and Random displays as a function of Element Proportion for the same test sessions as in the top panel. The results from the Uniform condition are not included in the bottom panel.

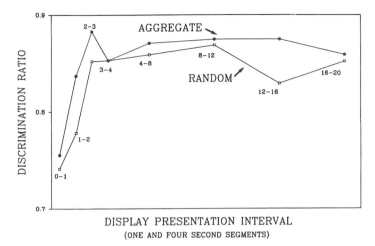

FIG. 14.4. Discrimination ratios for Aggregate and Random displays over 8 different temporal segments of the 20-s presentation interval. The left 4 data points are based on 1-s divisions of the first 4 s of the display interval. The right 4 data points are based on 4-s divisions of the last 16 s of presentation. Discrimination ratios were calculated by dividing the number of pecks to positive displays by the total pecks to both positive and negative displays during that interval segment.

relative proportions of the elements present in the displays suggests that the basic detection and discrimination of element mixture from uniformity was easy for the birds.

The second outcome of interest involved the two display configurations. Aggregate displays supported the superior discrimination, not the Random displays as predicted by the small aperture model. This finding is displayed in the lower panel of Fig. 14.3 and in Fig. 14.4. Fig. 14.3 shows that over the 20-s stimulus evaluation interval, the Aggregate displays received more pecks than the Random displays. Fig. 14.4 shows the time course of this difference, via a series of discrimination ratios calculated for different temporal segments of the display interval. The first four discrimination ratios on the left side of Fig. 14.4 are based on pecks emitted during the first 4 s following stimulus presentation. Pecks were grouped into separate 1-second divisions of that interval. The four discrimination ratios on the right side of the figure cover the remaining 16 s of presentation, and were divided into 4-s time segments. Fig. 14.4 shows that over the 20-s display interval the Aggregate displays were always pecked at higher rates than Random displays.

Overall, these results contradict the predictions of the small-aperture discrimination model described earlier. Both the minimal effects of element proportion and the overall inferiority of the Random displays suggest that these pigeons were visually processing the differences in these displays with only a few widely tuned scans. Further evidence of this is the clear above-chance discrimination of element mixture from uniformity within just the first second of texture presentation (see Fig. 14.4), suggesting that these features of the displays were detected quickly. Hence, the overall findings seem most consistent with the idea that the birds' viewing and comparison aperture permitted large-scale, parallel-like, evaluation of the visual differences in the display.

Such a large-aperture hypothesis, however, does not explain the superior discrimination of the Aggregate displays. Even in the completely parallel case, requiring only one scan of the display, there should be no performance differences between the two display types. The more rapid and higher rates of responding to the Aggregate displays suggests the presence of additional discriminative features over and above the color differences present in the displays. Again, the obvious features seem to be those emergent properties directly attributable to the grouping of the colored elements into regions. These include the long, continuous, contrasting edges emerging at the borders of the different regions, or perhaps the smaller square "figure" created by these contours in the "ground" of the surrounding elements.

One additional piece of evidence consistent with this reasoning comes from looking at where the birds pecked during Aggregate display presentations (responses to Random displays could not be analyzed for several reasons; see Cook, 1992c for details about the techniques used). All pigeons spontaneously pecked at either the small embedded group of elements in asymmetrically proportioned Aggregate displays, or directly at the extended borders present in the more evenly proportioned displays. Given the known propensity of pigeons to peck at the controlling attribute in discriminative stimuli (Hearst & Jenkins, 1974; Jenkins & Sainsbury, 1970), these peck location results suggest that the discontinuities between regions and the regions themselves were important features in controlling responding to these displays.

The Effects of Edge Discontinuities in Dot Texture Discrimination

Both lines of research discussed in the two previous sections have in different ways suggested that the emergent contours and related geometrical properties of textured stimuli are important in avian texture discrimination. The third and final line of research strengthens this conclusion. In these experiments texture stimuli constructed entirely from identical small dots, allowing only integrated global information to be used as the basis for any discrimination, were examined.

An example of a dotted texture stimulus is displayed in Fig. 14.5, the two regions of which are easily segregated by the human eye. It was formed by varying the probability of a dot being placed within the small target and larger surrounding distractor regions of the display. As the display is constructed entirely from one kind of element, the perceptual differences between the regions are

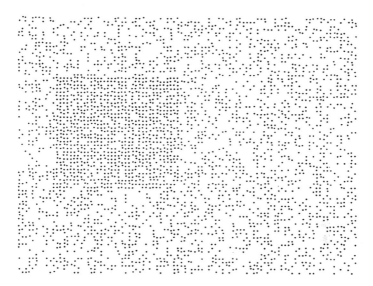

FIG. 14.5.   An example of a texture display differing in dot density. Variations of this type of display was used to train and test the pigeons tested in the third experiment described in this chapter.

determined exclusively by the emergent, global properties of the display (Julesz, 1975). Because local element differences are not present, dot textures have been important in isolating the grouping and integration mechanisms underlying texture perception. As a result, these kinds of displays have attracted a large amount of experimental attention from researchers of human perception (Barlow, 1978; Burgess, Wagner, Jennings, & Barlow, 1981; Julesz, 1975, 1981; Uttal, 1975, 1976).

In the experiments described next, the perception and discrimination of dotted texture displays was examined for the first time in pigeons. The pigeons were first trained to discriminate two types of dotted texture displays and were then tested to see which features of the display mediated this learned discrimination. They were tested using computer-generated random dot stimuli in a two-alternative forced-choice task. In this task, the pigeons were required to peck at the half of the display containing the target region in order to receive food. This target region was randomly located on the left or right half of the display between trials, and consisted of a 6-cm square area of dots that differed from its surrounding dotted context in either dot density or dot spacing.

Dot density displays were produced by either increasing or decreasing the probability of a dot occurring within the target region with respect to the surrounding region of background dots (Fig. 14.5 is an actual dot density display from these experiments). Such displays correspond to the first-order texture differences of Julesz's (1981) scheme for describing the statistical properties of texture displays. The regions of such displays differ on a number of properties, among which include dot density, overall luminance, and the average spacing between the dots. *Dot spacing* displays were produced by varying the average distance between the dots in each region, while holding the average dot density equivalent. The regions of such displays readily appear to the human eye as differing in "clumpiness." These displays correspond to the second-order texture differences of Julesz's scheme.

The pigeons easily acquired this type of discrimination with both kinds of dot displays, quickly learning to peck right or left depending on the target's location. Their rates of task acquisition and final levels of performance indicated that over the range of values tested, differences in dot density were more easily discriminated than those of dot spacing.

Of more specific concern to this chapter, however, were the subsequent experiments examining the role of edgelike discontinuities in these dotted texture discriminations. This was accomplished by varying the feature "profile" of the dotted stimuli through precise variations of dot density over the display. Examples of these profiles are in Fig. 14.6. For example, consider the dot probability profile of a standard dot density display like that in Fig. 14.5. If the plane of the display is imagined as being viewed from the side and in cross section, the dot probability of such a display starts out low and then suddenly increases for 6 cm (defining the target region) and then drops immediately back down to its initial value. The height of these "one-dimensional" profiles directly corresponds to the probability of a dot at any point in the display. A one-dimensional profile for the converse display is also portrayed in the top row of Fig. 14.6.

Below these two baseline dot density trials are the profiles of the test displays

Baseline Dot Density

Low Background Values   High Background Values

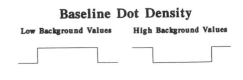

Test Conditions

FIG. 14.6. Target profiles for standard dot density displays and the four different test conditions used to test the contribution of edge-like features to dotted texture discrimination by pigeons.

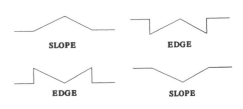

SLOPE   EDGE

EDGE   SLOPE

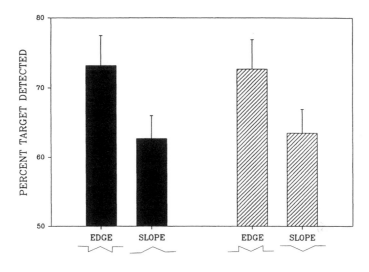

TARGET PROFILE

FIG. 14.7.   Mean target detection accuracy for the four different dotted test conditions displayed in Fig. 14.6.

examined in the next experiment. Because of the dot probability profiles of these displays, the resulting target regions have important differences in their global features. For instance, the profile of the first test display in Fig. 14.6 portrays a display having a target region with no edges. That is, the target region's dot probabilities are gradually increased from those of the background, peaking in the center of the target, and then just as gradually declining on the other side. This kind of profile produced a "pyramid" target of increasing density toward the center. The profile to the immediate right portrays the same target, but placed within a dotted background having an overall higher dot probability. This profile produces a sharp discontinuity at the target/ground border, creating an edge or visual contour around the target. As a result of this design, the presence and absence of edgelike features around what is otherwise the same target area could be changed as a simple function of the surrounding dot density. The lower row of Fig. 14.6 shows the converse set of conditions that also were tested in these experiments.

The outcome of transfer tests with these slope and edge test displays are shown in Fig. 14.7. The results showed that accuracy was clearly superior with the edge test displays. This was true whether the target was increasing or decreasing in its dot density toward the center. The implication of this result is like those of the two previous sections—the presence of implicit edges at the target/ground boundary, created this time by an abrupt difference in dot density, are important to the accurate discrimination of textured regions by pigeons.

SUMMARY AND IMPLICATIONS

All three lines of research converge toward a similar conclusion: The linear discontinuities at the borders of the textured regions are salient features and have a significant role in pigeon texture discrimination. In the first experiment, the linear arrangement of the distractors produced the greatest interference with target detection. In the second, aggregating the elements into a block resulted in the superior discrimination of color mixture rather than random distributions of these same elements. In the third, the presence of sharp dot density transitions forming an edge at the target boundary resulted in better discrimination than when this feature was absent. Hence, it appears from these experiments that the pigeons perceived the linear geometry of the textured regions by visually grouping the separate and disconnected local elements of the display. These data thus join other recent evidence suggesting that pigeons readily perceive the global properties of textured displays (Cook, 1992a, 1992b, 1992c).

Despite the large differences in the size and neural organization of the visual systems of people and pigeons, this line of comparative research suggests that the perceptual mechanisms for processing textured information are similar in both species. My earlier research suggested that the avian visual mechanisms for texture discrimination were capable of rapidly grouping separated local elements into global perceptions, and that this operation was done within, and not between, the color and shape dimensions.

The experiments in this chapter suggest a further similarity can be provisionally added to this description: specifically, that the early visual system of birds also functions to detect and highlight the edges and surfacelike characteristics of visual input. As mentioned earlier, the ability to rapidly generate and interpret edge and regional relations is critical to extracting the contour and surface information needed for discriminating objects within a visual image.

Are the avian rules for perceptually grouping visual information the same as those outlined for humans by modern visual theorists and the Gestalt psychologists? The experiments of this chapter were not specifically designed to evaluate this issue. However, it appears reasonably safe to say from my recent experiments that avian grouping processes are influenced by the proximity and similarity of the different component elements. As both of these factors are well correlated with the surface and contour properties of real objects, it should not be surprising that both species have taken advantage of these properties in representing their visual worlds. Evidence for the more intriguing grouping principles of closure and good continuity is as of yet undiscovered in birds, and await future investigation.

Closely related to the present set of issues is the phenomenon of subjective or illusory contours in human perception (e.g., Kanizsa, 1979). In this phenomenon, contours are perceptually induced in the complete absence of real lines or other luminance differences, indicating that the human visual system can self-generate contours from contextual nonluminance cues. So far, behavioral or neuronal responses indicative of subjective contour perception have been found in humans, monkeys (von der Heydt, Peterhans, & Baumgartner, 1984) and cats (Bravo, Blake, & Morrison, 1988; Orban, De Weerd, & Vandenbussche, 1990), suggesting that such contour-producing mechanisms are present in many mammals. Clearly an interesting area for future research is the examination of whether similar constructive mechanisms also operate in birds.

The multidimensional textures of these experiments are considerably more complex than the vast majority of visual stimuli used so far in testing pigeons. Yet in comparison to the real world they are exceedingly simple, with large regions of uniform, identical elements and large sharp contrasts between the hue, luminance, and shape differences of the regions. Real objects, their contours and surface textures, are considerably more complex and variable. For instance, moving objects constantly change their projected images, or disappear and reappear upon being occluded by other objects. Despite the complex computational problems associated with these realities, the highly mobile behavior of birds is a tantalizing and daily reminder that such visual problems can be readily solved by a nervous system that is small and presumably "simple" in nature.

ACKNOWLEDGMENTS

This research was supported by National Science Foundation Grant BNS–8909040. I would like to thank Kimberley Fulbright, Brian Cavoto, Mark Rilling, and Tom Zentall for their helpful comments on earlier drafts. Request for reprints should be sent to Robert G. Cook, Department of Psychology, Tufts University, Medford, Massachusetts 02155.

REFERENCES

Arbib, M. A., & Hanson, A. R. (1987). *Vision, brain and cooperative computation*. Cambridge, MA: Bradford.

Barlow, H. B. (1978). The efficiency of detecting changes of density in random dot patterns. *Vision Research, 18*, 637–650.

Beck, J. (1966). Effect of orientation and shape similarity on perceptual grouping. *Perception & Psychophysics, 2*, 491–495.

Beck, J. (1982). Textural segmentation. In J. Beck (Ed.), *Organization and representation in perception* (pp. 285–318). Hillsdale, NJ: Lawrence Erlbaum Associates.

Blough, D. S. (1982). Pigeon recognition of letters of the alphabet. *Science, 218*, 397–398.

Blough, D. S. (1985). Discrimination of letters and random dot patterns by pigeons and humans. *Journal of Experimental Psychology: Animal Behavior Processes, 11*, 261–280.

Blough, D. S., & Franklin, J. J. (1985). Pigeon discrimination of letters and other forms in texture displays. *Perception & Psychophysics, 38*, 523–532.

Bravo, M., Blake, R., & Morrison, S. (1988). Cats see subjective contours. *Vision Research, 28*, 861–865.

Broadbent, D. E. (1977). The hidden preattentive processes. *American Psychologist, 32,* 109–118.

Burgess, A. E., Wagner, R. F., Jennings, R. J., & Barlow, H. B. (1981). Efficiency of human visual signal discrimination. *Science, 214,* 93–94.

Cave, K. R., & Wolfe, J. M. (1990). Modeling the role of parallel processing in visual search. *Cognitive Psychology, 22,* 225–271.

Cohen, P. R., & Feigenbaum, E. A. (1982). *The handbook of artificial intelligence* (Vol. 3). Reading, MA: Addison-Wesley.

Cook, R. G. (1983). *The effects of compound stimulus structure on dimensional discrimination in pigeons and humans.* Unpublished doctoral dissertation, University of California, Berkeley.

Cook, R. G. (1992a). The acquisition and transfer of visual texture discriminations by pigeons. *Journal of Experimental Psychology: Animal Behavior Processes, 18,* 341–353.

Cook, R. G. (1992b). Dimensional organization and texture discrimination in pigeons. *Journal of Experimental Psychology: Animal Behavior Processes, 18,* 354–363.

Cook, R. G. (1992c). The visual perception and processing of textures by pigeons. In W. K. Honig & G. Fetterman (Eds.), *Cognitive aspects of stimulus control* (pp. 279–299). Hillsdale, NJ: Lawrence Erlbaum Associates.

Cook, R. G., Riley, D. A., & Brown, M. F. (1992). Spatial and configural factors in compound stimulus processing by pigeons. *Animal Learning & Behavior, 20,* 41–55.

Cook, R. G., Wright, A. A., & Kendrick, D. F. (1990). Visual categorization in pigeons. In M. L. Commons, R. Herrnstein, S. M. Kosslyn, & D. B. Mumford (Eds.), *Quantitative analyses of behavior: Behavioral approaches to pattern recognition and concept formation* (pp. 187–214). Hillsdale, NJ: Lawrence Erlbaum Associates.

Donovan, W. J. (1978). Structure and function of the pigeon visual system, *Physiological Psychology, 6,* 403–437.

Fischler, M. A., & Firschein, O. (1987). *Readings in computer vision.* Los Altos, CA: Morgan Kaufmann.

Fujita, K., Blough, D. S., & Blough, P. M. (1991). Pigeons see the ponzo illusion. *Animal Learning & Behavior, 3,* 283–293.

Grossberg, S., & Mingolla, E. (1985). Neural dynamics of perceptual grouping: Textures, boundaries, and emergent segmentations. *Perception & Psychophysics, 38,* 141–171.

Hearst, E., & Jenkins, H. M. (1974). *Sign tracking: The stimulus-reinforcer relation and directed action.* Monograph of the Psychonomics Society, Austin, TX.

Hoffman, J. E. (1979). A two-stage model of visual search. *Perception & Psychophysics, 4,* 319–327.

Honig, W. K. (1991). Discrimination by pigeons of mixture and uniformity in arrays of stimulus elements. *Journal of Experimental Psychology: Animal Behavior Processes, 17,* 68–80.

Jenkins, H. M., & Sainsbury, R. S. (1970). Discrimination learning with the distinctive feature on positive or negative trials. In D. Mostofsky (Ed.), *Attention: Contemporary theory and analysis* (pp. 239–274). New York: Appleton-Century-Crofts.

Julesz, B. (1975). Experiments in the visual perception of texture. *Scientific American, 212,* 34–43.

Julesz, B. (1981). Textons, the elements of texture perception and their interactions. *Nature, 290,* 91–97.

Kanizsa, G. (1979). *Organization in vision.* New York: Praeger.

Kohler, W. (1947). *Gestalt psychology.* New York: Liveright.

Lamb, M. R., & Riley, D. A. (1981). Effects of element arrangement on the processing of compound stimuli in pigeons (*Columba livia*). *Journal of Experimental Psychology: Animal Behavior Processes, 7,* 45–58.

Landy, M. S., & Movshon, J. A. (1991). *Computational models of visual processing.* Cambridge, MA: Bradford.

Leith, C. R., & Maki, W. S., Jr. (1975). Attention shifts during matching-to-sample performance in pigeons. *Animal Learning & Behavior, 3,* 85–89.

Marr, D. (1982). *Vision.* San Francisco: Freeman.

Neisser, U. (1967). *Cognitive psychology.* New York: Appleton-Century-Crofts.

Orban, G. A., De Weerd, P., & Vandenbussche, E. (1990). Cats discriminate orientations of illusory contours. In W. C. Stebbins & M. A. Berkley (Eds.), *Comparative perception: Complex signals* (pp. 157–186). New York: Wiley.

Pearson, R. (1972). *The avian brain.* New York: Academic.

Postman, L., & Riley, D. A. (1959). Degree of learning and interserial interference in retention. *University of California Publications in Psychology, 8,* 231–396.

Riley, D. A. (1962). Memory for form. In L. Postman (Ed.), *Psychology in the making* (pp. 402–465). New York: Knopf.

Riley, D. A. (1965). Stimulus generalization and transposition. *Gawein, 13,* 301–311.

Riley, D. A. (1968). *Discrimination learning.* Boston: Allyn & Bacon.

Riley, D. A., Cook, R. G., & Lamb, M. R. (1981). Classification and analysis of short term retention codes in pigeons. In G. H. Bower (Ed.), *The psychology of learning and motivation: Advances in research and theory* (pp. 51–79). New York: Academic Press.

Riley, D. A., Goggin, J. P., & Wright, D. C. (1963). Training level and cue separation as determiners of transposition and retention in rats. *Journal of Comparative and Physiological Psychology, 56,* 1044–1049.

Riley, D. A., & Leith, C. R. (1976). Multidimensional psychophysics and selective attention in animals. *Psychological Bulletin, 83,* 138–160.

Riley, D. A., & McKee, J. P. (1963). Pitch and loudness transposition in children and adults. *Child Development, 34,* 471–482.

Riley, D. A., & Roitblat, H. L. (1978). Selective attention and related cognitive processes in pigeons. In S. H. Hulse, H. Fowler, & W. K. Honig (Eds.), *Cognitive processes in animal behavior* (pp. 249–276). Hillsdale, NJ: Lawrence Erlbaum Associates.

Riley, D. A., & Rosenzweig, M. R. (1957). Echolocation in rats. *Journal of Comparative and Physiological Psychology, 50,* 323–328.

Treisman, A. (1986). Preattentive processing in vision. In A. Rosenfeld (Ed.), *Human and machine vision II* (pp. 313–334). New York: Academic.

Treisman, A., & Gelade, G. (1980). A feature-integration theory of attention. *Cognitive Psychology, 12,* 97–136.

Uttal, W. R. (1975). *An autocorrelation theory of form detection.* Hillsdale, NJ: Lawrence Erlbaum Associates.

Uttal, W. R. (1976). Visual spatial interactions between dotted line segments. *Vision Research, 16,* 581–586.

Uttal, W. R. (1981). *A taxonomy of visual processes.* Hillsdale, NJ: Lawrence Erlbaum Associates.

von der Heydt, R., Peterhans, E., & Baumgartner, G. (1984). Illusory contours and cortical neuron responses. *Science, 22,* 1260–1261.

Warren, L. R., Riley, D. A., & Scheier, M. F. (1974). Color changes of *Octopus rubescens* during attacks on unconditioned and conditioned stimuli. *Animal Behavior, 22,* 211–219.

Watt, R. (1988). *Visual processing.* London: Lawrence Erlbaum Associates.

15 Multidimensional Stimulus Control in Pigeons: Selective Attention and Other Issues

Diane L. Chatlosh
California State University, Chico

Edward A. Wasserman
The University of Iowa

Studies of discrimination learning have repeatedly demonstrated that differential responding in the presence of various compound stimuli does not guarantee that all aspects of the stimuli actually exercise control over behavior. Instead, stimulus control is sometimes selective, and the contributions of the individual components of the compounds to ultimate stimulus control are interdependent. A primary goal of discrimination-learning theories has been to predict which, to what extent, and under what conditions the various elements of compound stimuli will gain behavioral control.

Explanations of selective stimulus control often invoke attentional mechanisms. The fundamental assumption here is that organisms are limited-capacity processors of information (Broadbent, 1958; Riley, 1984; Sutherland & Mackintosh, 1971). When an organism is overloaded with information, it may not attend to all of the potential cues present in its environment; rather, its behavior may reflect the selective processing of only a subset of the available stimuli. Furthermore, according to the inverse hypothesis (Thomas, 1970), a competitive relationship may exist among stimuli, such that enhanced attention to some may be accompanied by reduced attention to others.

In spite of the hypothesis that selective attention is due to an overload of information, nearly all of the data pertaining to selective stimulus control have been obtained with discrimination tasks in which only two cues were experimentally manipulated. Certainly, it is highly probable that an organism's maximum processing capacity exceeds two bits of information. Indeed, Fink and Patton (1953) reported that rats are sensitive to changes in three stimulus dimensions and Nissen (1951) trained a chimp to make conditional discriminations based on five stimulus dimensions. Pigeons also have demonstrated skills that would seem to be much more cognitively demanding than two-dimensional compound discriminations, such as categorization (e.g., Bhatt, Wasserman, Reynolds, & Knauss, 1988; Cerella, 1979; Herrnstein, 1979; Herrnstein, Loveland, & Cable, 1976; Wasserman, Kiedinger, & Bhatt, 1988) and long-term memory for complex response sequences (Schwartz & Reilly, 1985) and for several hundred visual stimuli (Vaughan & Greene, 1984). Yet, only fairly recently has it been acknowledged that "experimenters attempting to investigate attention in animals have usually failed to structure their experiments to provide any *overload* of the animal's ability to take in all the available information within the allowed time" (Riley & Roitblat, 1978, pp. 249–250; also see Thomas, 1970). It would seem that precise specification of the conditions producing stimulus selection effects could be better accomplished with discrimination tasks that tax animals' capabilities; indeed, without a sufficiently demanding discrimination task, there is little reason to expect tradeoffs in the attention paid to the individual elements of compound stimuli.

An additional problem of prior investigations of stimulus selection pertains to the various tests employed to assess stimulus control. Usually, the tests are conducted after discrimination training and consist either of nonreinforced presentations of stimuli (e.g., Farthing & Hearst, 1970; Johnson & Cumming, 1968; Kamin, 1968; Reynolds, 1961; Seraganian & vom Saal, 1969; vom Saal & Jenkins, 1970) or of nondifferentially reinforced presentations of stimuli (e.g., Mackintosh, 1965; Sutherland & Holgate, 1966). Not only can different test procedures give rise to conflicting conclusions regarding stimulus control (see Wilkie & Masson, 1976), but also, one cannot be certain that postdiscrimination tests accurately reflect prior stimulus control (Johnson & Cumming, 1968). The latter concern cannot be resolved with standard discrimination procedures, because during training the measurement of stimulus control by each of the individual elements of the compounds is completely confounded. For example, with a typical compound discrimination task, the positive discriminative stimulus might be a red triangle and the negative discriminative stimulus might be a green circle. The discrimination can be solved by attending only to color, only to shape, or to both color and shape. Even if an animal comes to respond exclusively in the presence of the red triangle, we do not know whether one or both of the compound stimulus dimensions (color and shape) are controlling its behavior or whether each dimension is controlling a different aspect of the response (Jenkins, 1965). Thus, given this kind of discrimination procedure, the only way to compare the two dimensions with respect to stimulus control is to obtain postdiscrimination data.

A final shortcoming of most earlier research concerns the restricted choice of stimuli used for studying stimulus selection phenomena. Prior investigations of attention in instrumental discrimination learning have relied almost exclusively on visually presented elements and compounds that vary along static dimensions such as color, line orientation, or intensity. Whether and how the specific dimensions of cues influence attention are empirical questions that have been largely ignored. Factors that could produce stimulus dimension effects include innate attending hierarchies (Baron, 1965; cf. Kraemer & Roberts, 1987), the discriminability of stimulus values along a dimension (Johnson & Cumming, 1968), and the dimension of the other element(s) of a compound (DeLong & Wasserman, 1985).

An additional dimensional factor that could influence which stimuli contribute to behavioral control was suggested in an experiment that combined static visual cues with stimulus duration cues. DeLong and Wasserman (1985) found that although stimulus duration gained control over pigeons' responding in a simple discrimination task (also see Chatlosh & Wasserman, 1987), in a subsequent compound discrimination task in which stimulus duration and line orientation were equally reliable and redundant cues for reinforcement, there was a tendency for duration to lose some of its control over responding. DeLong and Wasserman pointed out that stimulus durations, unlike static visual stimuli, are cumulative, and their discrimination may require a memory process. Because of this cumulative property, the discrimination of stimulus durations may take more processing time than the discrimination of values along static stimulus dimensions. So, perhaps when other equally informative static cues that require less processing time are available, animals use them to guide their behavior, instead of the duration cues. The implication of these conjectures is that differences may exist among stimulus interactions depending on whether the cues are static or cumulative.

The present project was designed to circumvent the procedural problems of prior studies of selective attention. First, we attempted, if not to overload, certainly to strain pigeons' information-processing capacities by training the birds on a new discrimination task for which information about each of three stimulus dimensions was necessary for discriminative performance. Second, the nature of this procedure made it possible to obtain relative indices of discriminative control for each of the three independent stimulus dimensions as that control developed (i.e., during discrimination training). And third, because of the recent surge of interest in temporal control (e.g., Gibbon & Allan, 1984; Gibbon & Church, 1984), the paucity of selective attention research with temporal cues, and the suggestion that duration may function differently as a discriminative stimulus from static discriminative stimuli (DeLong & Wasserman, 1985), we used duration as one of the dimensions on which compound stimuli could vary.

In a procedure inspired by Heinemann and Chase (1970), pigeons were trained with multiple necessary cues (MNC). Eight line-color-time compounds were each defined by one of two values along each of the three dimensions. Specifically, each compound stimulus consisted of either a horizontal or a slanted line, presented on a blue or a purple background, with a duration of either 4 or 16 s. Pecking a red test stimulus was reinforced only if it had been preceded by a unique line-color-time compound. Thus, the task discouraged subjects from responding after the seven negative compound stimuli, a feat that requires the processing of information about all three independent stimulus dimensions (cf. Blough's two-dimensional discrimination tasks, 1969, 1972; Bourne's discussion of the "conjunctive rule" in the context of concept learning in humans, 1970; Lea & Harrison's research with "polymorphous stimulus sets," 1978). At issue was: (a) whether pigeons would ultimately learn to base their responding on information from all three stimulus dimensions, (b) whether pigeons would acquire control by these dimensions at different rates, and (c) whether the acquisition of control by one dimension would be accompanied by a reduction in control by one or more other dimensions, as the inverse hypothesis suggests.

METHOD

Subjects

Eight adult feral pigeons with prior experience in an unrelated key-pecking task were used. The birds were individually housed in a room with constant lighting, where they were given free access to water and daily provisions of grit. A vitamin supplement was added to the water supply two to three times a week. The subjects were maintained at approximately 85% of their ad lib weights by adjusting daily postsession food rations.

Apparatus

Four standard pigeon chambers were used (see DeLong & Wasserman, 1981, for specifications). Each box contained three clear Plexiglas keys, but only the center key was operative. Stimuli were presented via miniature readout projectors that were mounted behind the key. Six different stimuli were used, alone and/or in compounds: a yellow field, a red field, a blue field, a purple field, a white horizontal line, and a white line slanted 30° from vertical. When presented alone, each line was centered on a black field; when presented in compound, each was superimposed on the center of either a blue or a purple field.

Procedure

Pretraining. To minimize any response biases produced by prior key-pecking experience, all subjects were initially given five daily sessions of equivalence pretraining to the elemental stimuli. Each session entailed 48 trials; during each trial, one of the six single stimuli was equiprobably presented. When a subject met a specified response requirement, the stimulus was immediately terminated and the grain hopper was raised; if the response requirement was not met within 10 s, the stimulus was automatically terminated and the grain hopper was raised. Across the five sessions, the response requirement was increased from three to five key pecks and the hopper duration was gradually decreased from 4.5 to 3.0 s. By the end of pretraining, all but one of the birds met the response requirement on all trials (one bird failed to complete one trial), and no systematic differences in the time taken to complete each of the six trial types (corresponding to the six individual stimuli) were apparent.

MNC Training.   Each of 96 daily trials began with a variable inter-stimulus interval (ITI) averaging 20 s. Any response made during an ITI triggered the onset of another ITI; this cycle continued until an ITI elapsed without any responses. Next, a yellow (Y) stimulus was presented until a single key peck terminated it. This response requirement, to the yellow presample stimulus, ensured that subjects would be observing the key at the onset of the succeeding compound stimulus. One of eight line-color-time compounds was then imme-diately presented. These three-dimensional compounds were formed by superim-posing the horizontal- (H) or slanted-line (S) stimulus on either a blue (B) or purple (P) field, and by presenting them for either 4 (4) or 16 s (16). For each bird, a different, unique compound was positive (e.g., HB4), whereas all others were negative (e.g., SB4; HP4; SP4; HB16; SB16; HP16; SP16). A compound was automatically followed by a red (R) stimulus which, on negative trials, lasted 5 s and was followed by a 3-s blackout. On positive trials, the first key peck after 5 s extinguished R and produced 3-s access to food. Each of 100 daily sessions consisted of 12 randomized blocks of the eight trial types.

Behavioral Measures.   For each daily session, the numbers of key pecks during the first 5 s of R following each type of compound stimulus were re-corded. Response rates (per second) were obtained for successive 4-day blocks of training. In addition, separate discrimination ratios were calculated for each stimulus dimension, in 4-day blocks of training; this calculation was accom-plished by dividing the number of responses made during compound trials that entailed the positive value for a given stimulus dimension by the total number of responses made during all compound trials. Thus, for an individual subject, the denominator of the ratios remained constant for all three stimulus dimensions, but the numerators for each dimension ratio were obtained by summing response totals across different combinations of trial types. For example, if HB4 trials were positive, the numerators for each of the separate dimension ratios included response totals from the following combinations of trial types: HB4 + HP4 + HB16 + HP16 (line); HB4 + SB4 + HB16 + SB16 (color); HB4 + SB4 + HP4 + SP4 (time); in all cases, the denominator was the overall response output on all trials (HB4 + SB4 + HP4 + SP4 + HB16 + SB16 + HP16 + SP16). These three discrimination scores—which could range from .50 (indiscriminate perfor-mance) to 1.00 (perfect discrimination)—were used separately to assess the degree to which the stimuli from each of the dimensions controlled the pigeons' behavior.

RESULTS

Group Data

Group acquisition data are presented in Fig. 15.1. This figure illustrates the median number of 4-day training blocks needed for the individual line, color, and time discrimination ratios, and for all three discrimination ratios in the same block, to first meet successively stringent criteria for stimulus control (medians, rather than means, were plotted because not all birds reached the .90 criterion for all three stimulus dimensions). For each of the individual stimulus dimensions, performance became more discriminative as training progressed. Moreover, by the end of training, performance levels were high (discrimination ratios > .80) for each of the stimulus dimensions. Simultaneous discriminative control by all three stimulus dimensions in a single training block also increased as training progressed.

In addition, Fig. 15.1 shows that the speed of discrimination learning was not the same for each of the stimulus dimensions; rather, control by color emerged first, then control by time, and finally control by line (although not included here, a plot of mean discrimination ratios across blocks of training shows the same order of emerging control). The general order of the learning rates (color > time > line) was confirmed using two-tailed Mann–Whitney tests (Siegel, 1956). Specifically, fewer training blocks were required to reach the .6 criterion by color or time than by line ($U = 7.0$, $p < .01$ and $U = 11.0$, $p < .05$, respectively); fewer training blocks were required to reach the .7 criterion by color than by line ($U = 2.0$, $p < .001$) or time ($U = 8.0$, $p < .01$) and by time than by line ($U = 6.0$, $p < .005$); fewer training blocks were required to reach the .8 criterion by color than by line ($U = 1.5$, $p < .001$) or time ($U = 8.0$, $p < .01$); fewer training blocks were required to reach the .9 criterion by color than by line ($U = 4.0$, $p < .005$) or time ($U = 12.5$, $p < .05$).

Lest there be any concern about the discrimination ratios we used to represent dimensional control, we also examined subjects' absolute response rates during the tests that followed the various compound stimuli. Fig. 15.2 depicts the mean number of test key pecks per second for each of the eight line-color-time (LCT) compounds during the initial eight blocks of training; each element of the compounds is labeled positive (+) or negative (−), depending on whether its value was the same as or different from the value included in the uniquely positive compound (L+C+T+).

Clearly, the L+C+T+ compound maintained the highest level of responding across training blocks, although there was a slight decline in the response rate

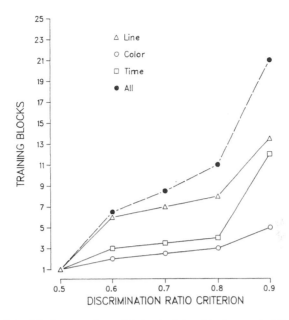

FIG. 15.1.   Median number of 4-day training blocks needed for discrimination ratios to first meet successively stringent criteria for stimulus control (*N* = 8). Data for each dimension individually are represented with open symbols: line (triangles), color (circles), time (squares). The filled symbols indicate when performance first met the successive criteria for all three dimensions in the same block.

after Block 4. In contrast, the response rates for all seven of the negative stimulus compounds quickly declined across training blocks, and to much lower levels than did responding for the positive compound stimulus. The fact that the differences between the response rates on L+C+T+ trials and on each type of negative compound trial tended to enlarge as training progressed, once again suggests that discriminative control by all three stimulus dimensions gradually emerged during acquisition.

Yet, subjects did not respond equally to all of the negative compounds. Note that the composition of the seven negative compounds differed in terms of the proportion of positive and negative elements. Three compounds entailed two positive elements and one negative element (L+C+T−, L+C−T+, and L−C+T+); three compounds entailed one positive element and two negative elements (L+C−T−, L−C+T−, and L−C−T+); and one compound entailed three negative elements (L−C−T−). If the line, color, and time discriminations were equally difficult, one would expect performance to be similar among compounds with equal numbers of positive and negative elements. With one possible exception (i.e., L−C+T−), there were few performance differences among the

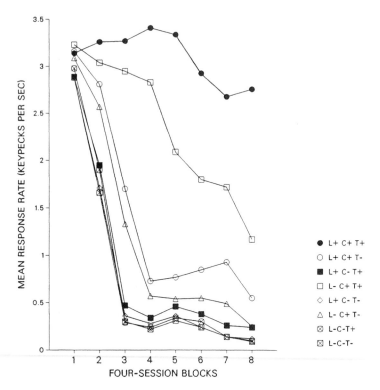

FIG. 15.2.   Mean test response rates (per second) for each of the eight line-color-time (LCT) compounds across successive 4-day training blocks ($N = 8$). Positive (+) and negative (−) signs indicate the stimulus value of each element of the compounds (see text for further explanation).

compounds with only one positive element; responding declined at similar rates. The general correspondence among these functions and their resemblance to the $L-C-T-$ function suggest that any clear evidence of dimensional differences would have to come from an examination of performance on compound trials with only one negative element. Responding on $L+C-T+$ trials decreased more rapidly than did responding on either $L-C+T+$ trials or $L+C+T-$ trials. In fact, the response decrement observed when only the color element was negative was very similar to that observed when two or three elements were negative. These comparisons suggest that among the three dimensions, differential responding developed most quickly for color. The previously noted exception among the compounds with only one positive element is consistent with this conclusion. Although the performance differences were not very pronounced, when color was the only positive element, response rates were somewhat higher than when either line or time was the only positive element. In addition, when only the time element was negative, responding declined faster than when only the line element was negative, suggesting that the time discrimination was more rapidly acquired than the line discrimination.

To better trace the development of stimulus control, the response rates for each training block represented in Fig. 15.2 were subjected to a $2 \times 2 \times 2$ factorial analysis of variance. The three factors were line, color, and time dimensions, each with a positive and a negative stimulus value. Thus, a main effect indicates differential responding along an individual stimulus dimension, whereas an interaction indicates conjoint discriminative control by two or three stimulus dimensions (c.f. the hierarchical analytic scheme of Roitblat, Scopatz, & Bever, 1987). Of particular interest is the order in which the various discrimination effects emerged. The main effect of color was statistically significant as early as Block 1 [$F(1,7) = 7.92, p < .05$]. Time was the next reliable main effect to appear, first in Block 2 [$F(1,7) = 17.55, p < .005$]. And, the main effect of line was not statistically significant until Block 3 [$F(1,7) = 5.99, p < .05$].

Conjoint control by two or more stimulus dimensions developed somewhat more slowly than did control by the individual stimulus dimensions. Although a reliable line $\times$ time interaction was found in Block 1 [$F(1,7) = 7.43, p < .05$], it did not recur until Block 6 [$F(1,7) = 10.13, p < .05$]; the color $\times$ time interaction was first reliable in Block 3 [$F(1,7) = 34.08, p < .001$]; and the line $\times$ color interaction was not statistically significant until Block 6 [($F(1,7) = 7.64, p < .05$]. Finally, it was not until Block 8 that the line $\times$ color $\times$ time interaction was statistically significant, demonstrating conjoint control by all three stimulus dimensions [$F(1,7) = 7.97, p < .05$].

In Fig. 15.3, the response rate data are reorganized to more clearly depict the differences among the individual stimulus dimensions. For each dimension, mean response rates are separately plotted for compounds that included the positive stimulus value and for those that included the negative stimulus value. For example, the L+ function represents performance on all of the compound trials that included the positive line value [i.e, (L+C+T+) + (L+C+T−) + (L+C−T+) + (L+C−T−)], whereas the L− function represents performance on all of the compound trials that included the negative line value [i.e., (L−C+T+) + (L−C+T−) + (L−C−T+) + (L−C−T−)].

Not surprisingly, all of the response rate functions in Fig. 15.3 declined as training progressed; after all, none of the trials included in the L−, C−, or T− functions entailed the positive compound and only one fourth of the trials in the L+, C+, and T+ functions did so. More important is the degree of differential responding within each stimulus dimension. The increasing divergence of the L+ and L− functions, the C+ and C− functions, and the T+ and T− functions, which occurred within the first few blocks, demonstrates that discriminative performance improved as training progressed. Furthermore, this figure clearly illustrates the order in which the individual dimensions gained behavioral control. Differential responding developed most quickly along the color dimension,

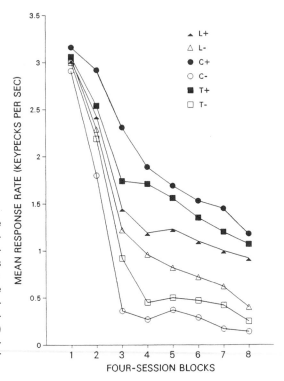

FIG. 15.3. Mean test response rates (per second) for compounds that included the positive (+) or negative (−) stimulus value (+ = closed symbols; − = open symbols) along the line (L), color (C), and time (T) dimensions (line = triangles; color = circles; time = squares) across successive 4-day training blocks (see text for further explanation).

somewhat less quickly along the time dimension, and most slowly along the line dimension.

In addition, to assess further the relative contributions of line, color, and time to stimulus control, we examined subjects' mean response rates across the initial eight blocks of training. These means were 3.10, 1.44, .87, 2.35, .76, 1.17, .79, and .73 for the L+C+T+, L+C+T−, L+C−T+, L−C+T+, L+C−T−, L−C+T−, L−C−T+, and L−C−T− stimulus compounds, respectively. Three sets of pairwise difference scores were used to compare the relative contributions of each dimension to the overall response rate. For example, the differences between mean response rates on L−C−T− trials and on L+C−T− trials (.03), L−C+T− trials (.44), and L−C−T+ trials (.06) reflect the relative contributions of line, color, and time, respectively, to the overall response rate on trials with a single positive stimulus value. Similarly, the average differences between response rates on trials with two positive stimulus values and one positive stimulus value [e.g., (L+C+T−) − (L−C+T−) and (L+C−T+) − (L−C−T+) for line] reflect the relative contributions of line (.18), color (1.12), and time (.64) to the overall response rate on trials with two positive stimulus values. Finally, the differences between response rates on L+C+T+ trials and on L−C+T+ (.75), L+C−T+ (2.23), and L+C+T− (1.66) trials, reflect the relative contributions of each dimension to the overall response rate on trials with three positive stimulus values. Notice that, in all three sets of comparisons, the difference scores were similarly ordered: Control by color was the greatest, followed by control by time, and then control by line. Moreover, the relative contributions of each dimension to stimulus control rose as the number of other positive stimulus values increased, suggesting a multiplicative relationship among the dimensions.

A series of computer simulations resulted in the following mathematical model, which defines the observed relationship among the stimulus dimensions: Mean response rate = a (L × C × T) + b. The parameters L, C, and T correspond to the line, color, and time dimensions, respectively. When the positive stimulus value for a dimension is presented, its corresponding parameter is set equal to 1.0. However, due to discriminability differences, the dimensional parameters diverge for negative stimulus values. Specifically, L = .669, C = .074, and T = .308 for compound stimuli that include the negative line, color, and/or time value. When the remaining parameters, a and b, are set equal to 2.428 and .690, respectively, the fit between the predicted response rates (3.12, 1.44, .87, 2.31, .74, 1.19, .81, and .73) and the obtained response rates (see aforementioned) is quite impressive. Notably, this multiplicative model accounts for a statistically significant portion of the observed variability in responding above that accounted for by an additive model.

Individual Data

Single-subject acquisition data are presented in Fig. 15.4. This figure separately depicts each subject's discriminative performance for all three dimensions in the 4-day training block on which the line discrimination first met the .70 criterion (Block 0); the three blocks prior to criterion (Blocks −3, −2, and −1); and the block following criterion (Block 1). For seven of the eight birds, as discriminative performance for line improved, discriminative performance for color and/or time worsened, at least temporarily. Birds 2, 3, 6, and 8 manifested decrements in both color and time discriminations during the acquisition of the line discrimination. However, by the end of the block following criterion, Bird 2 had already begun to recover both discriminations; Bird 3 exhibited only slight increments in both discriminations; Bird 6 had not begun to recover either discrimination; and Bird 8 showed a very slight increment in color discrimination only. Birds 1 and 4 showed a decrement in time discrimination only as line discrimination improved. Whereas Bird 1 quickly began to recover the time discrimination, Bird 4 showed no sign of recovery within the training period depicted in Fig. 15.4. Bird 7

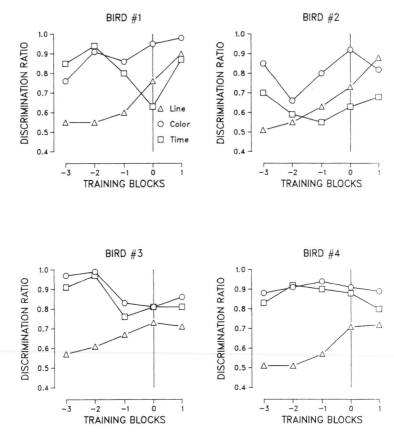

FIG. 15.4.   Single-subject discrimination ratios for each of the three dimensions (line = triangles; color = circles; time = squares) across successive 4-day training blocks. The block on which the line discrimination ratio first met the .70 criterion is designated Block 0 (Blocks 7, 7, 15, 18, 3, 8, 5, and 5 for Birds 1 through 8, respectively). Blocks −3, −2, −1, and 1 refer, respectively, to the three blocks prior to meeting this criterion and the block following criterion.

evidenced a decrement in color discrimination only during the acquisition of the line discrimination and no recovery until later in training. The performance of Bird 5 was atypical in that the emergence of discriminative control by line was not accompanied by a loss of discriminative control by either color or time; this bird also acquired the three-dimensional discrimination faster than any of the other subjects.

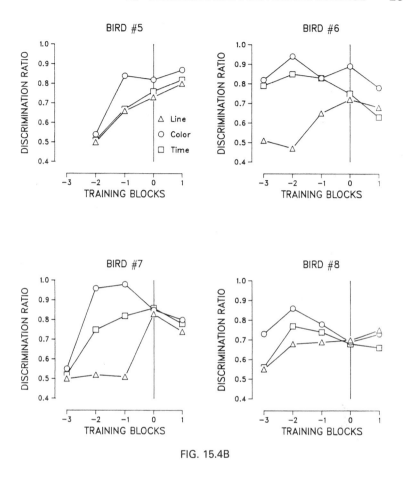

FIG. 15.4B

DISCUSSION

We examined pigeons' discriminative performance on a multiple necessary cue (MNC) task in which the compound stimuli that were presented varied along three separate dimensions. Typically, the compound stimuli that are used in research of attentional processes vary along only two dimensions and, usually, both dimensions are visual. In contrast, one of the three stimulus dimensions we manipulated was nonvisual (i.e., duration). Moreover, because discriminative performance on the MNC task required attention to all three dimensions (i.e., the cues were nonredundant), we were able to make relative comparisons of dimensional control during, rather than after, discrimination training. Because of these features, this project distinctively addresses several recurring issues in compound stimulus control and selective attention.

First, our data suggest that pigeons are very adept processors of information. Specifically, subjects ultimately did learn to base their key-pecking behavior on information about all three features of the compound stimuli. During acquisition, differential response rates for the positive compound stimulus and each of the seven negative compound stimuli gradually emerged (Fig. 15.2). And by the end of training, separate line, color, and time discrimination ratios all reached very high levels (Fig. 15.1). Yet, to demonstrate attention to all three dimensions requires that, over some period of time, the pigeons' behavior is controlled by all of the dimensions in question. The fact that, as training progressed, the pigeons met successively stringent criteria for control by all three dimensions in a single training block, provides more persuasive evidence for the processing of information about multiple stimulus dimensions (Fig. 15.1). However, one still might argue that the pigeons might have attended to one, two, or three dimensions on different trials within a training block.

To address this issue more directly, we inspected the individual records of subjects' behavior during the final 12 daily sessions. Within these 12 sessions, there were 144 cycles through the eight line-color-time compound stimuli. By setting a .95 confidence limit on the rate of key pecking on positive trials, it was possible to determine the proportion of cycles on which subjects demonstrated behavioral control by one, two, or three stimulus dimensions, and in the former two cases, to specify which dimension(s) controlled responding.

Representative single-session data from a subject trained with SB16 as the positive compound stimulus are presented in Table 15.1. During the last 12 sessions, this bird pecked the test stimulus seven or more times on 95% of the positive trials; thus, seven or more key pecks represented a "positive" report response (and 6 or fewer key pecks represented a "negative" report response). By examining the pattern of positive report responses, we were able to infer which dimension(s) of the compound stimuli received the subject's attention on each eight-trial cycle. For example, during the session shown in Table 15.1, on 10 of the 12 cycles (2, 4, 5, 6, 7, 8, 9, 10, 11, and 12) Bird 6 responded positively only on the trial with the positive compound stimulus, demonstrating discriminative control by all three stimulus dimensions. However, on the other two cycles (1 and 3), the bird's behavior was controlled by just two of the stimulus dimensions. On Cycle 1, this subject correctly responded "positive" on the SB16 trial, but incorrectly responded "positive" on the SB4 trial, demonstrating discriminative control by line and color, but not by time. On Cycle 3, Bird 6 again correctly responded "positive" on the SB16 trial, but now incorrectly responded "positive" on the SP16 trial, demonstrating discriminative control by line and time, but not by color. Thus, whenever subjects responded "positive" on compound trials that

TABLE 15.1
Single Session Test Keypecks (Bird 6, Day 94)

Randomized Cycles of 8 Trials	Individual Trials								Dimensional Control[a]
	HB4–	SB4–	HP4–	SP4–	HB16–	SB16+	HP16–	SP16–	
1	0	10	0	0	0	11	1	0	LC
2	2	2	0	0	0	8	0	1	LCT
3	0	1	0	0	0	9	1	8	LT
4	0	1	0	0	0	11	1	0	LCT
5	0	1	0	2	0	8	0	0	LCT
6	0	1	0	0	0	15	0	0	LCT
7	0	1	0	0	0	11	0	0	LCT
8	0	1	0	0	0	18	0	0	LCT
9	0	2	0	0	0	10	2	0	LCT
10	2	2	0	0	0	9	0	0	LCT
11	0	1	0	0	0	11	0	0	LCT
12	0	1	1	0	0	18	0	0	LCT
Total	4	24	1	2	0	139	5	9	

Note. .95 confidence limit on the number of keypecks on positive trials for the last 12 sessions $\geq$ 7 keypecks.
[a] L = Line, C = Color, T = Time

285

included either the positive or the negative value of a particular dimension, that dimension was assumed to have exerted no discriminative control during the cycle.

Over all eight subjects, 89.8% of the cycles from the final 12 sessions produced readily classifiable response patterns conforming to control by L, C, T, LC, LT, CT, or LCT (on 3.5% of the remaining cycles, the subject failed to respond "positive" on the trial with the positive compound stimulus; these trials were omitted from consideration because interpretational problems arise for cases in which a "positive" response is made exclusively on a trial with the negative value of a particular dimension—e.g., see Sidman, 1980; the residual 6.7% of the trials entailed uninterpretable patterns of responding). Using this procedure to infer attentional patterns, we found that over all subjects, 86.4% of the classifiable cycles involved attention to all three stimulus dimensions, 12.6% involved attention to two stimulus dimensions, and 1.0% involved attention to just one stimulus dimension. Therefore, on the vast majority of cycles (89.8% × 86.4% = 77.6% of all cycles), performance was controlled by all three stimulus dimensions.

Moreover, because the criterion for discriminative control by all three dimensions required a high relative response rate on the positive compound trial of a cycle and low relative response rates on all seven negative compound trials, we propose that the pigeons attended to all three stimulus dimensions, not only within most cycles, but also on most individual trials. Examination of single-subject data further substantiates this contention. Again setting a .95 confidence limit on the rate of key pecking on positive trials, we determined the proportion of all cycles during the final 12 sessions on which individual pigeons pecked on the positive compound trial more than on all seven negative compound trials. For the eight pigeons, the proportions were: .91, .94, .71, .33, .92, .85, .92, and .62. These scores greatly exceed the proportion expected by chance (7.42 × 10^{-10}). Thus, our data strongly suggest that pigeons can process information about three separate features of a compound stimulus on a single trial. It should be noted, however, that this conclusion does not necessarily imply that attention to multiple stimulus features occurs simultaneously; it is still possible that subjects shift their attention from one feature of a compound stimulus to another within a single presentation (Riley, 1984).

Our second conclusion is that learning about each of the stimulus dimensions did not occur at the same rate; rather, control by color emerged first, then control by time, and finally control by line (Figs. 15.1, 15.2, and 15.3). The reasons for this particular ordering are not yet certain (see the previous discussion of alternative explanations for differences among stimulus dimensions). However, our results do suggest that, despite the cumulative nature of stimulus duration cues, they function much like static visual cues in controlling behavior (cf. Bowers & Richards, 1990). Temporal control was acquired at a speed intermediate to the two visual dimensions and, for the most part, suffered a loss similar to that of color during acquisition of the more difficult line discrimination (Fig. 15.4). Thus, the cumulative or static nature of the stimulus dimension, per se, was not exclusively responsible for the rate of learning. Whether and how the nature of the stimulus dimension might interact with other factors—such as the ease of discriminating the particular values along each dimension—to determine relative acquisition rates, will be interesting issues to pursue in future investigations.

In any case, the mere fact that the speed of acquisition did vary among the dimensions implies that the birds decomposed the compound stimuli into separate components to solve the discrimination. Although this issue was not central to our study, it is clear that had the pigeons perceived the stimuli as holistic configurations, their discrimination scores for each of the dimensions would not have differed (see Baker, 1968; Kehoe & Gormezano, 1980; Razran, 1971, for reviews of configuration experiments).

In general, prior research suggests that organisms do not invariably process compound stimuli as unitary or holistic events, but do so only under certain conditions (Riley, 1984). For example, with pigeons, two factors that appear to promote elementistic, rather than holistic processing, are the spatial separation of the elements within the compound stimuli (Lamb, 1988; Lamb & Riley, 1981) and prior experience with the elements (Brown, 1987). In addition, research with humans suggests that the greater the number of compound stimuli presented during training, the less likely it is that processing will be holistic (Gibson, 1969). The extent to which these and other as yet unidentified factors may be responsible for the decomposition that occurred with the MNC procedure remains open to investigation. Plus, it is important to note that our elementistic interpretation of subjects' performance is valid only with regard to the acquisition of discriminative control; we are not yet certain about the basis of discrimination maintenance. Thus, further research is needed to see if asymptotic performance is the result of holistic or elementistic processing. This issue is particularly compelling in light of recent claims that the potential impact of configural cues on discrimination learning has been underestimated (e.g., Pearce & Wilson, 1990).

The occurrence of decomposition during acquisition of the MNC discrimination is consistent with the view that the attention paid to a compound stimulus is shared by its individual elements. It follows that the more elements that make up a compound stimulus, the greater the processing load. According to the limited capacity assumption, if the processing capacity of an organism is exceeded, selective attention may occur. The inverse hypothesis further suggests that the individual elements of a compound compete for attention, such that there are tradeoffs in the amount of stimulus control exerted by each. We attempted to maximize the likelihood of selective processing by requiring subjects to attend to three separate features of compound stimuli.

Thus, our third major finding is that, in general, the birds took longer to attain successive levels of discrimination mastery on all three dimensions than it took them to do so on the most difficult dimension of line orientation (Fig. 15.1). This result suggests that attention to the line orientation dimension may have come at the expense of attention to the color and/or time dimensions, as predicted by the inverse hypothesis. And, indeed, the individual subject data support this interpretation. Seven of the eight pigeons gave evidence that learning the line discrimination came at the sacrifice of the color and/or the time discriminations (Fig. 15.4).

Similar tradeoffs in attention have been reported in studies of human information processing (e.g., Kahneman, 1973). Nonetheless, animal researchers typically have been somewhat reluctant to accept the inverse hypothesis. This hesitation may be attributed to the availability of alternative explanations of selective attention when multiple cues are redundant, as in the case of blocking (e.g., Mackintosh, 1975; Rescorla & Wagner, 1972), and to the overwhelming lack of evidence for informational overload in most compound discrimination tasks (Riley & Roitblat, 1978). However, with the MNC procedure, the cues were not redundant and, although the fact that strong discriminative control was ultimately exerted by all three stimulus dimensions indicates that we did not succeed in completely overloading the pigeons' information processing system (Fig. 15.1), we did manage to strain it. As a result, control by a third dimension often came at the temporary expense of one or both of the other dimensions (Fig. 15.4). However, note that although these data suggest attentional competition, we cannot rule out the possibility that the temporary information overload we observed actually occurred during a later stage of processing (Brown & Morrison, 1990; Lamb, 1991). Specifically, because our sample and test stimuli were successively presented, it is conceivable that overload occurred during the memory stage rather than during encoding. Clearly, the locus of capacity limitations is yet another issue that requires further research attention. The eventual regaining of control by the other dimension(s) indicates that three may not be the upper limit on the number of stimulus features to which pigeons can attend. Indeed, von Fersen and Lea (1990) recently reported that pigeons successfully performed a five-feature polymorphous discrimination task, although only four of their eight birds did so in the absence of prior single-feature training. By expanding the number of stimulus dimensions within our MNC discrimination procedure, we hope more precisely to determine the pigeon's upper processing limit for simultaneous multidimensional discriminations.

Finally, we developed a computer-generated model that successfully predicts mean response rates for all eight of the compound trial types in our MNC discrimination task. More specifically, our preliminary attempt at model fitting suggests that simultaneous behavioral control by the three stimulus dimensions might best be described by a multiplicative rule, in which the parameters that correspond to each dimension are influenced by the relative discriminability of the stimulus values along the dimensions. Although multiplicative models of compound stimulus control are not unprecedented in the animal literature (e.g.,

Blough, 1972; Butter, 1963; Johnson, 1970), they also are not uncontested. For example, studies of configural conditioning have tended to favor an additive model in which the contributions of the individual stimulus elements of a compound sum with that of the unique stimulus compound, to determine discriminative control (e.g., Rescorla, 1972, 1973). Perhaps these discrepant conclusions are not surprising given the multitude of procedural differences among the various studies. Indeed, the mathematical form of the combination rule that best describes compound stimulus control might well be expected to depend on many of the same variables that influence the kind of processing that is assumed to occur (e.g., see prior discussion of holistic vs. elementistic processing). By manipulating some of these variables within the MNC discrimination procedure, and by simulating the development of behavioral control in individual subjects, we hope to glean a better understanding of the nature of complex discrimination learning in future projects.

In summary, using a new compound discrimination procedure, we found that: (a) Pigeons were quite capable of processing information about three independent features of a compound stimulus; (b) during acquisition of the MNC discrimination, the compound stimuli were decomposed into their separate elements; (c) duration functioned much like other discriminative cues in controlling behavior; (d) in support of the inverse hypothesis, when pigeons' processing abilities were sufficiently taxed, stimulus control by the most difficult to discriminate of three stimulus dimensions often came at the sacrifice of control by one or both of the other dimensions; and (e) our observations of simultaneous stimulus control by line, color, and time were quite accurately predicted by a nonadditive model. Additionally, this study demonstrates the utility of the MNC procedure for examining attentional processes. Until now, attempts to overload an organism's information processing system have typically involved limiting the available time to study the relevant information (Riley, 1984). Our alternative strategy was to tax the animal's information processing system via the number of relevant stimulus characteristics. Moreover, our method provided separate indices of discriminative control for each stimulus dimension as the control developed, as well as error profiles that were used to identify the controlling dimensions within successive training blocks (Table 15.1). Because of these procedural advantages, the MNC task seems especially well suited to the investigation of compound stimulus control; we hope that its further use will shed new light on such theoretically important stimulus selection phenomena as blocking and overshadowing.

ACKNOWLEDGMENTS

The authors thank Ramesh Bhatt and Greg Oden for their valuable help in conducting and analyzing this research.

REFERENCES

Baker, T. W. (1968). Properties of compound conditioned stimuli and their components. *Psychological Bulletin, 70*, 611–625.

Baron, M. R. (1965). The stimulus, stimulus control, and stimulus generalization. In D. I. Mostofsky (Ed.), *Stimulus generalization* (pp. 62–71). Stanford, CA: Stanford University Press.

Bhatt, R. S., Wasserman, E. A., Reynolds, W. F., Jr., & Knauss, K. S. (1988). Conceptual behavior in pigeons: Categorization of both familiar and novel examples from four classes of natural and artificial stimuli. *Journal of Experimental Psychology: Animal Behavior Processes, 14*, 219–234.

Blough, D. S. (1969). Attention shifts in a maintained discrimination. *Science, 166*, 125–126.

Blough, D. S. (1972). Recognition by the pigeon of stimuli varying in two dimensions. *Journal of the Experimental Analysis of Behavior, 18*, 345–367.

Bourne, L. E. (1970). Knowing and using concepts. *Psychological Review, 77*, 46–556.

Bowers, R. L., & Richards, R. W. (1990). Pigeons' short-term memory for temporal and visual stimuli in delayed matching-to-sample. *Animal Learning and Behavior, 18*, 23–28.

Broadbent, D. E. (1958). *Perception and communication.* London: Pergamon.

Brown, M. F. (1987). Dissociation of stimulus compounds by pigeons. *Journal of Experimental Psychology: Animal Behavior Processes, 13*, 80–91.

Brown, M. R., & Morrison, S. K. (1990). Element and compound matching-to-sample performance in pigeons: The roles of information overload and training history. *Journal of Experimental Psychology: Animal Behavior Processes, 16*, 185–192.

Butter, C. M. (1963). Stimulus generalization along one and two dimensions in pigeons. *Journal of Experimental Psychology, 65*, 339–346.

Cerella, J. (1979). Visual classes and natural categories in the pigeon. *Journal of Experimental Psychology: Human Perception and Performance, 5*, 68–77.

Chatlosh, D. L., & Wasserman, E. A. (1987). Delayed temporal discrimination in pigeons: A comparison of two procedures. *Journal of the Experimental Analysis of Behavior, 47*, 299–309.

DeLong, R. E., & Wasserman, E. A. (1981). Effects of differential reinforcement expectancies on successive matching-to-sample performance in pigeons. *Journal of Experimental Psychology: Animal Behavior Processes, 7*, 394–412.

DeLong, R. E., & Wasserman, E. A. (1985). Stimulus selection with duration as a relevant cue. *Learning and Motivation, 16*, 259–287.

Farthing, G. W., & Hearst, E. (1970). Attention in pigeons: Testing with compounds or elements. *Learning and Motivation, 1*, 65–78.

Fink, J. B., & Patton, R. M. (1953). Decrement of a learned drinking response accompanying changes in several stimulus characteristics. *Journal of Comparative and Physiological Psychology, 46*, 23–27.

Gibbon, J., & Allan, L. (Eds.). (1984). *Annals of the New York Academy of Sciences: Vol. 423. Timing and time perception.* New York: New York Academy of Sciences.

Gibbon, J., & Church, R. M. (1984). Sources of variance in an information processing theory of timing. In H. L. Roitblat, T. G. Bever, & H. S. Terrace (Eds.), *Animal cognition* (pp. 465–488). Hillsdale, NJ: Lawrence Erlbaum Associates.

Gibson, E. J. (1969). *Principles of perceptual learning and development.* New York: Appleton-Century-Crofts.

Heinemann, E. G., & Chase, S. (1970). Conditional stimulus control. *Journal of Experimental Psychology, 84*, 187–197.

Herrnstein, R. J. (1979). Acquisition, generalization, and discrimination reversal of a natural concept. *Journal of Experimental Psychology Animal Behavior Processes, 5*, 116–129.

Herrnstein, R. J., Loveland, D. H., & Cable, C. (1976). Natural concepts in pigeons. *Journal of Experimental Psychology: Animal Behavior Processes, 2*, 285–302.

Jenkins, H. M. (1965). Generalization gradients and the concept of inhibition. In D. I. Mostofsky (Ed.), *Stimulus generalization* (pp. 55–61). Stanford, CA: Stanford University Press.

Johnson, D. F. (1970). Determiners of selective stimulus control in the pigeon. *Journal of Comparative and Physiological Psychology, 70*, 298–307.

Johnson, D. F., & Cumming, W. W. (1968). Some determiners of attention. *Journal of the Experimental Analysis of Behavior, 11*, 157–166.

Kahneman, D. (1973). *Attention and effort.* Englewood Cliffs, NJ: Prentice-Hall.

Kamin, L. J. (1968). Attention-like processes in classical conditioning. In M. R.Jones (Ed.), *Miami symposium on the prediction of behavior, 1967* (pp. 9–32). Coral Gables, FL: University of Miami Press.

Kehoe, E. J., & Gormezano, I. (1980). Configuration and combination laws in conditioning with compound stimuli. *Psychological Bulletin, 87*, 351–378.

Kraemer, P. J., & Roberts, W. A. (1987). Restricted processing of simultaneously presented brightness and pattern stimuli in pigeons. *Animal Learning and Behavior, 15*, 15–24.

Lamb, M. R. (1988). Selective attention: Effects of cuing on the processing of different types of compound stimuli. *Journal of Experimental Psychology: Animal Behavior Processes, 14*, 96–104.

Lamb, M. R. (1991). Attention in humans and animals: Is there a capacity limitation at the time of encoding? *Journal of Experimental Psychology Animal Behavior Processes, 17*, 45–54.

Lamb, M. R., & Riley, D. A. (1981). Effects of element arrangement on the processing of compound stimuli in pigeons (*Columba livia*) *Journal of Experimental Psychology: Animal Behavior Processes, 7*, 45–58.

Lea, S. E. G., & Harrison, S. N. (1978). Discrimination of polymorphous stimulus sets by pigeons. *Quarterly Journal of Experimental Psychology, 30*, 521–537.

Mackintosh, N. J. (1965). Incidental cue learning in rats. *Quarterly Journal of Experimental Psychology, 17*, 292–300.

Mackintosh, N. J. (1975). A theory of attention: Variations in associability of stimuli with reinforcement. *Psychological Review, 82*, 276–298.

Nissen, H. W. (1951). Analysis of a complex conditional reaction in chimpanzee. *Journal of Comparative and Physiological Psychology, 44*, 9–16.

Pearce, J. M., & Wilson, P. N. (1990). Configural associations in discrimination learning. *Journal of Experimental Psychology: Animal Behavior Processes, 16*, 250–261.

Razran, G. (1971). *Mind in evolution.* New York: Houghton Mifflin.

Rescorla, R. A. (1972). "Configural" conditioning in discrete-trial bar pressing. *Journal of Comparative and Physiological Psychology, 79*, 307–317.

Rescorla, R. A. (1973). Evidence for "unique stimulus" account of configural conditioning. *Journal of Comparative and Physiological Psychology, 85*, 331–338.

Rescorla, R. A., & Wagner, A. R. (1972). A theory of Pavlovian conditioning: Variations in the effectiveness of reinforcement and nonreinforcement. In A. H. Black & W. F. Prokasy (Eds.), *Classical conditioning II: Current research and theory* (pp. 64–99). New York: Appleton–Century–Crofts.

Reynolds, G. S. (1961). Attention in the pigeon. *Journal of the Experimental Analysis of Behavior, 4*, 203–208.

Riley, D. A. (1984). Do pigeons decompose stimulus compounds? In H. L. Roitblat, T. G. Bever, & H. S. Terrace (Eds.), *Animal cognition* (pp. 333–349). Hillsdale, NJ: Lawrence Erlbaum Associates.

Riley, D. A., & Roitblat, H. L. (1978). Selective attention and related cognitive processes in pigeons. In S. H. Hulse, H. Fowler, & W. K. Honig (Eds.), *Cognitive processes in animal behavior* (pp. 249–276). Hillsdale NJ: Lawrence Erlbaum Associates.

Roitblat, H. L., Scopatz, R. A., & Bever, T. G. (1987). The hierarchical representation of three-item sequences. *Animal Learning and Behavior, 15*, 179–192.

Schwartz, G., & Reilly, M. (1985). Long-term retention of a complex operant in pigeons. *Journal of Experimental Psychology: Animal Behavior Processes, 11*, 337–355.

Seraganian, P., & vom Saal, W. (1969). Blocking the development of stimulus control when stimuli indicate periods of nonreinforcement. *Journal of the Experimental Analysis of Behavior, 12*, 767–772.

Sidman, M. (1980). A note on the measurement of conditional discrimination. *Journal of the Experimental Analysis of Behavior, 33*, 285–289.

Siegel, S. (1956). *Nonparametric statistics for the behavioral sciences.* New York: McGraw-Hill.

Sutherland, N. S., & Holgate, V. (1966). Two cue discrimination learning in rats. *Journal of Comparative and Physiological Psychology, 61*, 198–207.

Sutherland, N. S., & Mackintosh, N. J. (1971). *Mechanisms of animal discrimination learning.* New York: Academic.

Thomas, D. R. (1970). Stimulus selection, attention, and related matters. In J. H. Reynierse (Ed.), *Current issues in animal learning* (pp. 311–356). Lincoln: University of Nebraska Press.

Vaughan, W., Jr., & Greene, S. L. (1984). Pigeon visual memory capacity. *Journal of Experimental Psychology: Animal Behavior Processes, 10*, 256–271.

vom Saal, W., & Jenkins, H. M. (1970). Blocking the development of stimulus control. *Learning and Motivation, 1*, 52–64.

von Fersen, L., & Lea, S. E. G. (1990). Category discrimination by pigeons using five polymorphous features. *Journal of the Experimental Analysis of Behavior, 54*, 69–84.

Wasserman, E. A., Kiedinger, R. E., & Bhatt, R. S. (1988). Conceptual behavior in pigeons: Categories, subcategories, and pseudocategories. *Journal of Experimental Psychology: Animal Behavior Processes, 14*, 235–246.

Wilkie, D. M., & Masson, M. E. (1976). Attention in the pigeon: A reevaluation. *Journal of the Experimental Analysis of Behavior, 26*, 207–212.

16

From Elementary Associations to Animal Cognition: Connectionist Models of Discrimination Learning

William S. Maki
North Dakota State University

I began work in Al Riley's laboratory in the fall of 1970. At that time, discrimination learning was a theoretically central area in psychology, and Riley and his students were working on problems of discrimination and generalization in animals. *Discrimination Learning* had recently been published (Riley, 1968). Wagner, Logan, Haberlandt, and Price (1968), for example, had laid some empirical groundwork for the influential theory of Rescorla and Wagner (1972). And, the two-process theory of Sutherland and Mackintosh (1971) was imminent.

But discrimination learning per se has become a less popular topic than it once was. Consider, for example, the topic of reversal and nonreversal shifts (RS and NRS; Kendler & Kendler, 1968). Figure 16.1 shows the results of an electronic search of *Psychological Abstracts* from 1968 (the year of Riley's book) to 1990. Two sets of key words were used. One set contained RS and NRS (and related terms, like extradimensional and intradimensional shifts). The retrieved citations were then sorted by species (human vs. animal). Both human and animal literatures show parallel trends. The frequency of publications on RS/NRS peaked in 1973 and steadily declined thereafter. The current rate of publication is nearly zero. The other set of key words pertained to the current crop of computational models in cognitive science, referred to as connectionist, parallel distributed processing (PDP), or neural networks (McClelland & Rumelhart, 1988). This search set was limited to the literature on human learning or attention. The results of this search are also presented in Fig. 16.1. The trend is dramatically different from that in the RS/NRS literature. After a period of almost no activity through 1986, there was an explosive growth in publications on connectionist models of human learning.

I argue in this chapter that, in spite of the changes in the last 20 years, certain discrimination learning phenomena remain theoretically important. They are

293

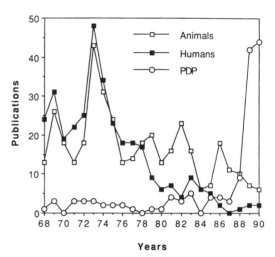

FIG. 16.1. Results of an electronic literature search for numbers of publications during 1968–1990. One search was for terms related to reversal and nonreversal shifts and the results are presented separately for humans and animals. The other search was for terms related to connectionist modeling (like PDP) of human learning.

important because they establish constraints that can and should guide the development of workable theories of learning. In particular, learning and transfer of relatively simple discriminations present serious challenges for the connectionist models. In this light, the trends displayed in Fig. 16.1 are most unfortunate; an important class of theoretical models of learning is not being sufficiently constrained by data. For illustration, I eventually focus on RS/NRS because I believe that such phenomena are especially instructive about connectionist models of both human and animal learning.

In the first of the ensuing three sections, I describe a connectionist model of acquisition and performance of the conditional discrimination known as delayed matching to sample (DMTS). I then describe an application of the model, by computer simulation, to certain experiments that were performed by Zentall and his collaborators on "common coding" (see chapter 12 of this volume). To anticipate, I have found that the connectionist model is accurate with respect to some findings but fails miserably on others. Interestingly, the failures occur on those transfer problems that resemble the prototypical RS/NRS experiment. In the second section, simulations studying the RS/NRS problem more directly reveal the cause of the failures. In the third section, a class of alternative models, more along the lines of attentional theories of discrimination learning (e.g., Sutherland & Mackintosh, 1971), are evaluated.

COMMON CODING: AN EXAMPLE OF CONNECTIONIST MODELING OF ANIMAL COGNITION

Like its parent discipline, animal learning has experienced something of a cognitive revolution during the last two decades, sparked at least in part by work that Riley and his students were doing (see, e.g., Riley & Roitblat, 1978). As the chapters in this book show (e.g., chapters 11 and 12), much of the modern field of animal *cognition* is concerned with cognitive processes examined in the context of relatively complex tasks like DMTS. Indeed, together with my lab-mates, Terry Leuin and Charles Leith, under Al Riley's mentorship, I pursued such research in Riley's lab (Maki & Leith, 1973; Maki & Leuin, 1972; Maki, Riley, & Leith, 1976).

During that period in Riley's lab, we read about the Rescorla–Wagner model of Pavlovian conditioning (Rescorla & Wagner, 1972). In spite of the commitment to an information processing approach to traditional issues in animal discrimination learning (Maki & Leuin, 1972), we were all impressed by the explanatory power of the Rescorla–Wagner model. During the next few years I continued to contemplate the possibility of extending the model to cover conditional discriminations like DMTS.

Riley and I were in the audience of a symposium at the 1986 Psychonomic Society on "Connectionist Models and Psychological Evidence." It was then that I realized how such models could be applied to animal cognition. Unlike the simple network of the Rescorla–Wagner model, these PDP models could learn new internal representations (McClelland, 1986; Rumelhart, 1986), so the models could solve conditional problems like DMTS. Moreover, we learned, the Rescorla–Wagner model turns out to be a special case of the "delta rule" (Gluck & Bower, 1986). Since then, part of my research has been focused on testing connectionist models of conditional discriminations, DMTS in particular, and initial results have been reported elsewhere (Maki & Abunawass, 1991).

Common Coding in DMTS

Zentall et al. (chapter 12 of this volume) report a series of experiments on common coding that motivated the simulations reported here. If two stimuli, A and B, share a common consequence X, then they might also share a common code C. Hence, the A-X and B-X associations are mediated by the code: A-C-X and B-C-X. Some interesting predictions follow. For example, if a new consequence Y is associated with stimulus A, then Y should also become associated with stimulus B because of the common code.

Much of the work on common coding has been done within the context of the DMTS procedure. In DMTS, the correct choice between two "comparison" stimuli depends on the identity of a preceding "sample" stimulus. The two comparison stimuli might be red and green disks (R and G), and the two sample stimuli might be vertical and horizontal orientations of a line (V and H). A set of contingencies relate the sample and comparisons. For example, choice of the red comparison would be rewarded following a vertical sample (V → R); choice of the green comparison would be rewarded following a horizontal sample (H → G).

A more complex DMTS problem is illustrated in Table 16.1. There are two sets of sample (R and G, and V and H) and two sets of comparisons (also R and G, and V and H). The correct comparison following each sample is given in the

TABLE 16.1
Conditions for Simulated Common Coding
of Both Sample and Comparison Stimuli

	Color Comparisons	Line Comparisons
Color Samples	R → R G → G	R → V G → H
Line Samples	V → R H → G	V → V H → H

Note. Simulated stimuli are red (R), green (G), vertical (V), and horizontal (H).

table. A separate DMTS problem is identified within each cell. Problems within rows have common samples. Problems within columns have common comparisons. In the work with pigeons, a single subject learns either the associations within a row or the associations within a column. Although this is a hypothetical experiment designed for a simulation study, the expectations about common codes are clear. The R and V samples should become treated as equivalent because of their common comparison stimuli (as should the G and H samples). Similarly, the R and V comparisons should become equivalent because of their common samples (as should the G and H comparisons).

A Connectionist Model of DMTS

The model employs the (now common) layered feed-forward network that learns by the "delta rule." The chapter in the PDP series by Rumelhart, Hinton, and Williams (1986) contains the mathematical derivation of the generalized delta rule, in which learning occurs by error backpropagation. Gluck and Bower (1988) showed the relationship of the delta rule to the Rescorla–Wagner model. These derivations are reviewed and applied in a form particular to the DMTS model in Maki and Abunawass (1991), so, instead of a complete mathematical treatment, only an overview of the functioning of the model is given here.

Figure 16.2 contains a diagram of the network used in simulations of common coding in DMTS. The "input units" are represented on the left of the figure. Each group of units represents the location (left, center, right) of one pecking key in an operant conditioning chamber designed for pigeons. The units assigned to each location represent one of four stimuli (R, G, V, H). (In other simulations, additional units were assigned to each key so as to represent more stimuli.) Each input unit is connected to each of eight "hidden units." Patterns of activity across these internal units are internal representations, or codes, that are learned by the network in response to environmental regularities. The rightmost pair of "output units" report the network's response to input patterns. In DMTS, two responses are possible: left or right.

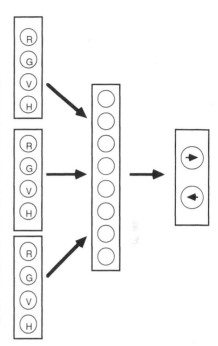

FIG. 16.2. Representation of delayed matching to sample by a connectionist network. Input units are displayed in the leftmost column and are identified by the external inputs that they receive: red (R), green (G), vertical (V), and horizontal (H). The three fields, from top to bottom, correspond to left, center, and right pecking keys in the conditioning chamber. The two output units are displayed in the right column, and represent the two choices—pecking the left or right keys. The middle column contains the hidden units. Each arrow indicates complete connectivity between fields of units.

In each training trial, a pair of input and output patterns is presented to the network. Each pattern consists of binary numbers, with "1" representing stimulus presence and "0" representing stimulus absence. Thus, the input pattern 0010 1000 0001 corresponds to the stimulus pattern VRH. The output pattern consists of the desired response from the network (sometimes called "teaching input"). A binary output pattern of 10 would indicate that the correct response is to choose the left stimulus, that is, R → V.

The input activation is propagated forward through the net, with each input being weighted by the strength of the connection between that input and a hidden unit. Connection strengths, or weights, can be negative (inhibitory) as well as positive (excitatory). The activation level of each hidden unit is the sum of the weighted inputs (net input) transformed by an activation function (a sigmoid) that constrains the output from the unit to fall between 0 and 1.

Hidden unit activations are then fed forward to output units, multiplied by appropriate (hidden-output) weights, and the activations of the output units are computed. The difference between each output activation and the desired output is a measure of error. That error is propagated backward through network and is used to adjust each weight according to the delta rule (Rumelhart et al., 1986). With repeated training, the network converges on a solution in which a set of weights is discovered that minimizes the output error.

These training procedures were used with the network diagrammed in Fig.

16.2 in all the common coding simulations. All simulation work was done using commercially available software (McClelland & Rumelhart, 1988). Each simulation reported consisted of many replicas of this network, each initialized with a unique set of small random weights. The results presented are averages, and conclusions were, without exception, confirmed statistically with appropriate analyses of variance. For other details of these kinds of simulations, see Maki and Abunawass (1991).

Analysis of Internal Representations

Common codes are not directly observed in experiments with real organisms. Rather, like any other mediator, inferences about codes are drawn from behavior. In contrast, common codes can be studied directly in connectionist models because the internal computational entities are open to inspection. For example, language learning has been studied in connectionist networks (e.g., St. John & McClelland, 1990). These networks are known to be sensitive to certain regularities in a corpus of sentences. For example, training of the sentences "ball breaks window" and "rock breaks window" causes the network to represent ball and rock as being equivalent. To put it another way, the network learns common codes for items that have common linguistic consequences. These internal codes are revealed by studies in which the responses of hidden units to various stimuli are directly measured and compared. Similarity (or distance) measures between stimuli are then analyzed using methods like hierarchical clustering in order to reveal the functional groups of items discovered by the network (e.g., Elman, 1989). An application of this technique to common coding in DMTS follows.

The network was trained on the DMTS problem illustrated in Table 16.1. However, unlike the work with pigeons, each simulated subject learned all of the associations concurrently. Training continued until each output for each type of trial was within 5% of the desired value. Then the network was presented with two sets of input patterns which acted as test trials. The first set of 16 patterns shown at the top right of Fig. 16.3 were the original training patterns. The second set of eight patterns, shown at the bottom of Fig. 16.3, had never appeared in training. This set consisted of the eight separate events within the DMTS task— the four samples and four pairs of comparison stimuli.

The hidden unit activations were recorded in response to each pattern in each test set. The activation values from each pattern in a test set were compared to those that resulted from every other pattern in that set by computing a difference measure, the Euclidean distance. For each test set, the distances were subjected to a hierarchical clustering analysis using Ward's minimum variance method (SAS, 1982). The results are presented in Fig. 16.3 as traditional tree diagrams. Short distances between nodes in the tree indicate similarity. In the training set, the analysis shows that samples with common comparisons clustered together (e.g., RRG, RVG) and that comparisons with common samples also clustered

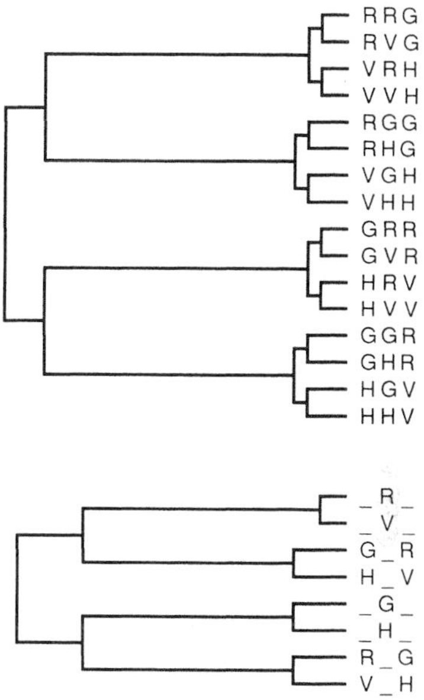

FIG. 16.3. Results of a hierarchical scaling study. After training on the problem in Table 16.1, the network was presented with the original training set (displayed at the top) and new test events (displayed at the bottom). Euclidean distances between pairs of hidden unit activation vectors formed the proximity matrix that was the input for the scaling program. Short routes through the tree represent similarity.

together, but at the next higher node. (Trace the path from RRG to VRH.) These relationships are shown more clearly in the other test set in the bottom of Fig. 16.3. These results show that the connectionist network learns common codes for samples with common comparisons, and learns common codes for comparisons with common samples.

Transfer of Common Codes

The most convincing evidence for common codes in pigeons comes from the transfer experiments. One example, the "substitution experiment," was originally reported by Urcuioli, Zentall, Jackson-Smith, and Steirn (1989), and is reviewed in chapter 12 of this volume. For convenience, the design of the experiment is presented in Table 16.2. In Phase 1, two comparison stimuli were associated with four sample stimuli. In Phase 2, two new comparison stimuli were associated with two of the samples. Finally, in Phase 3, the new comparison stimuli were tested with the other two samples. In this last phase, half the pigeons were rewarded for responding consistent with the common codes presumably learned in Phase 2, and for the other birds rewarded responding was inconsistent (sample-comparison mappings were reversed). Thus, if common coding occurred then positive transfer should be found for the consistent group but negative transfer for the inconsistent group. Urcuioli et al. obtained results during Phase 3 that confirmed the prediction. They found that the consistent

TABLE 16.2
Design of the Simulated Substitution Experiment

	Phase 1	Phase 2	Phase 3
Consistent	R → V	R → C	
	G → H	G → D	
	V → V		V → C
	H → H		H → D
Inconsistent	R → V	R → C	
	G → H	G → D	
	V → V		V → D
	H → H		H → C

Note. Six stimuli were represented: red (R), green (G), vertical line (V), horizontal line (H), a circle (C), and a dot (D).

group performed above chance (71.9% correct), and that the inconsistent group performed below chance (35.4% correct).

A network similar to that in Fig. 16.2 (but with 18 input units) was trained successively on the DMTS problems in Phases 1–3. Six input units were used to represent the stimuli on each pecking key (see Table 16.2). Separate simulations were conducted for the consistent and inconsistent conditions. The results, averaged over 30 simulations per condition, are displayed in Fig. 16.4. The dependent measure is the number of training epochs (trials) required to reach an error

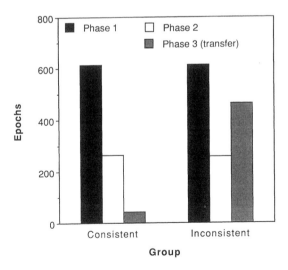

FIG. 16.4. Results of the simulation of the substitution experiment (Urcuioli et al., 1989). See Table 16.2 for the design of the experiment. Each "epoch" is a randomized block of simulated trials.

criterion of less than 5%. Reacquisition during Phase 3 was almost immediate in the consistent condition, but was severely retarded in the inconsistent condition. Thus, the connectionist model showed transfer results like those of pigeons.

Part Versus Whole Reversals

Another prediction of the common coding hypothesis has been tested in pigeons (Zentall, et al., in press; Zentall, Steirn, Sherburne, & Urcuioli, 1991). That research is also reviewed in chapter 12 of this volume. The designs of the relevant transfer experiments are displayed in Table 16.3. Two versions of the experiment were conducted ("many-to-one" and "one-to-many"). The findings are substantially the same in each case, so only the many-to-one experiment is discussed here.

Pigeons were trained on a DMTS task with two sets of samples (R/G and V/H) but only one set of comparisons (V/H). Different samples were associated with common comparisons: R → V, G → H, and V → V, H → H. Then the pigeons were transferred to one of two reversal conditions. In the part reversal, the correct comparisons for one set of samples was reversed. In the whole reversal, the correct comparisons were reversed for both sets of samples. The prediction from common coding is that reversing the comparison responses (whole reversal) would be easier than learning new codes (part reversal). Zentall et al. (in chapter 12 of this volume) describe the evidence suggesting that pigeons show exactly this effect (as do rats; see Nakagawa, 1986).

Does the connectionist model also learn whole reversals faster than part reversals? To answer this question, different networks were trained on either many-to-one or one-to-many problems. Each network was first taught the DMTS training problem diagrammed in Table 16.3. Then, the network was retrained on

TABLE 16.3
Design of the Simulated Part–Whole
Reversal Experiment

	Training	Part	Whole
Many-to-One	R → V	R → H	R → H
	G → H	G → V	G → V
	V → V	V → V	V → H
	H → H	H → H	H → V
One-to-Many	V → R	V → G	V → G
	H → G	H → R	H → R
	V → V	V → V	V → H
	H → H	H → H	H → V

Note. Four stimuli were represented: red (R), green (G), vertical line (V), and horizontal line (H).

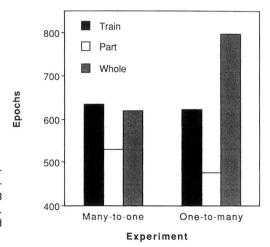

FIG. 16.5. Results of the simulation of the part-whole reversal experiment. See Table 16.3 for design of the experiment. Each "epoch" is a randomized block of simulated trials.

either the part or the whole transfer problem. The results are shown in Fig. 16.5. Both many-to-one and one-to-many versions show the same pattern. The whole reversal is learned slower than the part reversal, and this result is opposite that found with pigeons.

Summary

The connectionist model illustrated in Fig. 16.2 does learn DMTS problems, and it shows numerous effects found in experiments with pigeons (Maki & Abunawass, 1991). The list now includes transfer of common codes (cf. Urcuioli et al., 1989). However, the simulations reported here have shown that the model learns whole reversals more slowly than part reversals, whereas pigeons learn whole reversals faster than part reversals. This discrepancy motivated the explorations described next.

AN ANALYSIS OF REVERSAL LEARNING BY
THE CONNECTIONIST MODEL

I now have run many simulation studies attempting to discover some set of conditions under which the connectionist model of DMTS will show faster learning during whole-reversal transfer than during part-reversal transfer (Maki, 1990a, 1991). The conclusion at this point is quite clear. Under no conditions whatsoever has the model shown such an effect. Inevitably, the model produces the reverse effect displayed in Fig. 16.5.

The simulation work on this problem followed two parallel tracks. One line of work focused on simulation of the RS/NRS experiment (Kendler & Kendler,

1968) in which college students show faster learning of RS than NRS. Figure 16.6 illustrates the approach. The two-dimensional stimuli and the contingencies were adapted from the Kendlers' chapter. In this particular set of simulations, Gluck and Bower's (1988) configural cue model was examined. This model works just like the one in Fig. 16.2, but it has no hidden units. Instead, additional input units code the conjunctions of stimuli.

The second line of work focused on a related phenomenon reported by Nakagawa (1986). Nakagawa's experiment is summarized in Table 16.4. Rats were trained on concurrent, simultaneous discriminations in a Y-maze. For example, some rats were trained to choose black over white, and (separately) a vertical

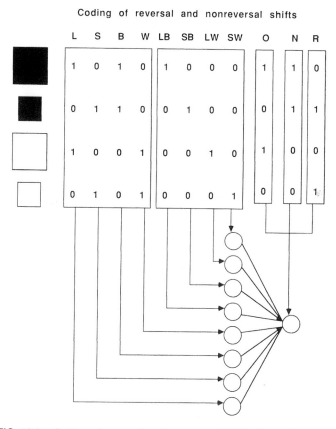

FIG. 16.6. Coding of reversal and nonreversal shifts for a single layer network with configural units. The stimulus values are large (L), small (S), black (B), and white (W) and their pairwise conjunctions (e.g., LB). The desired output activations (teaching inputs) are given for original learning (O), nonreversal (N), and reversal (R) discrimination problems.

TABLE 16.4
Summary of Nakagawa's (1986) Experiment

	Original	Transfer	Degree of Learning	
			Criterion	Overtraining
Whole-Reversal Group	B+W− V+H−	B−W+ V−H+	18.9	8.5
Part-Reversal Group	B+W− V+H−	V−H+	8.5	17.3

Note. Stimuli were black (B) or white (W) panels, or vertical (V) or horizontal (H) stripes. Data are average days to criterion on the orientation (V, H) discrimination.

pattern over a horizontal pattern. Then the rats were transferred to a whole reversal or a part reversal. In the part reversal, only one set of stimuli was presented (e.g., the patterns) with reversed contingencies. Nakagawa observed that whole reversals were learned faster than part reversals, but only with overtraining. When the original discriminations were learned to criterion, the part-reversal problem was learned fastest.

In the course of simulating both the RS/NRS and the part–whole reversal experiments, network architecture and learning parameters were systematically varied. In one simulation of Nakagawa's (1986) part–whole reversal experiment, the effects of overtraining were investigated. The network was trained to criterion (95% correct), and then run for an additional 0, 500, 1,000, 2,000, 4,000, or 8,000 blocks of trials. Then the network was switched to either the part or the whole reversal. Figure 16.7 shows the results. Part reversals were learned faster than whole reversals independently of degree of original learning. In fact, overtraining tended to retard learning during transfer regardless of the type of transfer problem.

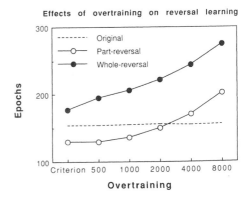

FIG. 16.7. Effects of overtraining on part- and whole-reversal learning. The dashed line is the average number of epochs required to learn the original problem. See Table 16.4 for the design of the simulation.

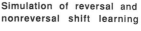

Simulation of reversal and
nonreversal shift learning

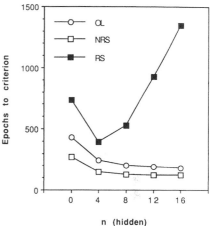

FIG. 16.8.   Effects of number of
hidden units on nonreversal
and reversal shift learning.

Varying the number of hidden units proved just as fruitless. In one simulation
of the RS/NRS experiment, the number of hidden units were varied from none (a
network like that in Fig. 16.6) to 16 (obtaining a network structure more like Fig.
16.2). Figure 16.8 shows the result. The networks learned NRS faster than RS
always. Adding more hidden units actually amplified the difference. It thus
appeared that the more obvious structural or parametric modifications of the
connectionist model were not going to be helpful in obtaining whole-reversal (or
RS) superiority.

The answer to this problem comes from a computational analysis of the
RS/NRS transfer design. The RS/NRS experiment is reviewed in a somewhat
different form in the top part of Fig. 16.9. For consistency, the terms appropriate
for the Kendler and Kendler (1968) example are used (see Fig. 16.6). So, in
original learning the response R1 is appropriate for the large black object (LB),
and the response R2 is appropriate for the small white object (SW). The appropriate responses for the NRS and RS transfer problems also are listed.

The bottom part of Fig. 16.9 represents the same information in a different
form. Each 2 × 2 panel is composed of the two-dimensional stimulus space. The
appropriate responses appear at each of the four points in each space. The
shading is provided so as to mark the responses in each category, and, for
example, might be taken to identify the reinforced responses. The line separating the space into the two response categories rotates 90° counterclockwise
from original learning to NRS, and again from NRS to RS. This is a somewhat
fancy way of pointing out that only two responses must change in NRS, but all
four must change in RS. The delta rule (Rumelhart et al., 1986) is an error–

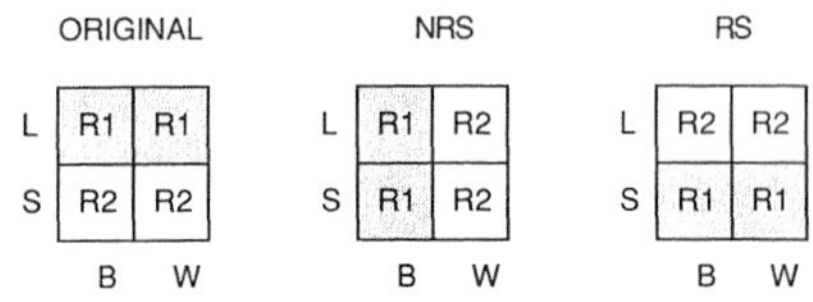

FIG. 16.9. Analysis of reversal/nonreversal shift problems in terms of classification of a two-dimensional stimulus space. The top panel shows the stimuli and appropriate responses (R1 or R2); see Fig. 16.6 for the stimuli and a representative network. The bottom panel shows the two-dimensional (size × color) spaces. The shaded areas mark the region of each space in which R1 is the appropriate response.

| DIMENSIONS | | RESPONSES | | |
SIZE	COLOR	ORIGINAL	NRS	RS
L	B	R1	R1	R2
L	W	R1	R2	R2
S	B	R2	R1	R1
S	W	R2	R2	R1

ORIGINAL

	B	W
L	R1	R1
S	R2	R2

NRS

	B	W
L	R1	R2
S	R1	R2

RS

	B	W
L	R2	R2
S	R1	R1

driven learning method. As such, the connectionist model explored here must correct more error in RS than in NRS, and we might expect that it will take longer to do so.

The real utility of the representation in Fig. 16.9 is that it casts the problem in spatial terms. If the RS transfer requires a larger movement through this kind of category space, then it might well demand that the network move farther in "weight space" (e.g., McCloskey & Cohen, 1989). The simplest connectionist model that can learn the discrimination problems considered here consists of two input units and one output unit (and therefore two weights). The two weights make up a two-dimensional weight space. Any discrimination problem then can be considered in terms of the region of weight space that will produce appropriate behavior.

Movement through weight space during RS and NRS transfer was studied in networks with and without hidden units. The weights were saved after original learning and were saved again after RS and NRS transfer. In all cases, the Euclidean distance in weight space between original learning and NRS was shorter than the distance between original learning and RS.

Summary

The connectionist model investigated in this chapter is a feedforward network that learns by error backpropagation. The model learns whole reversals (and RS) more slowly than part reversals (NRS). In contrast, experiments with rats, pigeons, and humans show exactly the opposite effect. The reason for the discrepancy is now known. RS and NRS discriminations present different computational demands. This is a difficult problem that needs to be solved by if we are to have a reasonable connectionist model of DMTS.

ATTENTIONAL LEARNING BY CONNECTIONIST MODELS

Two-process theories of discrimination learning incorporate constructs like selective attention and dimensional learning, topics of great concern in Riley's lab in the 1970s. Sutherland and Mackintosh (1971), for example, proposed that animals learn to switch in analyzers for (attend to) relevant dimensions, and also learn to attach specific responses to analyzer outputs. Response attachments happen relatively rapidly, but the attentional learning lags behind. Thus, overtraining improves the attentional learning. Can such a theory be cast in connectionist form? Would such an implementation exhibit faster RS learning? The simulation work described in the next two sections was designed to answer those questions.

A Connectionist Model with Attentional Learning (ALCOVE)

ALCOVE stands for *A*ttentional *L*earning with a *COVE*ring map. ALCOVE was designed by Kruschke (1990, 1992) as an improved connectionist model of human category learning. ALCOVE shares the same basic structure with the more common connectionist model that has been the focus of this chapter so far—a multilayered feedforward network with error backpropagation (BP). There are three layers of units—input, hidden, and output. Both are feedforward. And both learn by error backpropagation. ALCOVE, however, differs from BP in two important respects.

First, the methods of computing hidden unit activations are different in ALCOVE and BP. In BP, at least as used in this chapter, all the activation functions are sigmoids, converting the linear net input for each unit into an S-shaped function ranging from 0 to 1 by the logistic function (Rumelhart et al., 1986). In ALCOVE, the hidden units function more like neural feature detectors and have relatively well-defined receptive fields. ALCOVE's hidden units thus respond more selectively than those in BP to inputs located within a constrained area of dimensional space. Each such unit reports stimulus input to the extent that the conjunction of the two inputs falls at or near the center of its receptive field.

The second way in which ALCOVE differs from BP is in the functioning of the input layer and the first layer of weights. In BP, the weights in the first layer are treated as any other weight and thus subject to modification by the delta rule. In ALCOVE, the weights in the first layer are fixed and not usually modified. This first layer of weights in ALCOVE functions as the set of coordinates in dimensional space that defines the center locations of the receptive fields of the hidden units. In BP, each input unit assumes the value of its external stimulus input. But in ALCOVE, the external stimulus input is weighted by a dimension-specific attentional parameter. The attentional parameters are then treated as weights and are modified in proportion to backpropagated error.

These two differences combine to define different approaches in ALCOVE and BP to the internal representations at the level of hidden units (the site of common codes in connectionist models). In BP, all internal representations are acquired by modification of the first layer of weights. Thus, the "coordinates" of each hidden unit change during training. In ALCOVE, however, the hidden layer is initialized with a population of narrowly defined detectors, so the potential internal representations are predefined by fixed coordinates. ALCOVE then learns how much attention should be paid to each dimension.

I have performed extensive simulation of transfer experiments using AL-COVE.[1] The simulations show that in spite of its ability to do attentional learning, and in spite of its success in modeling certain other phenomena, ALCOVE does not learn RS faster than NRS. The problem faced by ALCOVE is exactly the same faced by BP. The RS transfer problem presents twice the output error as the NRS problem. The backpropagated error causes substantial changes in the attentional parameters. Thus, even after learning to pay attention to the relevant dimension during original learning, ALCOVE's attention will shift away from that dimension during RS as well as during NRS transfer. ALCOVE will rediscover the relevant dimension during RS, to be sure, but the longer distance that must be moved through weight space in the RS relative to NRS then causes a longer time to solution. So ALCOVE does not handle the RS/NRS phenomena in its published form (Kruschke, 1992).

Adaptation of Attentional Rates

The solution for ALCOVE, and maybe BP as well, was suggested in a different context by Gluck and Bower (1988). They observed that single-layer models like that in Fig. 16.6 would be incapable of generating superior RS transfer and could not, in general, mimic those findings that motivate a two-process theory of discrimination learning like that of Sutherland and Mackintosh (1971). Gluck and Bower suggested that a two-layer model should learn codes (weights between input and hidden units) and then "lock in (clamp) these intermediate codes while learning different responses to these stimuli in a second problem" (p. 190). In the remainder of this section, I suggest a method for accomplishing this consolidation of attentional learning.

The general idea is to let the attentional parameters become less modifiable as learning progresses. So, as the backpropagated error diminishes during original training, the rate of change of the attentional parameters diminishes also. The attentional learning rate becomes the focus of this extension to ALCOVE. If

---

[1] I thank John Kruschke for providing the ALCOVE program and documentation.

the rate at which an attentional parameter is modified approaches zero, then the learned codes would effectively be clamped. By itself, this is inadequate because NRS learning would be impossible, so some means to redirect attention must also be provided. My extension follows ALCOVE in letting attention be modified as a function of backpropagated error. The rate at which attention is modified, the attentional learning rate, is decreased by autonomous decay toward zero. The attentional learning rate is increased when two conditions are satisfied: (a) There exists some discrepancy between the maximum rate and the current value of the learning rate, and (b) the error is nonzero.

ALCOVE was modified to reflect the attentional learning rate modifications, and the RS/NRS experiment was simulated twice more. In one simulation, original learning was terminated when the criterion (95% correct) had been reached. In the second simulation, the network was given overtraining after having reached the criterion. The results are shown in Fig. 16.10. When original learning was terminated at criterion, ALCOVE learned NRS faster than RS. However, after overtraining, RS started at a disadvantage but ended up being learned faster than NRS. Overtraining allowed more time for the attentional learning rates to decay, so the "clamping" of attentional weights was more complete following overtraining than at criterion.

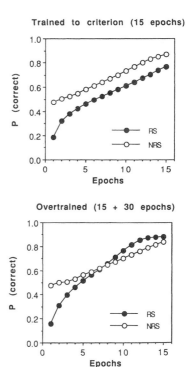

FIG. 16.10. Performance of the modified ALCOVE on reversal (RS) and nonreversal (NRS) shift problems when trained to criterion and with overtraining. Data are from NRS and RS transfer phases.

These results are encouraging and support Gluck and Bower's (1988) suggestion about how attentional learning might be implemented by connectionist models. The work on this problem, however, is far from over. Kruschke (1992) suggested a different method by which attentional weights might be rendered less plastic during the course of training. It remains to be seen which proposal better enables ALCOVE to cope with a full range of attentional learning phenomena.

The addition of attentional learning, and some clamping process, still leaves us quite a distance from an acceptable model, but with a sense of some directions that future work can take. Because dimensional space is populated with fixed feature detectors (the covering map), ALCOVE in its current version is incapable of learning new representations. Moreover, ALCOVE is specialized for the categorization of single stimuli. Additional work needs to be done to determine whether ALCOVE can be adapted to the domain of conditional discriminations involving choice among multiple objects (such as in DMTS). If so, the question will be whether a newer ALCOVE will acquire common codes and use them during transfer problems in the way that real organisms do. Also, we should pursue the present lead and search for a principled way of keeping the BP model (Fig. 16.2) intact but letting first-layer weights become relatively stable during the course of learning, using either a mechanism such as that proposed here or the one proposed by Kruschke (1992).

CONCLUDING REMARKS

This chapter resulted from the convergence of three intellectual themes that have (for me) their origins in Al Riley's lab—discrimination learning, animal cognition, and connectionist models. The theme of this chapter was displayed graphically in Fig. 16.1. Research on certain topics in human (and animal) discrimination learning have all but extinguished during the past two decades, whereas connectionist theories of human category learning have enjoyed an enormous increase in popularity. One could even draw the conclusion from Fig. 16.1 that the two literatures have passed each other by without much contact. It is important that such contact be renewed. A reunification of conditioning and cognition is overdue, and connectionism offers the tools by which constructive interactions can occur (Maki, 1990b). I have argued in this chapter that connectionist models of learning need to be applied to results of discrimination learning experiments, and especially to results from transfer designs. I suspect that substantial changes in the models will result. In the absence of the constraints on theorizing that are presented by the full range of discrimination learning phenomena from both humans and animals, connectionism risks becoming a science of what can be computed instead of what ought to be computed.

ACKNOWLEDGMENTS

Portions of this chapter were presented at the 1990 Psychonomic Society meeting, and as an invited paper at the 1991 meeting of the Midwestern Psychological Association. Preparation of the chapter was facilitated by the computing resources of the Intelligent Systems Cluster (ISC) of North Dakota State University. ISC is supported by the National Science Foundation's Experimental Program to Stimulate Competitive Research.

REFERENCES

Elman, J. L. (1989). Structured representations and connectionist models. *Proceedings of the Eleventh Annual Conference of the Cognitive Science Society* (pp. 17–23). Hillsdale, NJ: Lawrence Erlbaum Associates.

Gluck, M. A., & Bower, G. H. (1986, November). Category learning, judgement, and the Rescorla–Wagner model (aka the Delta-Rule). In *Connectionist models and psychological evidence.* Symposium conducted at the meeting of the Psychonomic Society, New Orleans.

Gluck, M. A., & Bower, G. H. (1988). Evaluating an adaptive network model of human learning. *Journal of Memory and Language, 27,* 166–195.

Kendler, H. H., & Kendler, T. S. (1968). Mediation and conceptual behavior. In K. W. Spence & J. T. Spence (Eds.), *The psychology of learning and motivation: Advances in research and theory* (Vol. 2, pp. 197–244). New York: Academic.

Kruschke, J. K. (1990). *ALCOVE: A connectionist model of category learning* (Cognitive Science Research Report No. 19). Bloomington: Indiana University.

Kruschke, J. K. (1992). ALCOVE: An exemplar-based connectionist model of category learning. *Psychological Review, 99,* 22–44.

Maki, W. S. (1990a, November). *Exploring common coding with a connectionist network.* Paper presented at the meeting of the Psychonomic Society, New Orleans.

Maki, W. S. (1990b). Toward a unification of conditioning and cognition. *The Behavioral and Brain Sciences, 13,* 501–502.

Maki, W. S. (1991, May). *Connectionist networks as models of discrimination learning.* Paper presented at the meeting of the Midwestern Psychological Association, Chicago.

Maki, W. S., & Abunawass, A. M. (1991). A connectionist approach to conditional discriminations: Learning, short-term memory, and attention. In M. Commons, S. Grossberg, & J. Staddon (Eds.), *Neural network models of conditioning and action* (pp. 241–278). Hillsdale, NJ: Lawrence Erlbaum Associates.

Maki, W. S., & Leith, C. R. (1973). Shared attention in pigeons. *Journal of the Experimental Analysis of Behavior, 19,* 345–349.

Maki, W. S., & Leuin, T. C. (1972). Information processing by pigeons. *Science, 176,* 535–536.

Maki, W. S., Riley, D. A., & Leith, C. R. (1976). The role of test stimuli in matching to compound samples. *Animal Learning & Behavior, 4,* 13–21.

McClelland, J. L. (1986, November). A thumbnail sketch of connectionist modeling. In *Connectionist models and psychological evidence.* Symposium conducted at the meeting of the Psychonomic Society, New Orleans.

McClelland, J. L., & Rumelhart, D. E. (1988). *Explorations in parallel distributed processing: A handbook of models, paradigms, and exercises.* Cambridge, MA: MIT Press.

McCloskey, M., & Cohen, N. J. (1989). Catastrophic interference in connectionist networks: The sequential learning problem. In G. H. Bower (Ed.), *The psychology of learning and motivation: Advances in research and theory* (Vol. 24, pp. 109–165). New York: Academic.

Nakagawa, E. (1986). Overtraining, extinction and shift learning in a concurrent discrimination in rats. *Quarterly Journal of Experimental Psychology, 38B*, 313–326.

Rescorla, R. A., & Wagner, A. R. (1972). A theory of Pavlovian conditioning: Variations in the effectiveness of reinforcement and nonreinforcement. In A. H. Black & W. F. Prokasy (Eds.), *Classical conditioning II: Current theory and research* (pp. 64–99). New York: Appleton–Century–Crofts.

Riley, D. A. (1968). *Discrimination learning.* Boston: Allyn & Bacon.

Riley, D. A., & Roitblat, H. L. (1978). Selective attention and related cognitive processes in pigeons. In S. H. Hulse, H. F. Fowler, & W. K. Honig (Eds.), *Cognitive processes in animal behavior* (pp. 249–276). Hillsdale, NJ: Lawrence Erlbaum Associates.

Rumelhart, D. E. (1986). A parallel distributed account of the acquisition of past tense in English. In *Connectionist models and psychological evidence.* Symposium conducted at the meeting of the Psychonomic Society, New Orleans.

Rumelhart, D. E., Hinton, G. E., & Williams, R. J. (1986). Learning internal representations by error propagation. In D. E. Rumelhart, J. L. McClelland, & the PDP Research Group, *Parallel distributed processing: Explorations in the microstructure of cognition. Vol. 1: Foundations* (pp. 318–362). Cambridge, MA: MIT Press.

St. John, M. F., & McClelland, J. L. (1990). Learning and applying contextual constraints in sentence comprehension. *Artificial Intelligence, 46*, 217–257.

SAS Institute, Inc. (1982). *SAS user's guide: Statistics, 1982 edition.* Cary, NC: SAS Institute, Inc.

Sutherland, N. S., & Mackintosh, N. J. (1971). *Mechanisms of animal discrimination learning.* New York: Academic.

Urcuioli, P. J., Zentall, T. R., Jackson-Smith, P., & Steirn, J. N. (1989). Evidence for common coding in many-to-one matching: Retention, intertrial interference, and transfer. *Journal of Experimental Psychology: Animal Behavior Processes, 15.* 264–273.

Wagner, A. R., Logan, F. A., Haberlandt, K., & Price, T. (1968). Stimulus selection in animal discrimination learning. *Journal of Experimental Psychology, 76*, 171–180.

Zentall, T. R., Sherburne, L. M., Steirn, J. N., Randall, C. K., Roper, K. L., & Urcuioli, P. J. (in press). Comnmon coding in pigeons: Partial versus total reversals of one-to-many conditional discriminations. *Animal Learning & Behavior.*

Zentall, T. R., Steirn, J. N., Sherburne, L. M., & Urcuioli, P. J. (1991). Common coding in pigeons assessed through partial versus total reversals of many-to-one conditional discriminations. *Journal of Experimental Psychology: Animal Behavior Processes, 17*, 194–201.

17

A Comparative, Hierarchical Theory for Object Recognition and Action

Mark Rilling,
Luke LaClaire,
and Mark Warner
Michigan State University

"It seems likely that when animals, such as pigeons, view the world about them, they perceive objects and discrete events in that world as such, and respond appropriately to them" (Riley, 1984, p. 333).

The goal of research in comparative perception is an experimental analysis of the behavior of organisms with respect to the complexity of the environmental events that form the basis of action (Stebbins & Berkley, 1990). Such research is greatly facilitated by a theoretical framework. As the date of the aforementioned quote indicates, Al Riley was one of the first investigators in animal cognition to recognize the importance of theories originally developed to explain human perception as a conceptual framework for guiding for research in animal perception. This chapter is an effort to heed Riley's call by implementing a research program on object recognition in pigeons. Simply put, this chapter uses theories that have been developed to study object recognition with human subjects as a framework for comparative research on object recognition with pigeons.

A central problem for research in comparative perception is to identify a psychological unit for the meaningful experimental analysis of complex stimuli. Skinner (1935) identified the problem when he described the task of the experimental analysis of behavior as breaking the environment into parts that retain their identity across time. Riley (1984) chose the object as the unit of analysis at a time when most investigators were biased toward smaller units such as simple features. This chapter assumes that contemporary theories of object recognition provide a comparative framework for identifying the units that animals extract from the visual environment.

Biederman (1990) defined object recognition as the activation in memory of a representation of a stimulus class—for example, a cube or birds—from an image projected on the retina by an object. A key criterion for object recognition is that the stimulus input is extremely variable, as when an object is viewed from different orientations or when different exemplars of a stimulus class are employed, yet the indicator response for object recognition remains invariant. As Hummel and Biederman (1992) pointed out, the most striking feature of human visual object recognition is that the input to the retina is extremely variable, yet the output response of object recognition from the subject is invariant. In fact, the task of a theory of human object recognition is essentially to account for these invariances.

Our strategy for comparative object recognition is based on the assumption that object recognition is organized hierarchically. The hallmark of a hierarchical system is that there is no single representation of an object. A hierarchical system is a series of levels of increasing complexity—for example, simple features, an intermediate representation, with the final level a representation or representations of a whole object. Species may vary in whether the final level of the system is a feature or whole object.

Our view was inspired by Palmer (1977) who rejected the dichotomy between analytic and holistic processing in favor of a hierarchical approach. An hierarchical approach makes neurophysiological sense, because it now appears that the complexity of the representation of visual scenes increases progressively across the visual system and that visual information is processed simultaneously in several visual areas (McClurkin, Optican, Richmond, & Gawne, 1991).

One alternative to a hierarchical framework is a dichotomy between simple features and whole objects. The question is whether the final stage of the perceptual process yields objects as wholes or constituent features selected from a more complex whole. It is not necessary to assume that each species' system for object recognition actually ends with the psychological representation of object as the highest level of the hierarchy. For example, fish may only extract features without necessarily integrating these features into whole objects. When faced with a complex visual stimulus, does the organism parse the array into objects as wholes, not features; or does the organism parse constituent features, not objects? In the quotation that begins this chapter, Riley (1984) placed comparative perception squarely in the holistic camp. Until very recently, researchers of animal behavior favored analytic mechanisms for perception, such as the sign stimuli of the ethologists, so Riley's view runs against a long tradition of parsimony in animal behavior.

If our thesis of hierarchical object recognition is correct, then the distinction between analytic and holistic processing for object recognition is somewhat overdrawn. Therefore, Riley's (1984) hypothesis that pigeons perceive objects remains a viable theoretical possibility.

MIDSEGMENTS AND VERTICES AS FEATURES

Our technique for identifying features in pigeons was inspired, in part, by an unlikely source, Lewis Carroll's *Alice in Wonderland*. Lewis Carroll gave the Cheshire cat the ability to vanish at will. The grin was the last feature remaining before the representation of the cat completely vanished from Alice's head. Similar to Carroll, contour deletion was the first technique employed to identify the features by which pigeons represent line drawings in memory. The procedure varied the percentage of contour deleted.

Pigeons were exposed to simple line drawings that were presented on a video screen. First, the pigeon learned a simple successive discrimination between one line drawing that was associated with reinforcement and another line drawing that was associated with extinction. Second, after reaching a criterion of discrimination, various percentages of contour were deleted, either from the midsegments or from the vertices, during a stimulus generalization test. The number of pixels from which the vertices and midsegments were constructed was held constant across each percentage of deletion. Therefore differences that might emerge between the two conditions could not be attributed to the number of pixels used to construct the contour elements.

The dependent variable was the rate of responding to each part during the stimulus generalization test. If midsegments exerted more control over responding than vertices, then the rate of responding to midsegments should be higher than the rate of responding to vertices. If vertices were represented in memory as a stronger feature than midsegments, then the rate of responding to vertices should be higher than the rate of responding to midsegments.

The assumption underlying this approach is that during the acquisition of the discrimination, a representation or multiple representations of the contour elements used to construct the line drawings are formed in memory. The assumption tested is that the representation is not iconic, but composed of a primitive unit or units less complex than the whole object. The primitive units can be identified from those parts of the object that produce the highest rates of responding during the stimulus generalization test. The weight assigned to each primitive unit or feature is a function of the rate of responding to each component during the stimulus generalization test.

The technique of contour deletion as a tool for identifying features was employed by Blickle (1989) in a dissertation carried out in Biederman's laboratory. Line drawings such as a square or a cube, can be conceived as constructed from two different classes of contour elements. In turn these contour elements form two different potential features: midsegments and vertices. A feature theory usually assumes that some contours are represented in memory more strongly than other contours. Consider the predictions from a feature model. If the pigeons assign more weight to midsegments than vertices, then deleting the midsegments, which leaves the vertices intact, produces a larger decrement in responding than deleting the vertices. If the pigeons assign more weight to vertices than midsegments, then deleting the vertices, which leaves the midsegments intact, produces a larger decrement in responding than deleting the midsegments.

As a null hypothesis, consider the idea that the pigeon encodes a line drawing as a holistic template in which each pattern element or set of pixels is assigned an equal weight. A holistic template model with equally weighted pattern elements predicts no differences in the rates of responding for vertex and midsegment deletion as long as each test stimulus is constructed from the same number of pixels. Therefore, a holistic template theory predicts equal generalization decrements for responding to pattern elements constructed from contours deleted either at midsegments or vertices, provided the number of pixels deleted is the same for each type of contour. For example, a pigeon could represent a square either as a holistic template of the entire shape or as a set of features that represent the pattern elements of the shape. In a holistic template model, the square is represented in memory as a replica of the retinal stimulation projected by the square. In a feature model, there is a correspondence between the retinal stimulation provided by the contours of the square and features such as vertices or midsegments.

Two conditions with different line drawings were employed. In Condition 1, the positive stimulus was a square associated with reinforced responding on a fixed-interval schedule, whereas the negative stimulus was an isosceles triangle associated with extinction. The purpose of Condition 2 was to extend the results of Condition 1 to slightly more complex forms. The S+ was a cube and the S− was a planar projection of a three-dimensional truncated pyramid. The stimuli are presented in Fig. 17.1. Seven pigeons served as subjects for Condition 1, whereas six pigeons served as subjects for Condition 2.

After acquisition of the discrimination, 25%, 50%, or 75% of the contours for the positive forms were deleted either at midsegments or vertices during a stimulus generalization test in which responding to the test stimuli was extinguished. At each percentage of contour deletion, the number of pixels remaining for each of the two contour-deleted rectangles was identical. For midsegment deletion, these pixels were removed from the midpoint of each line, whereas for vertex deletion, these pixels were removed from the corners. Because the vertex and midsegment elements were each constructed from the same number of pixels, the two line segments that met to form the vertex were half the length of the midsegment. In order to determine the relative weight assigned to midsegments and vertices, one feature remained intact after contour deletion of the other.

The upper and lower panels of Fig. 17.2 shows the results of the experiment for conditions 1 and 2 respectively. Contrary to the prediction from holistic template matching theory, the data in Fig. 17.2 demonstrate that the amount of

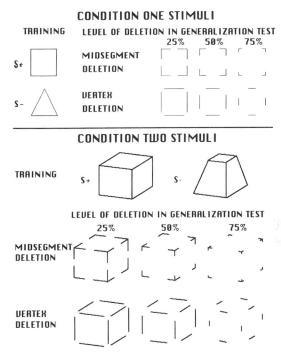

FIG. 17.1.   Upper panel: stimuli employed during discrimination training, a square (S+) versus a triangle (S−), and generalization testing for
Condition 1. Lower panel: stimuli employed during discrimination
training, a planar projection of a cube (S+) versus a truncated pyramid
(S−), and generalization testing for Condition 2. Generalization testing
involved contour deletion of either midsegments or vertices.

stimulus generalization depends on the locus of contour deletion. As compared
with the number of responses to the unoccluded S+, the square, each percentage
of contour deletion produced a decrement in the number of responses per trial.
Therefore, both vertices and midsegments acquired stimulus control as features
of the square during the acquisition of the discrimination. Midsegments acquired
more stimulus control than the vertices in both Condition 1 and Condition 2 as
demonstrated by more responding to midsegments than vertices at each level of
contour deletion in the top and bottom panels of Fig. 17.2. Therefore, as we
expected, a holistic template matching theories fails to account for the data. The
data do support a feature theory because the rate of responding to midsegments
was higher than the rate of responding to vertices.

Contour deletion was employed as the first technique for identifying the
features by which pigeons represent objects. Converging evidence from different
operations provides stronger evidence for cognitive concepts than reliance on a

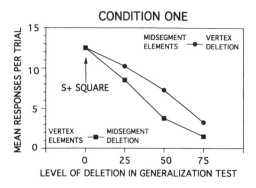

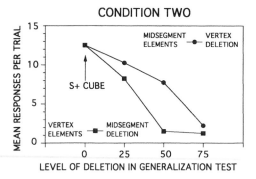

FIG. 17.2. Upper panel: average number of responses per trial during generalization testing as a function of the percentage of contour deletion in Condition 1. Lower panel: average number of responses per trial during generalization testing as a function of the percentage of contour deletion in condition 2. The point at 0% represents responses to S+ with no contour deletion, whereas other points represent varying percentages of contour deletion.

single technique. Therefore, following a suggestion from Ed Wasserman, a second technique for identifying features in pigeons was breaking and expanding or "exploding" the line drawings either at midsegments or vertices. This technique holds constant the number of pixels or contour between the training and the novel stimuli introduced during probe testing, but introduces a gap between parts. The size of the gap was varied. The question of interest was whether introducing a gap at each of the midsegments would produce a greater stimulus generalization decrement than introducing the gap at each of the vertices.

Contour breaking was designed to control for several possible alternative interpretations of the results of the experiment described previously. Contour deletion holds constant the area of the screen occupied by contour during training and testing, but it introduced a brightness difference because some black pixels present during acquisition were replaced by white pixels during the stimulus generalization test. Therefore some of the absolute, but not the relative, decrements in responding for vertex and midsegment deletion could be attributed to brightness differences between training and testing. Another variable that might influence the representation of S+ is the selection of S−. Spence's (1937) well-known model of stimulus generalization predicts that the rate of responding to S+ will be strongly influenced by the similarity of positive and negative stimuli.

Therefore, the third experiment introduced a circle as the negative stimulus. The contour of the circle is different from the straight lines from which the negative stimuli were constructed in Experiment 1. The positive stimulus was the two-dimensional square employed in Experiment 1. During the stimulus generalization test, the rectangle was broken into four parts by breaking the object either at the four midsegments or the four vertices. Also, the fixed-interval schedule was increased from 10 to 20s. In other respects the procedure for the control experiment was the same as the procedure for Experiments 1 and 2.

The results of the second technique for identifying features are presented in Fig. 17.3. This figure shows higher rates of responding to forms containing

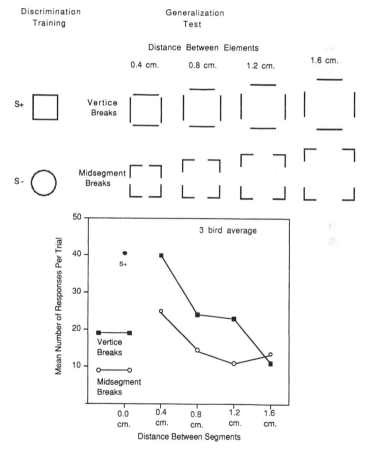

FIG. 17.3.   Upper panel: stimuli employed during discrimination training. In generalization testing contour is broken and expanded either at each vertex or at each midsegment. Lower panel: mean number of responses per trial as a function of distance between segments for breaks between lines at vertices or midsegments.

midsegments than to forms containing vertices. The absolute rate of responding decreased as a function of distance between contours. Evidence from two different techniques for identifying features shows that midsegments are represented more strongly than vertices. Although it is possible that the stimulus selected as S− may influence the representation of S+, midsegments were represented more strongly than vertices independently of whether S− was a triangle (Experiment 1) or a circle (Experiment 2). Therefore, the results appear to reflect the basic cognitive processes by which pigeons represent objects, rather than differences between S+ and S− or procedural differences between training and testing. The data from both experiments imply that models with a provision for features provide a more attractive interpretation of the data than template matching models. In Conditions 1 and 2 of the first experiment, the absolute decrement in responding may be attributed, in part, to the overall loss of contour. However, the second experiment held contour constant between training and testing. Taken together these two experiments demonstrate that pigeons extract features from line drawings.

In the second experiment, the absolute decrement in responding as a function of the distance between contour elements requires explanation. Breaking and expanding the contour reduces the proximity between the contour elements that was present during training. Proximity may be one of the organizing principles by which pigeons integrate contour elements of line drawings. Steele (1990) also obtained evidence for proximity as one of the mechanisms underlying pattern recognition in pigeons. After training pigeons to discriminate one letter from another, an interference task was introduced by adding additional letters as distractors. The proximity manipulation varied the distance between the letters. Steele found support for the proximity mechanism with the finding that the interference was greatest when the distance between letters was least. Loss of proximity may explain part of the absolute decrement in the rate of responding in Experiment 2. Thus our data support Steele's interpretation that proximity is one of the variables underlying object recognition in pigeons.

In a speeded naming task where the stimulus duration was less than 1 s, Biederman (1987) found that vertices were represented more strongly than midsegments. Biederman's data demonstrating the superiority of vertices over midsegments were obtained in a dissertation by Blickle (1989) conducted in Biederman's laboratory. The pigeons in our experiments showed the opposite outcome. It is not clear why. This difference is not due to the greater complexity of the forms employed by Biederman, because Wasserman, DeVolder, Van Hamme, and Biederman (1990) in an experiment with pigeons obtained results similar to ours with the more complex forms employed by Biederman for human subjects. Species differences in visual information processing between people and pigeons also might account for these differences; but we agree with Blough (in press) that conclusions about species differences in object recognition are premature because of the limited availability of data on comparative object recognition.

In Condition 2 of Experiment 1, the stimulus was a planar projection of a cube, an object that people perceive as three-dimensional. This raises the interesting question of whether pigeons perceive such objects as three-dimensional. From our data, there is no evidence that pigeons perceive such line drawings as three-dimensional. Although Condition 2 of Experiment 1 confirmed Cerella's (1990a, 1990b) finding that pigeons discriminate between different three-dimensional orientations of the same object, our experiment was not designed to address the question of whether the pigeon perceived depth. Herrnstein (1984) conservatively interpreted one of Cerella's experiments in which the pigeon learned to discriminate between planar projections of a cube as evidence that pigeons discriminate between different patterns without recovering the three-dimensional structure from the line drawings. Two separate questions arise from work in which pigeons discriminate among planar projections of three-dimensional objects. The first is, has the pigeon acquired an object concept independently of whether the structure represented is two- or three-dimensional? The second is, given that the pigeon has acquired a concept, has the pigeon recovered the three-dimensional structure? Given the findings that pigeons readily acquire concepts when presented with wide ranges of different objects (e.g. Wasserman, Kiedinger, & Bhatt, 1988), we think it likely that given sufficient exemplars and adequate training conditions, pigeons will acquire an object concept when planar projections of three-dimensional objects are employed as discriminative stimuli in concept learning experiments with pigeons. Therefore, when pigeons discriminate between planar projections of three-dimensional objects, we conclude that pigeons may acquire a concept of the object without necessarily recovering the three-dimensional structure.

EXTRAPOLATED MOVEMENT AND "CATCHING"

Our research on extrapolated movement and "catching" in the pigeon may strike the reader as an abrupt jump from our previously described research on feature extraction. The conventional wisdom is to investigate object recognition separately from action. Because perception and the biologically appropriate action presumable evolved simultaneously, we have long been interested in the link between perception and action.

By sitting still and scanning the visual environment an organism can assess the risk/reward potential of various actions. Suppose an organism detects a predator moving about its environment. The organism has the choice of sitting still or engaging in various defensive behaviors such as running away. Field research suggests that a wide range of species have the ability of extrapolate the movement of a predator and respond appropriately. For example Burger, Gochfeld, and Murray (1992) examined the ability of basking iguanas to discriminate the risk from a person walking directly toward the iguanas compared with the risk from a person walking tangentially by them. Similar research on anti-predatory behavior was conducted by Ristau (1991) with the piping plover. The piping plover is a shorebird that typically nests on sandy beaches in the eastern United States. During the breeding season, there is a period of about 2 months during which the eggs are incubated and the young have not yet learned to fly. During this time, the nest is especially susceptible to predation. The piping plover has evolved a broken-wing display that is an adaptive specialization that distracts predators from the nest. The behaviors in this display are so dramatic that they have caught the attention of ethologists. In the full display, the plover stretches its wings so that they drag along the ground. This display occurs while the animal is moving forward along the sand away from its nest. When the plover has reached a point far away from the predator, it flies away. The adaptive value of this display is that it reduces predation by distracting the predator from the nest site.

In a field experiment, Ristau (1991) varied the path of a predator and demonstrated that the probability of the display is highest when a predator is on a trajectory in which the nest falls on its path. The surrogate predators were human. For the "safe" path, the predator moved along a trajectory tangent to the nest. For the "dangerous" path, the observer moved along a path directly toward the nest. The dependent variable was the percentage of trials during which the plover left the nest. The plover left the nest on 52% of the trials when the approach was along the dangerous path.

For an experimental psychologist, the question raised by this antipredatory behavior is, what are the mechanisms underlying this antipredatory behavior? This kind of experimental analysis requires a trip to the laboratory. One possibility from Ristau's (1991) and Burger et al. (1992) data is that the plover and iguana extrapolate the path of the predator. If the path intersects with the nest, then the plover leaves the nest and the injured wing display is presented to the predator. If the path does not intersect the nest, then the plover remains on the nest incubating the eggs.

Something is always lost during the trip from the field to the laboratory. In our case we lost the iguana and the plover and substituted our experimental organism, the pigeon. Pigeons do not appear to exhibit the adaptive specialization of the broken wing display, but they may have evolved an ability to extrapolate the path of a moving object. Our question was, can the pigeon acquire a concept of extrapolating movement?

Figure 17.4 illustrates the stimulus array for the experimental procedure. Eight circles were arranged in a diamond configuration and enclosed by lines. The width of the outer frame of the diamond was 11.5 cm and the invisible frame within which the circles were drawn on the screen on the Macintosh computer was 10 pixels square. In our simulation, these eight circles corresponded to eight potential nest sites and were potential goals for a visually moving target. The task of the pigeon was to peck at one of the eight goals that lay on the linear path of

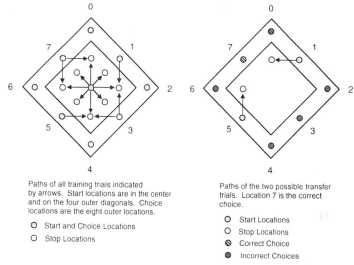

Paths of all training trials indicated by arrows. Start locations are in the center and on the four outer diagonals. Choice locations are the eight outer locations.

O   Start and Choice Locations

O   Stop Locations

Paths of the two possible transfer trials. Location 7 is the correct choice.

O   Start Locations

O   Stop Locations

◎   Correct Choice

●   Incorrect Choices

All circles, when visible, were solid black. Arrows and numbers were never shown.

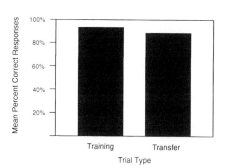

FIG. 17.4.   Upper panel: paths of the moving target during acquisition of the discrimination (left panel) and the test for transfer (right panel) for the extrapolated movement experiment. Lower panel: mean percentage of correct responses for two pigeons during discrimination training and transfer.

the moving target. In terms of a concept formation experiment, the experiment was designed to determine to what extent the pigeon could master a concept of whether a point lay on a straight line. The basic task was to visually track a moving computer generated circle through half of its linear path and then peck at the location at which the circle would have arrived had it completed its path.

At the beginning of the trial, the eight goals were not visible. A trial began with the presentation of a single circle. A peck at the circle started movement of the circle along the linear path toward one of the eight invisible goals indicated by the arrows in Fig. 17.4. The velocity of movement was 0.83 cm/s. When the circle reached the midpoint of its path at which time it was equidistant from three possible goal locations, it disappeared, and the eight potential goal circles were now illuminated simultaneously. A peck at the correct circle produced food reinforcement, whereas a peck at any other locations ended the trial without reinforcement.

A series of successive approximations was employed in order to develop final accurate performance on the discrimination. At first, the entire trajectory from start to goal was visible, so the circle stopped within the goal, at which time a peck at this goal produced food. Then, the stop location was gradually backed off from the correct goal, until the circle moved through 50% of its entire path.

In order to encourage the acquisition of a concept of extrapolated movement, we eliminated the start and stop locations as unique cues for the correct goal location. The experiment was designed so that the pigeon could solve the task by representing at least two points, but not solely by representing the beginning or end points.

Fourteen paths were used for training. Eight started at the center and moved outward and 6 started at one of four outer circles and moved toward another outer circle. After reaching a criterion of about 90% correct on two successive sessions, the two pigeons were tested for transfer to two novel paths not previously presented. Acquisition required about 60 sessions. These two paths are indicated by the right panel of Fig. 17.4.

The bottom panel of Fig. 17.4 shows the results of the experiment. The percentage of correct responses on the transfer test with novel trajectories is almost as high as the percentage correct during acquisition.

What strategy are the birds employing to master this task? Have the birds learned a concept of extrapolated movement? If the pigeon solves the problem with a concept, which is perhaps equivalent to an algorithm for determining whether a point lies on a straight line, then it is necessary for the pigeon to represent at least two points along the line. At least the start and stop locations were eliminated as sole cues for solving the problem. Fig. 17.4 shows that each start location led to several different stop locations. Each stop location, the point at which the circle disappeared, was equidistant from three potential goal locations, so the birds could not use the stop location alone as a cue for the correct choice. The high percentage of correct responses on the transfer test demonstrates that the pigeons were not relying on the start or stop locations as sole cues; rather the bird was basing its choice on some information associated with the path of movement. Based on this narrow set of conditions, definitive conclusions about the mechanisms employed by the pigeons to solve this problem are premature. Nevertheless, we entertain as a hypothesis worth further testing that the pigeons in our experiment learned to represent the path of a straight line and used this information to select the goal where pecks were reinforced. At least, we think that this type of experiment asks investigators to consider how object recognition serves action.

A HIERARCHICAL THEORY FOR COMPARATIVE OBJECT RECOGNITION

Comparative behavioral research on object recognition is now a topic attracting attention in several laboratories. These investigations are aimed at identifying the mechanisms underlying object recognition in a variety of species. Our approach to object recognition is a model with a hierarchical architecture. The input to a hierarchical model is a visual array composed of stimulus elements and the output of the model is an indicator response that provides an index of object recognition. The core of the model is a series of layers. Successive layers reflect enhancement of the complexity of the representation from individual features to the object. The essence of the idea is that object recognition consists of a series of stages and the goal of research is to identify which stages make sense for a particular species. For simplicity, the focus of the model in this chapter is object recognition in pigeons. Somewhat surprisingly, even at this early stage of re-search, a rather clear picture has emerged about which stages are necessary for explaining object recognition in pigeons. The model is intended to provide a framework for integrating data from different laboratories, rather than to test quantitative predictions. Fig. 17.5 provides an overview of the model.

Hierarchical theories, in which the complexity of the representation increases from one level to the next, are employed to explain object recognition in humans (Hummel & Biederman, 1992; Palmer, 1977; Treisman & Gormican, 1988). As stated in the introduction to this chapter, a hierarchical approach has been implic-it in much of the recent work on object recognition in pigeons, so this chapter makes an implicit idea explicit.

An organism faces a complex visual environment. Attention is the process by which the organism concentrates its information-processing capacity on a subset of the visual field. Following Kinchla (1992), attention refers to the processing tradeoffs that occur when an organism is simultaneously presented with a pattern containing multiple stimulus elements. Attention may operate throughout the hierarchical object recognition system proposed here. For example, attention may operate as suggested by Treisman and Gelade (1980) to bind features togeth-er at the same spatial location in the representation. Alternatively, attention may operate in the opposite fashion, as suggested by Hummel and Biederman (1992), as an inhibitory process preventing the accidental binding together of features not part of the same object.

Evidence that attention plays a role in object recognition in pigeons has been obtained by P. Blough (1991). In a visual search task using reaction time as the

Stages in Comparative Object Recognition

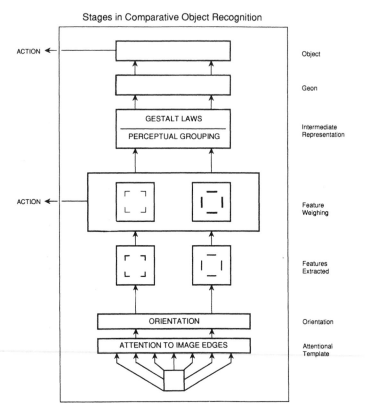

FIG. 17.5. A comparative hierarchical model for object recognition and action.

dependent variable, P. Blough obtained convincing evidence for sequential priming in pigeons. Reaction time was faster when the target was preceded by an uninterrupted series of trials with the same target than when the prior occurrence of the target was random. Priming may provide the mechanism for the concept of "search images" employed by the ethologists. With evidence for selective attention in hand, additional experiments can be designed to determine if the mechanisms for selective attention in pigeons include feature binding and/or inhibitory processes.

One of the first problems facing any visual system is segregating the figure, or object of interest, from the ground. By attending to changes in the texture of the visual array, an organism can subject appropriate subsets of this array to further processing. In a series of convincing experiments, Cook (chapter 14 of this volume) demonstrates that edge extraction via texture is an early perceptual mechanism for pigeons. For example, in one of Cook's experiments the pigeon viewed a video display constructed entirely from small dots. An "edge" was produced by an abrupt change in the density of dots within a target region as compared with the density of dots in the surrounding area. The task of the pigeon was to learn a discrimination in which a peck at the target region, which randomly alternated between the left and right halves of the display, produced food. The target was a square of dots. After learning to peck the target, the density of the dots was varied on transfer tests. In one condition, the edge was eliminated gradually by increasing the density of the dots on either side of the square. In the other condition, the edge was eliminated by abruptly increasing the density of the dots between the square and the surround. The results of a transfer test showed that the percentage of trials when the target was detected were much higher for the target profile with a sharp edge than for the target profile with a gradual slope. Cook's data strongly support the view that pigeons attend to edges and use this information to attend to psychologically relevant contours and direct attention away from psychologically irrelevant aspects of the visual array.

The next stage of the model is feature extraction. Differential attention to features produces feature extraction and differential feature weighting. The determination of feature weighting is an empirical matter. The experiments with two-dimensional line drawings, previously described in this chapter, show that pigeons weight midsegments more strongly than vertices. Feature extraction is the best established of the stages in comparative object recognition. Von Fersen and Lea (1990) demonstrated that pigeons may use as many as five features for object recognition. Other recent reviews of research that supports feature extraction for pigeons are those of D. Blough (1991), Blough and Blough (1990), Cerella (1990a, 1990b), and Lea and Ryan (1990). Species may differ in the features extracted from the visual patterns and in the weight assigned to these features.

The model also specifies that the orientation of features is represented in memory. As De Valois and De Valois (1990) pointed out, orientation selectivity is a characteristic of many of the cells within the visual system. In the behavioral literature for pigeons, Honig, Boneau, Burstein, and Pennypacker (1963) demonstrated orientation sensitivity in the pigeon by showing that following discrimination training at one orientation, a generalization decrement is obtained when the orientation of the test stimulus is rotated from that of the training stimulus.

Following feature weighting, the next level is obtained by grouping the weighted features according to the laws of perceptual organization. Although the Gestalt psychologists pioneered the study of perceptual grouping in human perception, very little is known about such perceptual grouping in other species such as the pigeons employed in our research. Among the gestalt laws are proximity, closure, and texture. See Pomerantz and Kubovy (1986) for a contemporary discussion of these gestalt laws.

Originally, the gestalt laws were formulated qualitatively. In order to employ perceptual organization as an independent variable in an object recognition experiment with pigeons, it is necessary to quantify the gestalt law before the experimenter can manipulate the perceptual organization in the stimulus. An article by Palmer (1977) advocates the kind of hierarchical approach proposed here and extends this approach to the gestalt law of proximity. Proximity refers to the closeness of the elements of a form. Palmer defined proximity quantitatively. In a dissertation recently completed at Michigan State University, La Claire (1992) extended Palmer's technique for quantifying proximity to the planar representation of a three-dimensional object, such as the cube depicted in the bottom panel of Fig. 17.1.

La Claire (1992) reinforced the pigeon for pecking at one planar projection of a cube and extinguished the pigeon for responding to a different three-dimensional projection of a cube. A planar projection of a cube was constructed from nine line segments. Proximity was varied by breaking the cube into parts constructed from subsets of the nine lines and introducing a gap between the parts. Following Palmer's (1977) formula, high- and low-proximity values were employed, each of which consisted of the nine line segments from which the cube was constructed. Therefore, each test stimulus was constructed from all nine line segments, but the line segments that were connected varied so as to produce test stimuli of high and low proximity according to Palmer's formula. The two proximity levels were presented as probe stimuli during a stimulus generalization test. Also varied was the separation between the parts, which was either 2 mm or 4 mm. La Claire's dissertation demonstrates that the gestalt law of proximity can be quantified and introduced into an experiment as an independent variable.

For five of the seven subjects in the experiment, the rate of responding to the high-proximity parts was higher than the rate of responding to the low-proximity parts. Unfortunately, the statistical significance of the results was marginal. The rate of responding to each of the four probe stimuli was only slightly reduced from the rate of responding to the reinforced training stimulus, the original cube, so a ceiling effect may have blocked the potential effects of the proximity variable. Fig. 17.3 shows that in this procedure the distance between the parts of the test stimuli is a major variable controlling the rate of responding. The separation between the parts may have been too small to reduce the rate of responding sufficiently to permit the emergence of a proximity effect. It is possible that proximity might emerge as an organizing principle for pigeons when larger distances separate the contour elements. These speculations notwithstanding, perceptual grouping by gestalt laws such as proximity or closure remains a stage of the model that requires further research for confirmation.

The geon level of the model refers specifically to a theory of object recognition developed by Biederman (1987) and expanded into a connectionist network by Hummel and Biederman (1992). Roughly speaking, a geon is a primitive representation intermediate between a feature and the representation of the whole object. A geon corresponds to representations of the parts of an object, components constructed from simple volumetric primitives.

Van Hamme, Wasserman, and Biederman (1992) recently obtained evidence that pigeons represent line drawings via the representation of component parts or geons. First, pigeons were trained to discriminate among four drawings consisting of a single exemplar of four different pictures from which half of the contour was deleted. Each object was associated with reinforced responding at four choice keys mounted around the screen on which the pictures were displayed. The original training picture was constructed by deleting every other edge and vertex from each line of the intact image so that approximately only 50% of the contour remained. This produced two stimuli: an original training stimulus and its complement. The complement became a stimulus for measuring transfer from original training. The pigeon never saw the complete stimulus.

If pigeons represent line drawings in terms of components, then complete transfer to the complementary line drawing should be obtained. Although the complement contains no edges or vertices in common with the original training stimulus, the theoretical assumption is that the complement activates the same geons as the original drawing. The other possibility, advocated over the years by Cerella (1990a, 1990b) and others, is that the pigeon represents line drawings by extracting features. A feature theory leads to the prediction that the pigeons should fail to recognize the complementary images because none of the edges or vertices in the complement were present in the original training stimuli. The results of three experiments showed substantial transfer to the complementary images with performance well above the 25% or chance level. The substantial transfer to the complements, stimuli never seen by the pigeon until the generalization test, supports the idea of an intermediate representation. However in each experiment, a substantial generalization decrement was obtained in which the percentage of correct choice responses to the complementary stimuli was less than the percentage of responses to the original training stimuli. The incomplete transfer supports the idea that pigeons also extract features, such as vertices or midsegments, from line drawings and that these features contribute to object recognition in pigeons. However, in the Van Hamme et al. (in press) experiment, the weight assigned to the component or geon representation was higher than the weight assigned to the local features. The complementary images were discriminated with an accuracy of 70%, much higher than the accuracy when the images were scrambled between training and testing where accuracy ranged between 36% and 42%. The scrambled image holds constant local features but makes the representation of the geon nonrecoverable.

The final stage of the model is the object. Of course, people see a world of objects and events and it is reasonable to agree with Riley (1984) that pigeons do as well. However, in terms of comparative cognition, the object is a psychological construct that requires experimental validation. How can we determine if features, geons, or intermediate representations are assembled or combined into an interpretation of a single object? At present this final stage, the object, is the stage for which the evidence is weakest.

A major problem in the comparative study of form recognition is providing a criterion for identifying perception of a single whole object as opposed to perception of a set of nonintegrated features. A visual search experiment conducted with human subjects by Donnelly, Humphreys, and Riddoch (1991) provides an example of a criterion for "objectness" as a graded characteristic, which means that sets of independent perceptual structures vary in the degree to which they can be selected as single objects according to the gestalt laws of closure, good continuation, and so on. They predict that the features of a "good" pattern are more likely to bind into a single whole object because of gestalt organizing principles. The gestalt organizational principles fail to bind the contour elements of a "bad" pattern into a single form.

Donnelly et al. (1991) tested their theory with human subjects in a visual search task. One example of a good pattern was four L vertices aligned inward so that they appear to form a square by the operation of gestalt principles. A corresponding bad pattern was four L vertices aligned outward or randomly aligned so that they do not appear to form a square. In a task in which the subject searched for a misaligned vertex, parallel processing was obtained when the forms pointed outward. Using the types of visual search tasks developed for pigeons by D. Blough (1991, in press), one could readily determine if pigeons process these features similarly to humans or whether the processing is fundamentally different. This is the type of experiment that would move the construct of an object in animal perception out of the armchair of anthropomorphism and into the laboratory.

In general it appears that object recognition in birds and humans occurs in similar stages. For example, Brown and Dooling (1992) have discovered that Budgerigars, a species of parrot, readily discriminate differences among budgerigar faces. One of their most interesting findings is that their ability to discriminate differences among conspecific budgeriar faces is greater than their ability to discriminate differences among nonconspecific zebra finch faces. The finding that the discrimination is based on features of the faces is clearly consistent with the view of this chapter that birds may organize a complex form, such as a conspecific face, as a whole object. These similarities notwithstanding, one might also look for differences between birds and people in face perception. The adult human has a lifetime of practice in differentiating human faces. Our naive pigeons have very little if any practice differentiating faces. Carey (1992) reported that young children do not form representations of newly encountered faces with the efficiency of adults. Therefore, with very complex and biologically significant stimuli such as those involved in kin recognition one might confidently expect both species similarities and species differences in object recognition.

CONCLUSIONS

Although the problem of how animals see the world has been a subject for research and speculation from the inception of research on comparative perception, it has recently received attention from investigators who study object recognition with pigeons. The thesis of this chapter is that object recognition in pigeons is hierarchical, which means that information processing occurs in a series of stages in which the complexity of the representation increases. Object recognition and action presumably evolved simultaneously. Some of our data suggests that pigeons may have an ability to act upon a moving object by extrapolating its direction of movement.

Attention is the process that directs an organism's attention to biologically relevant aspects of the environment. Cook's research (chapter 14 of this volume) shows that pigeons use texture to discriminate a figure from its surrounding ground. When pigeons view on a video monitor a simple line drawing such as a cube or planar projection of a three-dimensional object, our research demonstrates that midsegments and vertices are extracted as features. Midsegments control a higher rate of responding than vertices. After feature extraction, features are bound together according to the organizational principles of Gestalt psychology such as proximity and closure. Van Hamme et al. (1992) found that pigeons represent line drawings hierarchically as parts or geons in a representation that is intermediate between the simple features and the whole object. Geons support a theory proposed by Biederman (1987). The object is the final level of the theory. Somewhat surprisingly, evidence that the object is the final stage of the representation is weak. The problem that remains unsolved is an experimental criterion that separates the perception of objects from the separation of multiple features that are not bound as objects. The advantage of the hierarchical approach is that it integrates data collected from different laboratories. Clearly this type of model could extend to species other than people and pigeons. Even though renewed interest on object recognition in pigeons is a recent development, the data collected so far fit rather neatly within a hierarchical framework.

REFERENCES

Biederman, I. (1987). Recognition-by-components: A theory of human image understanding. *Psychological Review, 94*, 115–147.

Biederman, I. (1990). Higher-level vision. In D. N. Osherson, S. M. Kosslyn, & J. M. Hollerbach (Eds.), *Visual cognition and action: An invitation to cognitive science* (Vol. 2, pp. 41–72). Cambridge, MA: MIT Press.

Blickle, T. W. (1989). *Recognition of contour deleted images* (Tech. Rep. No. 89–2). Minneapolis: University of Minnesota, Image Understanding Laboratory.

Blough, D. S. (1991). Perceptual analysis in pigeon visual search. In G. R. Lockhead and J. R. Pomerantz (Eds.), *The perception of structure: Essays in honor of Wendell R. Garner* (pp. 213–225). Hillsdale, NJ: Lawrence Erlbaum Associates.

Blough, D. S. (in press). Features of forms in pigeon perception. In W. K. Honig & G. Fetterman (Eds.), *Cognitive aspects of stimulus control*. Hillsdale, NJ: Lawrence Erlbaum Associates.

Blough, D. S., & Blough, P. M. (1990). Reaction time assessments of visual perception in pigeons. In W. C. Stebbins & M. A. Berkley (Eds.), *Comparative perception: Vol. 2. Complex signals* (pp. 245–276) New York: Wiley.

Blough, P. M. (1991). Selective attention and search images in pigeons. *Journal of Experimental Psychology: Animal Behavior Processes*, 7, 292–298.

Brown, S. D., & Dooling, R. J. (1992). Perception of conspecific faces by Budgerigars (Melopsittacus undulatus): I. Natural faces. *Journal of Comparative Psychology, 106*, 203–216.

Burger, J., Gochfeld, M., & Murray, B. G. (1992). Risk discrimination of eye contact and directness of approach in black iguanas (*Ctenosaura similis*). *Journal of Comparative Psychology*, 106, 97–101.

Carey, S. (1992). Becoming a face expert. *Philosophical Transactions of the Royal Society of London*, 335(B), 95–102.

Cerella, J. (1990a). Pigeon pattern perception: Limits on perspective inavriance. *Perception, 19*, 145–159.

Cerella, J. (1990b). Shape constancy in the pigeon: The perspective transformations decomposed. In M. L. Commons, R. J. Herrnstein, S. M. Kosslyn, & D. B. Mumford (Eds.), *Quantitative analyses of behavior: Behavioral appraches to pattern recognition and concept formation* (Vol. 8, pp. 145–163). Hillsdale, NJ: Lawrence Erlbaum Associates.

De Valois, R. L., & De Valois, K. K. (1990). *Spatial vision*. New York: Oxford University Press.

Donnelly, N., Humphreys, G. W., & Riddoch, M. J. (1991). Parellel computation of primitive shape descriptions. *Journal of Experimental Psychology: Human Perception and Performance*, 17, 561–570.

Herrnstein, R. J. (1984). Objects, categories, and discriminative stimuli. In H. L. Roitblat, T. G. Bever, & H. S. Terrace (Eds.), *Animal cognition* (pp. 233–261). Hillsdale, NJ: Lawrence Erlbaum Associates.

Honig, W. K., Boneau, C. A., Burstein, K. R., & Pennypacker, H. S. (1963). Positive and negative generalization gradients obtained under equivalent training conditions. *Journal of Comparataive and Physiological Psychology, 56*, 111–116.

Hummel, J. E., & Biederman, I. (1992). Dynamic binding in a neural network for shape recognition. *Psychological Review, 99*, 480–517.

Kinchla, R. A. (1992). Attention. In M. R. Rosenzweig & L. W. Porter (Eds.), *Annual review of psychology* (pp. 711–742) Palo Alto, CA: Annual Reviews Inc.

LaClaire, T. L. (1992). *The use of features in object recognition in pigeons*. Unpublished doctoral dissertation, Michigan State University, East Lansing.

Lea, S. E. G., & Ryan, C. M. E. (1990). Unnatural concepts and the theory of concept discrimination in birds. In M. L. Commons, R. J. Herrnstein, S. M. Kosslyn, & M. B. Mumford (Eds.), *Behavioral approaches to pattern recognition and concept formation: Quantitative analyses of behavior* (Vol. 8, pp. 165–185). Hillsdale, NJ: Lawrence Erlbaum Associates.

McClurkin, J. W., Optican, L. M., Richmond, B. J., & Gawne, T. J. (1991). Concurrent processing and complexity of temporally encoded neuronal messages in visual perception. *Science, 253*, 675–677.

Palmer, S. E. (1977). Hierarchical structure in perceptual representation. *Cognitive Psychology, 9*, 441–474.

Pomerantz, J. R., & Kubovy, M. (1986). Theoretical approaches to perceptual organization: Simplicity and likelihood principles. In K. R. Boff, L. Kaufmann, & J. P. Thomas (Eds.), *Handbook of perception and human performance: Vol. 11. Cognitive processes and performance* (pp. 1–46). New York: Wiley.

Riley, D. A. (1984). Do pigeons decompse stimulus compounds? In H. L. Roitblat, T. G. Bever, & H. S. Terrace (Eds.), *Animal cognition* (pp. 333–350). Hillsdale, NJ: Lawrence Erlbaum Associates.

Ristau, C. A. (1991). Aspects of the cognitive ethology of an injury-feigning bird, the piping plover. In C. A. Ristau (Ed.), *Cognitive ethology: The minds of other animals* (pp. 91–126). Hillsdale, NJ: Lawrence Erlbaum Associates.

Skinner, B. F. (1935). The generic nature of the concepts of stimulus and response. *The Journal of Psychology, 12*, 40–65.

Spence, K. W. (1937). The differential response in animals to stimuli varying within a single dimension. *Psychological Review, 44*, 430–444.

Stebbins, W. C., & Berkley, M. A. (1990). *Comparative perception: Vol. 2. Complex signals.* New York: Wiley.

Steele, K. M. (1990). Configural processes in pigeon perception. In M. L. Commons, R. J. Herrnstein, S. M. Kosslyn & D. B. Mumford (Eds.) *Quantitative analyses of behavior behavioral approaches to pattern recognition and concept formation* (Vol. 8, (pp. 111–125) Hillsdale, NJ: Lawrence Erlbaum Associataes.

Treisman, A. M., & Gelade, G. (1980). A feature integration theory of attention. *Cognitive Psychology, 12*, 97–136.

Treisman, A. M., & Gormican, S. (1988). Feature analysis in early vision: Evidence from search asymmetries. *Psychological Review, 95*, 15–48.

Van Hamme, L. J., Wasserman, E. A., & Biederman, I. (1992). Discrimination of contour-deleted images by pigeons. *Journal of Experimental Psychology: Animal Behavior Processes, 18*, 387–399.

von Fersen, L., & Lea, S. E. G. (1990). Category discrimination by pigeons using five polymorphous features. *Journal of the Experimental Analysis of Behavior, 54*, 69–84.

Wasserman, E. A., DeVolder, C. L., VanHamme, L. J., & Biederman, I. C. (1990, November). *Recognition by components: Comparataive evaluations of visual descriminations by pigeons.* Paper presented at the meeting of the Psychonomic Society, New Orleans.

Wasserman, E. A., Kiedinger, R. E., & Bhatt, R. S. (1988). Conceptual behavior in pigeons: Categories, subcategories, and pseudocategories. *Journal of Experimental Psychology: Animal Behavior Processes, 14*, 235–246.

18

Absolutes and Relations in Acoustic Perception by Songbirds

Stewart H. Hulse
Johns Hopkins University

The distinction between relational and absolute processing is the major theoretical focus of this chapter. This issue is studied in the context of the capacity of animals to perceive certain complex sounds. Sometimes, animals code auditory information by representing *relations* among sounds. Other times, however, they ignore relations and code sounds on the basis of *absolute features*. Why should this be so? The issue is not unique to auditory perception. The question of how we perceive relations and absolutes pervades much of experimental psychology—from sensory perception to abstract concept formation (Herrnstein, 1990; Medin, 1989; Nosofsky, 1991; Premack, 1978, 1983; Rosch & Mervis, 1975).

As a preview, animals use absolute features to process and remember serial patterns of pitches, that is to learn and remember patterns in which sounds of different pitch come one after another. On the other hand, animals prefer to use relations among sounds to process frequency information that occurs simultaneously. In this case, the relations lie in a sound's organized harmonic structure. In particular, animals show perceptual invariance over changes in pitch for sounds with the same harmonic structure.

Auditory streaming, and new work that is underway on the phenomenon provides the next topic. Auditory streaming is a complex process, but the general idea is that—for humans at any rate—there is a predisposition to hear sounds that hang together on some acoustic basis as if they are a single, coherent auditory object that stands out and moves against a background. Usually, auditory objects are sounds that move in time or space.

Certain things characterize all the work to be covered. First, it has been done with operant techniques using artificial, synthesized, acoustic stimuli. Second, it has been done with songbirds, chiefly the European starling (*Sturnus vulgaris*). However, some work has been done with other species, the mockingbird (*Mimus polyglottos*), and the cowbird (*Molothrus ater*) to assure at least some measure of comparative generality among songbirds.

Third, the work has been stimulated extensively by using human music perception as a metaphor for cognitive principles of auditory perception in animals. The visual world, and the photograph or computer display, are the target and the tools for studies of visual perception and concept formation based on visual stimuli, and that is no less true of research with animals than it is of research with people. For the research to be reviewed here, however, the auditory world is most important, and realizations of that world are the organized formalisms that are characteristic of musical structure and the sounds coming from the computer music laboratory. It is virtually inconceivable that starlings or any other animal have much sense of music as humans know it (although see Porter & Neuringer, 1984). However, as a composer or anyone interested in the cognitive psychology of music perception understands, music depends profoundly on the fact that people hear relations among the acoustic events from which music is built: pitch, timbre, loudness, and rhythm and tempo. Most important, music is based on the fact that people hear these relations as invariant over many types of acoustic transformation. In fact, people hear music much as if it is a collection of auditory concepts based on organized sound relations. Perhaps animals also hear things that way. The guiding hunch in the work described has been that music embodies general principles of auditory perception, principles that might apply to information processing in many species. If so, comparative studies of complex auditory perception should yield important new information about all natural forms of auditory perception, processing, and communication.

Pitch Relations and Their Invariance

To illustrate some things about absolutes and relations and their expression in music, let me describe pitch relations and their invariance. Consider a simple familiar melody, such as *Yankee Doodle*. Suppose, in musical parlance, it is played in the key of C major, and then played again in another key, the key of F# major. All this means, literally, is that the frequencies characteristic of the notes in the first melody have been multiplied by a constant—1.414 in fact—to produce the frequencies of the notes in the second melody. Observe that this is a *ratio transformation*—a simple example of a constant stimulus relation. All this is implied, of course, in the fact that the tones of music are drawn from pitches that are placed at regular intervals on a log scale of frequency.

Ratio transformations are to be distinguished from, for example, *additive transformations*, where notes in one melody might be changed by adding some constant frequency to each of them—such as 200 Hz. Additive transformations do maintain one aspect of a melody, the melody's *pitch contour*—the ups and downs in pitch that occur from one note to the next. But an additive transformation distorts the percept of the melody substantially. To the ear, small pitch changes from note to note in the original melody appear exaggerated in the transposition, and conversely, large pitch changes appear diminished.

For present purposes, there are two important things about the ratio transformations represented by key changes—from C major to F# major or to any other major key. First, the original and the transformed melodies remain perceptually identical with respect to the note-to-note changes in pitch. That is, the melodies are perceptually invariant. They are still the same tune. Second, memory for the pitch relations of familiar melodies is, by definition, excellent and would show a virtually limitless memory span. However, memory for the specific, or absolute, pitches involved in familiar melodies is generally quite poor. Performance on a same–different task testing memory for the pitch of specific notes would show a rapid decrement with time for most people. In other words, most people have good memories for the pitch relations that comprise melodies, but very poor memories for the actual pitches involved. Some people, however—about 3% to 6% of the population—do have perfect pitch (Takeuchi & Hulse, in press; Ward & Burns, 1982). They would show little, if any, decrement with time in their memory for the actual pitches of a melody.

SERIAL PITCH PERCEPTION

Let us turn then to serial pitch perception. This work is older, and it is merely summarized. However, the work lays some groundwork about serial pitch perception that must be understood for future purposes. The research began initially with a very simple question: Could starlings detect whether or not a sequence of tones went up or down in pitch? This is the simplest pitch relation imaginable. Furthermore, it is a pitch relationship that, for humans at least, is independent of the specific pitches involved. That is, rising and falling pitch relations show perceptual invariance. The relation is independent of the absolute pitches in the sequences themselves.

Starlings had little difficulty learning a discrimination between a set of rising and a set of falling pitch sequences (Hulse & Cynx, 1985, 1986; Hulse, Cynx, & Humpal, 1984). This clearly suggested that birds could process and discriminate rising and falling pitch relations. A transfer test for perceptual invariance of the rising/falling pitch relationship over a transformation in frequency was used to see if this was so (Cynx, Hulse, & Polyzois, 1986; Hulse & Cynx, 1985). The test simply halved the frequencies of all the tones in the original baseline set. This procedure lowered the pitches of the rising and falling patterns by one octave. If the birds had learned the rising/ falling pitch relationship, and if they showed perceptual invariance for pitch relations, they should readily transfer the rising/falling discrimination to the new frequencies.

On the day in which the transfer test took place, the birds lost the discrimination immediately and completely. They failed once again when they were shifted to novel frequencies an octave above the original baseline frequency range. However—and this is important—the starlings successfully transferred to rising and falling patterns built with new frequencies within the original baseline range. That transfer was done by shifting the baseline frequencies just a semitone instead of a whole octave, a shift that is easily detectable by starlings. Apparently, the birds' ability to discriminate new rising and falling patterns was restricted to the frequency range in which the discrimination had been trained originally. This restriction was termed a *frequency range constraint* to emphasize the fact that it was a failure to generalize the discrimination outside a range of familiar frequencies. It was not a failure to generalize to any new frequency at all.

Since its original discovery, the frequency range constraint has proven rather general. Besides starlings, it also holds for other songbirds such as mockingbirds (although only one has been tested to date) and cowbirds. D'Amato (1988) found it for Cebus monkeys, so the constraint holds for at least one primate species. Work with Suzanne Page and Richard Braaten showed that the critical range for starlings could be as small as a few hundred hertz or so large as to span virtually the entire frequency range to which starlings are sensitive (Hulse, Page, & Braaten, 1990a). In short, animals in serial pitch experiments appeared to process pitch sequences and show perceptual invariance only within a range of specific frequencies with which they were trained and highly familiar. This observation was the first clue that *absolute pitch* might play a role in frequency perception by songbirds. To this point, however, the evidence was indirect because it was based on a failure to respond on the basis of pitch relations.

Other evidence was quick to follow. Given the frequency range constraint, it appeared possible that the birds were not responding on the basis of whether pitches were going up or down in the pitch-relation experiments at all. Perhaps in the initial rising/falling pitch discriminations they were memorizing the pitch of isolated tones in the patterns and discriminating on that basis. That possibility proved to be correct. A number of experiments have now shown that starlings—and other songbirds—are capable of memorizing the pitch of isolated tones in serial sound patterns. Indeed, they appear to prefer to do so. Given a choice between learning a serial pitch sequence on the basis of pitch relations or absolute pitch features, the birds will opt for absolute features (Hulse, Page, & Braaten, 1990b; Page, Hulse, & Cynx, 1989).

It would be unwarranted to conclude that absolute pitch perception accounts for all the data obtained from discriminations based on serial pitch sequences. For one thing, birds like the starling do transfer successfully to novel rising and falling patterns within a familiar pitch range (Hulse & Cynx, 1985). And a number of experiments have shown that starlings can learn to use pitch relations in serial patterns (Page et al., 1989). In fact, the evidence suggests that songbirds—and perhaps other animals—have a hierarchy of perceptual strategies for auditory perception of serial pitch patterns. Absolute pitch processing lies at the bottom of the hierarchy, and relative pitch processing at the top. Animals learn the hierarchy from the bottom up (Hulse et al., 1990b). The striking new discovery, however, is the powerful role that absolute pitch appears to play in the auditory world of some animals.

PITCH RELATIONS IN SIMULTANEOUS PITCH PATTERNS: PERCEPTUAL INVARIANCES IN THE PERCEPTION OF SPECTRAL STRUCTURE

Unlike humans, then, songbirds do not appear to be adept at processing relations in serial pitch patterns. There is another form of frequency organization that is also a candidate for relational perception, however. That organization is based on the *spectral structure* of acoustic stimuli. A brief reminder about complex sounds facilitates the discussion.

A *complex sound* is a sound composed of a fundamental frequency and one or more harmonics, or overtones, where each harmonic is a *multiple* of the fundamental frequency. Any complex sound can be analyzed into a fundamental and a set of harmonics of varying energies (Fourier analysis), and any complex sound can be synthesized by reversing the process. For example, as Fig. 18.1 demonstrates, pick a fundamental frequency—say 500 Hz. Then add sound energy simultaneously at frequencies two, three, and four times the fundamental. That is, add energy at the second, third, and fourth harmonics—1,000, 1,500, and 2,000 Hz. Note that this is, once again, a stimulus constructed on the basis of a

Fundamental Frequency (F)	Harmonic Relations (Integer Ratios)	Spectral Frequencies
500 Hz	1 X F =	500 Hz
500	2 X F =	1000
500	3 X F =	1500
500	4 X F =	2000
700	1 X F =	700
700	2 X F =	1400
700	3 X F =	2100
700	4 X F =	2800

FIG. 18.1.   The frequencies that would be included in a spectrum with energy at a fundamental frequency (F) and at harmonics with frequencies two, three, and four times the fundamental. Spectra are shown for fundamentals of 500 Hz (top) and 700 Hz (bottom). Absolute frequency differences between successive harmonics in the two spectra (right column) change substantially, but both spectra are based on a common integer-ratio relation among the harmonics (middle column).

set of different frequencies, but this time the frequencies are arranged simultaneously, not serially, in time.

Now suppose the fundamental changes to 700 Hz, but the formal harmonic structure remains the same. That is, there is a new fundamental frequency, but the same set of harmonics. Harmonic frequencies are once again two, three, and four times the fundamental frequency: 1,400, 2,100, and 2,800 Hz. The frequencies of the component tones have changed in this new spectrum, but the multiple relation among them has not.

Complex tones with the same harmonic structure have an interesting perceptual property. They all continue to have the same quality as their fundamental frequency changes. In other words, given a common frequency spectrum, tonal quality remains perceptually invariant over changes in fundamental frequency.

In the parlance of a musical metaphor, once again, all the foregoing is a description of an acoustic factor that is a primary determinant of the *timbre* of a tone. Any given musical instrument—an oboe or a piano—has its own unique timbre, that is, its own unique sound quality. Furthermore, each instrument largely retains its own unique sound quality no matter what pitch it plays.

Given the idea of spectral structure and timbre, perhaps the next question is obvious. Might it be that songbirds—like humans—would show perceptual invariance for sounds of the same spectral structure? That is, might birds perceive sounds with a common harmonic structure as having the same sound quality regardless of the sounds' fundamental frequencies?

Suzanne Page first tackled this problem in her doctoral dissertation (Page, 1989). In a task in which birds had to judge similarities among sound qualities, she was able to show that starlings could indeed classify sounds on the basis of spectral structure. For example, they could readily tell the difference between sine waves and square waves regardless of fundamental frequency. Sine waves and square waves differ solely in their harmonic spectra. A sine wave only contains energy at a single fundamental frequency. Square waves contain energy at the fundamental frequency and at frequencies that are odd integer multiples above the fundamental.

Braaten and Hulse (1991) examined starlings' perception of synthesized harmonic spectra in another way. They trained starlings to discriminate between one positive spectral pattern for which responses were reinforced, and seven other negative spectral patterns for which responses were not reinforced. Some representative harmonic structures that were used as stimuli are shown in Fig. 18.2.

The S+ spectrum shown in the figure always contained equal energy at the fundamental and the fourth harmonic. The seven S− spectra (three examples are shown in the figure) contained equal energy at the fundamental and at none, or one or more of the second, third, and fourth harmonics. The S+ spectral pattern and the seven S− spectral patterns were combined with five fundamental frequencies to generate 35 different discrimination problems. Each day, the starlings

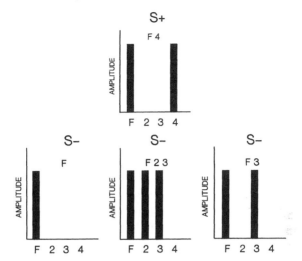

FIG. 18.2. Sample spectra from the spectral discriminations studied by Braaten and Hulse (1991). Spectra were generated with equal energy at the fundamental (F) and at one or more of the first, second, and third harmonics. The spectrum marked F was a sine tone with energy only at the fundamental.

faced a new discrimination problem involving the S+ spectrum and one of the S− spectra. From one day to the next, the S+ spectrum was always the same, but the S− spectrum changed. The fundamental frequency of the S+ and the S− spectral patterns was the same on any given day, so the birds could not discriminate the patterns on that basis. However, the fundamental frequency changed from day to day.

Of course, because fundamental frequencies changed from problem to problem during the course of the experiment, the starlings faced a discrimination between the S+ spectrum and a particular S− spectrum, say F23, when the two spectra sometimes involved one fundamental and sometimes another. This was a direct test for the perceptual invariance of the sound qualities associated with the spectra over changes of the fundamental. If perceptual invariance obtained, there should be immediate transfer of the S+/F23 spectral discrimination from one fundamental frequency to another.

There was a simple procedure for each daily problem. The bird started a trial by pecking on one key in the experimental chamber. Either the S+ or the S− spectrum then played. If S+ played, the bird was rewarded with food if it pecked a key to the right of the first key. If S− played, the bird was required to withhold a peck from the right key for 4 s. Errors produced a brief time out. The discrimination emerged as the bird learned to respond for the S+ spectrum and to withhold responses for the S− spectrum. Errors were recorded when the bird failed to respond to the S+ spectrum (which rarely occurred) or responded to the S− spectrum. A daily session lasted 2 hr.

It is important to remember that the starlings in the experiment got a brand new problem each day for 35 days regardless of how they performed on any given day.

The results showed that the starlings became remarkably adept at spectral discriminations. Mean percent correct was above chance (50%) within the first seven problems. By this time the starlings had discriminated between the S+ spectrum and all seven S− spectra. Over the last block of 7 problems in the set of 35 problems, mean performance was almost 80% correct.

Performance within a problem also became very good. By the last seven problems, the birds were reliably above chance within the first 20 trials on the problem of the day, and performance improved steadily to 80% correct or more by the final 20 trials of the day.

The data critical for a demonstration of perceptual invariance for sound quality associated with spectral structure are displayed in Fig. 18.3, which is slightly revised from Braaten and Hulse (1991). All the birds were tested with one S+ structure and all seven S− structures on three fundamental frequencies, 700, 1,000, and 1,400 Hz. (Other fundamental frequencies were also used, which were not common for all birds, on some problems.) If perceptual invariance were

S-SPECTRA	FUNDAMENTAL FREQUENCY (Hz)			
	700	1000	1400	MEAN
F	60.0	75.0	73.0	69.3
F2	77.2	80.2	77.4	78.3
F23	81.6	79.2	75.0	78.6
F234	81.0	65.8	69.6	72.1
F24	69.2	78.8	78.8	75.6
F3	75.4	62.8	68.2	68.8
F34	73.0	54.8	54.2	60.7
MEAN	73.9	70.9	70.9	71.9

FIG. 18.3.   Discrimination performance (mean percent correct) for five birds for seven S− spectra at three common fundamental frequencies. The S+ spectrum was F24 played with the same fundamental frequency as the S− spectrum for any given problem.

to hold, the birds should do equally well on all three fundamental frequencies for each of the seven problems outlined in the figure.

Just as they did on all problems, the starlings did equally well on the critical problems involving a common fundamental, as the figure shows. From one problem to the next, performance across the three fundamentals varied somewhat, but there was no consistent variation associated with problems, and mean performance across all problems was virtually identical and not statistically significant. This result means that the starlings were transferring the spectral discrimination readily across a change in frequency of the fundamental, and that is direct positive evidence for relational perception of frequency structures. If robust relational perception for serial patterns of pitches was difficult to demonstrate, relational perception was easy to show for formally organized patterns of simultaneous harmonics.

It is also worth noting in Fig. 18.3 that performance on problems in which the S− stimulus contained the second harmonic (F2, F234, and F24) was better than on the other problems, significantly so. As noted later, this observation may have some ecological significance.

It is possible, of course, that the starlings in the experiment were simply developing a learning set and becoming very skillful at solving spectral discrimination problems. To test for this possibility one new problem was presented after the main experiment was done that reversed the stimulus-reinforcement contingencies for the S+ and S− spectral structures. Under these circumstances, the birds performed at chance. Clearly, in the main experiment, they had become skilled at solving spectral discrimination problems on the basis of spectral relations. They had not just become rapid problem solvers.

It is important to remember that the starlings were tackling a new problem every day. Therefore, the 80% level of accuracy reached within each problem was not necessarily an accurate reflection of just how well starlings could do. To assess this, one bird was transferred at the end of the main experiment to a new problem in which it had to discriminate six exemplars of the S+ structure from six exemplars of the S− structures. The starling was correct on 89% of the trials in the first session. It had reached virtual perfection by the 10th session when it was correct 97% of the time.

Summary

What does the foregoing have to say about the issue of relative and absolute perception? To sum up, songbirds—and some other animals so far tested—have substantial difficulty learning organized relations among acoustic frequencies when that information is presented serially. However, they have little or no difficulty processing relations among frequencies when that information is presented simultaneously in the form of an organized acoustic spectrum. Hold the frequency relations in the spectrum constant, and the spectrum keeps the same quality across substantial changes in the specific frequencies involved. The quality of the spectrum is perceptually invariant over changes in pitch. To put this once again in terms of our musical metaphor, starlings are constrained in their ability to perceive and remember the rising and falling pitch relations in a melody to the frequency range in which they learned the melody originally. However, they are very skillful when it comes to identifying the acoustic qualities associated with complex spectra or timbres. In effect, they can easily learn to discriminate a piano, say, from seven other musical instruments. Furthermore, they can generalize the discrimination to whatever pitch the piano or the seven other instruments might play.

The fact that starlings can process the relations that define auditory spectra is fundamentally important for the ideas addressed next. Accordingly, we turn now to the topic of *auditory objects* and *auditory streaming*.

AUDITORY STREAMING

Everyone knows how the sound of someone's voice can be singled out at a noisy party. Attention can also be switched easily from the voice of one person at the party to the voice of someone else who might even be standing out of sight. Although computers are not yet able to do that job, human perception is extraordinarily good at picking specific acoustic signals out of noisy backgrounds and switching attention selectively from one acoustic signal to another. Albert Bregman (1990) has written a marvelous book, *Auditory Scene Analysis*, which describes how sound sources like voices and musical instruments can acquire the characteristic of unique, cohesive *objects* in the environment. Auditory objects are sounds that hang together as unitary perceptual wholes. Sounds become coherent objects to the extent that they have the same quality, come from the same source in space, and move together in time. In the latter sense particularly, they may be said to *stream*—against one another and against a "noisy" auditory background. Familiar gestalt principles are very important for the organization and unique identification of acoustic objects.

Auditory objects are often perceptually invariant, to use a now-familiar idea, for many different kinds of stimulus transformations. The voice of a particular friend, for example, remains qualitatively identical whether they speak normally or raise or lower their voice, or whether they are close at hand or far away. By the same token, the sound quality of an oboe remains the same no matter what pitch it plays.

Another relevant aspect of human perception and auditory streams is that perception tends to actively *partition* the ongoing auditory world into auditory objects. In this sense, one voice in a crowded room emerges from a babble of others. An oboe—or whatever other musical instrument we might choose to attend to—emerges from the rich mixture of other instruments in the ongoing performance of a symphony orchestra. The ongoing auditory world becomes partitioned into separate streams of auditory objects that flow in time with a characteristic rhythm or syntax. Attention is important in this process, because attention can be focused on just one ongoing stream at a time.

Perhaps a visual analogy is useful. Recall one of the well-known reversible figures that are so common in psychology textbooks. For example, there is one that can be seen globally as a black and white configuration or design. However, the configuration also can be broken down and seen as a vase or, by changing attentional focus, as two faces in profile. In the latter case part of the configuration becomes foreground whereas the other becomes background, and attention shifts back and forth between the two. The visual scene is analyzed by the perceptual system into two distinct objects, each having its own characteristic form.

There is now an extensive literature demonstrating that the same thing happens with sounds (Bregman, 1990; Bregman & Campbell, 1971; Bregman & Pinker, 1978; McAdams, 1989). For a simple example, imagine a sequence of tones that can assume one of two pitches, X and O, where X is a high pitch and O is a low pitch. Assume that the sequence is organized as X O X O X X O O, and that it repeats continuously and indefinitely, that is, X O X O X X O O / X O X. . . . If the pitches of X and of O are discriminably different, and if the sequence unfolds at a slow rate or tempo (less than, say, two events per second), one hears a sequence of tones that moves successively up and down in pitch; that is, one perceives the tone-to-tone pitch contour. However, if the tempo of the sequence increases, a rate is eventually reached at which the sequence *splits* or *divides* into two streams, one composed of the high pitches (X - X - X X - - / X - X - X X . . .), and one composed of the low pitches (- O - O - - O O / - O - O - - . . .). Each stream has its own unique rhythm or temporal structure. Like the reversible vase/faces, the listener is aware that both streams are part of a larger melodic line that includes both high tones and low tones (Xs and Os), but attention focuses on a stream of either the Xs or the Os. Attention can shift from one stream to the other, but attention can not focus on both streams at the same time.

Streaming is also easy to demonstrate for other acoustic events, such as sounds that differ in spectral structure or timbre. If X now represents a sound with one spectral structure and O represents a sound with another spectral structure, and if the structures are discriminably different, there is a tempo at which the structures will divide and stream as separate figures, each with its own quality and rhythmic pattern.

What makes sound patterns divide and cohere into separate auditory objects? Bregman (1990), in summarizing this literature (see also, Handel, 1989), pointed out that sounds tend to hang together as streams or objects when they share a common quality or timbre, when they share a common pitch space or pitch range, and when they move together in location and in time. Time is especially important. Sounds become auditory objects to the extent they move or change systematically with the passage of time.

The Perception of Auditory Streams by Starlings

The next question, obviously enough, is whether animals also perceive auditory objects. This question was first put to test with animals by Richard Braaten (1991) in part of his dissertation. The original motivation behind the work was an analysis of serial sound patterns modeled after some data and theory published by W. R. Garner (1974) some time ago. But in the light of what happened, Braaten's work seems at least as relevant to auditory objects as it does to the Garnerian analysis.

Using a standard two-alternative, forced-choice operant technique, Braaten (1991) had starlings learn to discriminate between two serial patterns of sounds much like the ones used as examples earlier. His eight-element patterns were made from two timbres, which are designated by X and O in Fig. 18.4.

On each trial, the starling started an acoustic pattern by pecking a key located in the center of the response panel. The patterns repeated over and over again for 25 s or so—the time varied from trial to trial. Then, after waiting through the listening period, the bird was free to indicate a choice between the patterns by pecking one of two side keys. Responses to the left key were correlated with one pattern and food, responses to the right key were correlated with the other pattern and food. Errors led to a time out.

Learning proceeded without incident, but the task was, in general, difficult. Two birds reached stable performance within 40 daily sessions, but two others required almost 160 sessions before performance stabilized. After performance reached asymptote, transfer tests were undertaken to see if the starlings heard the larger patterns segregate and stream into subpatterns composed separately of the X and O timbres.

The first question was whether or not the starlings had learned something about the temporal structure of the patterns that was independent of the quality of their specific elements—the X timbres and the O timbres. If so, that would satisfy a condition known to be important for streaming to occur. To answer this question, Braaten (1991) substituted entirely new timbres for the original Xs and

X O X O X X O O / X O X O...

X X X X X O O O / X X X X ...

FIG. 18.4. Stimulus patterns and their parameters used to study streaming by Braaten (1991). The patterns, which repeated, were composed of tones with two spectra, X and O, arranged as shown.

X = O = 100 MSEC

Intertone Interval = 100 msec

Pattern Rate = 5 tones/sec

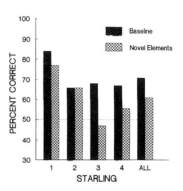

FIG. 18.5. Discrimination performance for baseline patterns (filled bars) and for transfer patterns with novel elements (hatched bars). The discrimination is well maintained with the novel elements. (Redrawn from Braaten, 1991.)

Os. Now, the patterns were made from Y timbres and Z timbres—but the new timbres had the same temporal arrangement as the Xs and Os. As Fig. 18.5 shows, performance dropped somewhat for all starlings with the new patterns. One starling (marked "3" in the figure) performed below chance, and another (marked "4" in the figure) was not significantly above chance. But there was clear evidence that the discrimination was maintained with the new spectral structures for the remaining birds.

This transfer suggests that two of the starlings encoded a representation of the original patterns that was independent of the quality of pattern elements. It follows that they must have been differentiating the patterns on the basis of their temporal structure because that was the only common property of the X-O patterns and the Y-Z patterns. Given this, the next question was obvious.

Assume that the birds had heard the main patterns stream into the two subpatterns defined by the X timbres and the O timbres as they learned the initial discrimination. If so, would they maintain the discrimination if one pattern element—say the O element—was eliminated and the pattern was played with the X elements alone? That should happen because the unique temporal structure of the X elements alone should still suffice for the main discrimination. The same reasoning would hold, of course, if the X elements were removed, and the birds heard the O elements by themselves.

As Fig. 18.6 shows, discrimination performance deteriorated only slightly compared with baseline when the X elements were played alone. Although the data are not shown in Fig. 18.6, virtually identical results were obtained with patterns composed of the O elements alone. Interestingly, the two starlings whose performance had fallen to chance when novel timbres were substituted (Fig. 18.5) did well on these transfers to X-alone and O-alone.

These results suggest that in the baseline discrimination, the starlings were not just listening to a large serial mixture of X timbres and O timbres. Sometimes they were listening just to Xs and sometimes just to Os, and they learned to solve the discrimination on the basis of either subpattern. If so, this is strong evidence that streaming took place during training. That is, the original pattern of Xs and

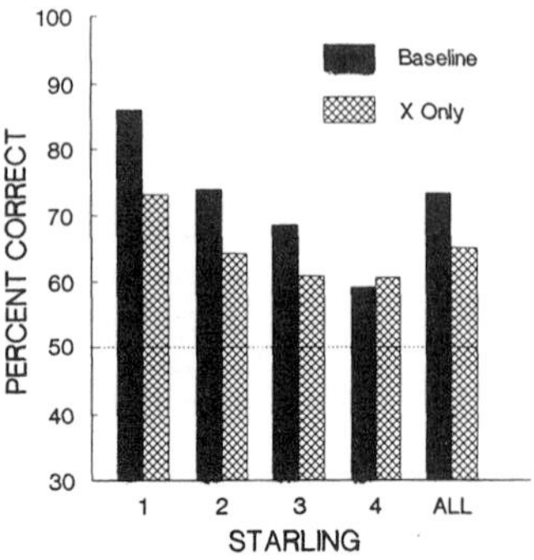

FIG. 18.6. Discrimination performance for baseline patterns (filled bars) and for patterns composed of the X elements only (hatched bars). (Redrawn from Braaten, 1991.)

Os was perceived not as a unitary auditory object composed of both Xs and Os, but as two separate objects, one composed of Xs and the other composed of Os.

Finally, Braaten (1991) varied the speed or tempo of the patterns over a range that both doubled and halved the baseline tempo of 5 tones per second. His intent was to see if there would be perceptual invariance for the temporal structure of the X - O patterns with the tempo change, that is whether the starlings would continue to hear the patterns as the same rhythmically organized structures despite the change in the patterns' presentation rate. This is again analogous to a phenomenon in music perception where the rhythmic structure of a piece of music is perceptually invariant—within broad limits—over faster or slower tempos.

As the 6 data points at the left in Fig. 18.7 show, the discrimination remained above chance over a broad range of pattern rates. There is some suggestion that performance deteriorated at tempos faster than the baseline rate of 5 events per second and at the slowest rate that Braaten (1991) tested of 2.5 events per second. Apparently, the starlings did show perceptual invariance for the patterns involved in the discrimination over this range of tempos.

However, there is another prediction that can be made about pattern discrimination when the tempo of the pattern changes. If the starlings were using streamed patterns to help solve the discrimination, then performance might deteriorate if pattern speed became slow enough to halt the streaming percept. This possibility emerged after Braaten (1991) had finished his dissertation, but his work was a necessary preliminary. Daniel Bernard used Braaten's X - O patterns and the same subjects for an additional test in which the patterns were presented at tempos slower still than the tempos Braaten had used, namely, 1.25 and 0.67 events per second. Under these conditions, as the 2 points at the far right in Fig. 18.7 show, there is no question that the discrimination broke down as rate slowed below 2.5 events per second. Somewhere between 2.5 and 1.5 events per second, the starlings lost the discrimination completely as would be expected if at the faster tempos the starlings perceived the patterns as two distinct streams rather than as a single coherent stream.

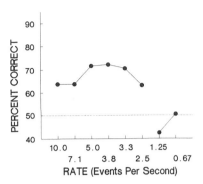

FIG. 18.7. Discrimination performance for X and O spectral patterns played at various rates. The baseline rate was five tones per second. The discrimination was maintained for all rates above two per second, but was lost at lower rates.

There is reason for a certain measure of caution in interpreting these data because the discrimination also could fail if the starlings were unable to integrate pattern information in general. The slowest tempos might have strained short-term memory for successive pattern elements. Only future research can dismiss that possibility.

Nevertheless, to sum up, Braaten's (1991) data are well described by the proposition that starlings are capable of perceiving auditory objects—patterns of streaming timbres in this case. This is an exciting new observation that ought to lead to interesting new developments in comparative perception.

ECOLOGICAL IMPLICATIONS

Perhaps some of those developments can be suggested by turning to the possible ecological significance of the research described in this chapter. To be sure, all the work has been done with artificial stimuli synthesized in the laboratory. That was a deliberate strategy to assure precise control over the physical properties of the acoustic stimuli. The skeptic may say it is all well and good to show that animals can process synthesized sounds this way or that way. But are these sounds of any real significance to the animals? Or are we simply dealing with laboratory curiosities?

Without doubt, starlings in the wild have rarely, if ever, heard ascending and descending patterns of sine tones, or spectral structures composed of just a fundamental and three harmonics. The important thing is that artificial stimuli like the ones we have used incorporate general cognitive principles that could be functionally important in processing the sounds that enter into natural communication. In fact, at least some songbird species do use absolute and relative pitch perception in their natural song—as Weisman and his colleagues have shown (Weary & Weisman, 1991; Weisman & Ratcliffe, 1989; Weisman, Ratcliffe, Johnsrude, & Hurly, 1990). Chickadees, for example, sometimes shift parts of their natural song (a two-pitched *fee bee* phrase) from one pitch level to another with a *ratio* transposition—just like a perceptually invariant musical key change. They do so with synthesized *fee* and *bee* syllables in an operant task, however, only within a set range of frequencies with which they are familiar (Weary & Weisman, 1991)—strongly reminiscent of the frequency range constraint. Recently, Cynx, Williams, and Nottebohm (1990) showed that zebra finches are extraordinarily sensitive to alterations in the frequency structure of certain syllables in their complex song. For some syllables, the second harmonic seems especially important in discriminating among the acoustic components used for natural communication. For zebra finches, at least, spectral structure appears to be an important cue for acoustic communication. It is interesting that starlings in the research reviewed here found spectral discriminations between artificial spectra easier if the second harmonic was included in the discriminative stimuli than if it was not. This seems a parallel worth exploring further for starlings—to say nothing of other avian species.

Animals certainly vary the acoustic spectra of their natural sound production, and they hear and use spectral differences, too. Crows do not sound like larks. Hawks do not sound like mice. Alarm calls are loud with a frequency and spectral structure that makes them hard to localize (Brown, 1982; Marler, 1955). Birdsong is based on a *syntax* of changing spectra in syllables and phrases. It may be that nonhuman animals use absolute pitch more than people under some circumstances. However, many species may share a capacity to use relative acoustic information to construct and perceive distinctive auditory objects that stand out from the acoustic background. Certainly acoustic sources that move or loom systematically in space are cases in point—such as the unique vocalizations of mates moving from one location to another or the sounds of a rapidly approaching predator. Furthermore, these auditory objects show surprising perceptual invariance over many acoustic transformations—the natural transformations that occur with movement, with echoes, with changes in distance, or with other kinds of acoustic modulation introduced by the natural environment (Brown, 1989; Marten & Marler, 1977; Waser & Brown, 1986). This chapter has discussed invariances associated with pitch and spectral structure, but there are many others—such as tempo and rhythm (Hulse, Humpal, & Cynx, 1984). Here are echoes of music perception once again.

Finally, the world in which we all live is a noisy place. It must be of adaptive significance for all species to segregate important acoustic objects from the cacophony of a natural environment. Like people, animals must distinguish among voices of their own species, and they must distinguish sounds of their own species from those of other animals. Sounds that predators make are no doubt of special significance. And all these must somehow stand out from the other irrelevant noises of city, meadow, forest, or jungle. Of course, most of the research remains to be done to prove the existence in nonhuman animals of auditory object formation and streaming. Animals may be the same as people in some respects. They are probably different in others. But the problems are important, and the prospects for new discoveries are promising.

OVERVIEW

This chapter has been organized around the study of some phenomena that depend fundamentally on the perception and processing of relations among acoustic events for their reality. Most of the work has depended on the principle of *perceptual invariance* as evidence for processing on a relational basis. That is, perceptual events are presumed to be based on relational processing if the events remain identifiably the same after one or another stimulus transformation. Typically, the transformations leading to perceptual invariance are based on ratio changes of stimulus values.

Starlings and, most likely, other songbirds show unquestioned relational perception of sound spectra. Sounds with a given spectrum (that is, a constant harmonic structure) remain perceptually invariant despite large changes in the absolute frequency of the fundamental. This is to be contrasted with starlings' perception of simple sequences of single frequencies that appears to be based preferentially— and surprisingly, from a human perspective—on memorization of the absolute frequencies involved.

However, if distinctive serial patterns of spectra are intermixed with those of another as they unfold in time, starlings apparently can attend to each of the subpatterns independently, a phenomenon known as auditory streaming. That process, too, is based fundamentally on stimulus relations: those that comprise the spectral elements themselves, and those that determine the unique temporal structures of the subpatterns.

This chapter honors Donald A. Riley. It is fair to say that much of Professor Riley's work has been based on attention and concept formation in animals, albeit mostly in the visual world. As I have tried to show, attention and the perception of stimulus relations is an important, indeed fundamental, part of the auditory as well as the visual world. Animals, human and otherwise, learn to pay attention to sounds, and to hear them out as coherent, classifiable objects from the surrounding environment in which they occur. Although Professor Riley and I have worked with different sensory systems in recent years, his work has been a constant inspiration. Those of us in the field of animal cognition, broadly defined, owe him a great deal. Some of his inspiration—and that debt—is reflected in this chapter.

ACKNOWLEDGMENTS

The research was supported by National Science Foundation Research grants. I gratefully acknowledge the ideas and the efforts of all those—especially Richard Braaten, Jeffrey Cynx, Suzanne Page, and Annie Takeuchi—who have contributed to the research in the past. Similar acknowledgments go to Daniel Bernard and Cynthia Gray who are helping to assure that the research continues. David Olton, Daniel Bernard, Cynthia Gray, and Annie Takeuchi read the manuscript and provided many helpful comments.

REFERENCES

Braaten, R. F. (1991). *Auditory perceptual organization by a songbird, the European starling (Sturnus vulgaris)*. Unpublished doctoral dissertation, Johns Hopkins University, Baltimore.

Braaten, R. F., & Hulse, S. H. (1991). A songbird, the European starling (*Sturnus vulgaris*), shows perceptual constancy for acoustic spectral structure. *Journal of Comparative Psychology, 105,* 221–231.

Bregman, A. S. (1990). *Auditory scene analysis: The perceptual organization of sound.* Cambridge, MA: MIT Press.

Bregman, A. S., & Campbell, J. (1971). Primary auditory stream segregation and perception of order in rapid sequences of tones. *Journal of Experimental Psychology, 89,* 244–249.

Bregman, A. S., & Pinker, S. (1978). Auditory streaming and the building of timbre. *Canadian Journal of Psychology, 32,* 19–31.

Brown, C. H. (1982). Ventriloquial and locatable vocalizations in birds. *Zeitschrift fur Tier Psychologie, 59,* 338–350.

Brown, C. H. (1989). The acoustic ecology of East African primates and the perception of vocal signals by Grey-cheeked mangabeys and Blue monkeys. In R. J. Dooling & S. H. Hulse (Eds.), *The comparative psychology of audition: Perceiving complex sounds* (pp. 201–239). Hillsdale, NJ: Lawrence Erlbaum Associates.

Cynx, J., Hulse, S. H., & Polyzois, S. (1986). A psychophysical measure of pitch discrimination loss resulting from a frequency range constraint in European Starlings (*Sturnus vulgaris*). *Journal of Experimental Psychology: Animal Behavior Processes, 12,* 394–402.

Cynx, J., Williams, H., & Nottebohm, F. (1990). Timbre discrimination in zebra finch (*Taeniopygia guttata*) song syllables. *Journal of Comparative Psychology, 104,* 303–308.

D'Amato, M. A. (1988). A search for tonal pattern perception in Cebus monkeys: Why monkeys can't hum a tune. *Music Perception, 5,* 453–480.

Garner, W. R. (1974). *The processing of information and structure.* Hillsdale, NJ: Lawrence Erlbaum Associates.

Handel, S. (1989). *Listening.* Cambridge, MA: MIT Press.

Herrnstein, R. J. (1990). Levels of stimulus control: A functional approach. *Cognition, 37,* 133–166.

Hulse, S. H., & Cynx, J. (1985). Relative pitch perception is constrained by absolute pitch in songbirds (*Mimus, Molothrus,* and *Sturnus*). *Journal of Comparative Psychology, 99,* 176–196.

Hulse, S. H., & Cynx, J. (1986). Interval and contour in serial pitch perception by a passerine bird, the European starling (*Sturnus vulgaris*). *Journal of Comparative Psychology, 100,* 215–228.

Hulse, S. H., Cynx, J., & Humpal, J. (1984). Absolute and relative pitch discrimination in serial pitch perception by birds. *Journal of Experimental Psychology: General, 113,* 38–54.

Hulse, S. H., Humpal, J., & Cynx, J. (1984). Discrimination and generalization of rhythmic and arrhythmic sound patterns by european starlings (*Sturnus vulgaris*). *Music Perception, 1*(4), 442–446.

Hulse, S. H., Page, S. C., & Braaten, R. F. (1990a). Frequency range size and the frequency range constraint in auditory perception by European starlings (*Sturnus vulgaris*). *Animal Learning and Behavior, 18,* 238–245.

Hulse, S. H., Page, S. C., & Braaten, R. F. (1990b). An integrative approach to auditory perception by songbirds. In M. Berkley & W. C. Stebbins (Eds.), *Comparative perception* (Vol. 2, pp. 3–34). New York: Wiley.

Marler, P. (1955). Characteristics of some animal calls. *Nature, 176,* 6–8.

Marten, K., & Marler, P. (1977). Sound transmission and its significance for animal vocalizations: I. Temperate habitats. *Behavioral Ecology and Sociobiology, 2,* 272–290.

McAdams, S. (1989). Segregation of concurrent sounds: I. Effects of frequency modulation coherence. *Journal of the Acoustical Society of America, 86,* 2148–2159.

Medin, D. L. (1989). Concepts and conceptual structure. *American Psychologist, 44*, 1469–1481.

Nosofsky, R. M. (1991). Tests of an exemplar model for relating perceptual classification and recognition memory. *Journal of Experimental Psychology: Human Perception and Performance, 17*, 3–27.

Page, S. C. (1989). *The perception of pitch and timbre by a songbird (Sturnus vulgaris)*. Unpublished doctoral dissertation. Johns Hopkins University, Baltimore.

Page, S. C., Hulse, S. H., & Cynx, J. (1989). Relative pitch perception in the European starling (*Sturnus vulgaris*): Further evidence for an elusive phenomenon. *Journal of Experimental Psychology: Animal Behavior Processes, 15*, 137–146.

Porter, D., & Neuringer, A. (1984). Music discrimination by pigeons. *Journal of Experimental Psychology: Animal Behavior Processes, 10*, 138–148.

Premack, D. (1978). On the abstractness of human concepts: Why it would be difficult to talk to a pigeon. In S. H. Hulse, H. Fowler, & W. K. Honig (Eds.), *Cognitive processes in animal behavior* (pp. 423–451). Hillsdale, NJ: Lawrence Erlbaum Associates.

Premack, D. (1983). The codes of man and beasts. *Behavioral and Brain Sciences, 6*, 126–167.

Rosch, E., & Mervis, C. B. (1975). Family resemblances: Studies in the internal structure of categories. *Cognitive Psychology, 7*, 573–605.

Takeuchi, A. H., & Hulse, S. H. (in press). Absolute pitch. *Psychological Bulletin.*

Ward, W. D., & Burns, E. M. (1982). Absolute pitch. In D. Deutsch (Ed.), *The psychology of music* (pp. 431–451). New York: Academic.

Waser, P. M., & Brown, C. H. (1986). Habitat acoustics and primate communication. *American Journal of Primatology, 10*, 135–154.

Weary, D. M., & Weisman, R. G. (1991). Operant discrimination of frequency and frequency ratio in the black-capped chickadee (*Parus atricapillus*). *Journal of Comparative Psychology, 105*, 253–259.

Weisman, R. G., & Ratcliffe, L. (1989). Absolute and relative pitch processing in black-capped chickadees (*Parus atricapillus*). *Animal Behavior, 38*, 685–692.

Weisman, R. G., Ratcliffe, L., Johnsrude, I. S., & Hurly, T. A. (1990). Absolute and relative pitch production in the song of the black-capped chickadee (*Parus atricapillus*). *Condor, 92*, 118–124.

Author Index

Numbers in *italics* denote pages with complete bibliographic information.

Subject Index

Printed and bound by CPI Group (UK) Ltd, Croydon, CR0 4YY

17/10/2024

01775687-0010